Digitalization and Future of Digital Society

Suat KOLUKIRIK (ed.)

Digitalization and Future of Digital Society

PETER LANG

Bibliographic Information published by the Deutsche Nationalbibliothek
The Deutsche Nationalbibliothek lists this publication in the Deutsche Nationalbibliografie; detailed bibliographic data is available online at http://dnb.d-nb.de.

Library of Congress Cataloging-in-Publication Data
A CIP catalog record for this book has been applied for at the Library of Congress.

This publication was printed with the support of InGlobe Academy.

Cover illustration: © Pixabay@pexels.com

ISBN 978-3-631-83489-3 (Print) · E-ISBN 978-3-631-84231-7 (E-PDF)
E-ISBN 978-3-631-84232-4 (EPUB) · E-ISBN 978-3-631-84233-1 (MOBI)
DOI 10.3726/b17856

Peter Lang – Berlin · Bern · Bruxelles · New York · Oxford · Warszawa · Wien

This publication has been peer reviewed.

www.peterlang.com

Contents

List of Contributors

Editor
Prof. Dr. Suat KOLUKIRIK
https://orcid.org/0000-0003-0399-666X
Akdeniz University
Department of Sociology
suatkolukirik@akdeniz.edu.tr

Chapter 1
Assoc. Prof. Dr. Gül AKTAŞ
https://orcid.org/0000-0001-6424-4256
Pamukkale University
Department of Sociology
gaktas@pau.edu.tr

Chapter 2
Dr. Abdilcelil KOÇ
https://orcid.org/0000-0002-9952-9830
Isparta University of Applied Sciences
Department of Management and Organization
acelilkoc@gmail.com

Chapter 3
Daniel NGUYEN
https://orcid.org/0000-0002-3992-541X
Missouri State University
Benefits Advisor
nguyen597@live.missouristate.edu
Assoc. Prof. Dr. Subhasree Basu ROY
http://orcid.org/0000-0002-2989-6416
Missouri State University
Department of Economics
sbasuroy@missouristate.edu

Chapter 4
Dr. Bahar MURATOĞLU PEHLİVAN
https://orcid.org/0000-0001-8977-822X
baharmuratoglu@gmail.com

Chapter 5
Assist. Prof. Sevilay Ece GÜMÜŞ ÖZUYAR
https://orcid.org/0000-0002-1957-3648
Necmettin Erbakan University
Faculty of Political Sciences
sevilayecegumus@gmail.com

Chapter 6
Lect. Emre Kadir ÖZEKENCİ
https://orcid.org/0000-0001-6669-0006
Çağ University
Faculty of Economics and Administrative Sciences
ekadirozekenci@cag.edu.tr

Chapter 7
Dr. Mehmet ÖZYÜREK
https://orcid.org/0000-0002-7980-4655
Yalova University
Faculty of Law
ozyureksez@yahoo.com.tr

Chapter 8
Assist. Prof. Seda Gökçe TURAN
https://orcid.org/0000-0002-6563-1378
Bahçeşehir University
Department of Child Development
sedagokce.turan@vsh.bau.edu.tr

Chapter 9
PhD Candidate Eren KESER
https://orcid.org/0000-0003-3810-0376
Akdeniz University
Department of Communication
erenkeserpr@gmail.com
Assoc. Prof. Dr. Fatih USLU
https://orcid.org/0000-0003-3192-4145
Akdeniz University
Faculty of Education
fatihuslu@akdeniz.edu.tr

Chapter 10
Dr. Eyüp ÖZ
https://orcid.org/0000-0003-4485-2902
Haliç University
Department of Political Science and International Relations eyupozum@hotmail.com

Chapter 11
Dr. Esra Ebru MAVİ
https://orcid.org/0000-0002-8734-2630
Council of Higher Education
esraebrumavi@gmail.com

Chapter 12
Lecturer Hatice OĞUZ
https://orcid.org/0000-0001-5901-3488
Isparta University of Applied Sciences
Therapy and Rehabilitation Department
haticeoguz@isparta.edu.tr
Prof. Dr. Metin ÖZKUL
https://orcid.org/0000-0003-2511-9780
Süleyman Demirel University
Department of Sociology
metinozkul@sdu.edu.tr

Chapter 13
Dr. Esra DOĞAN
https://orcid.org/0000-0002-9886-4186
Eskisehir Osmangazi University
Department of Public Finance
edogan82@hotmail.com

Chapter 14
MA Student Defne Nazlı BAŞ
https://orcid.org/0000-0002-1301-1740
İzmir Bakırçay University
Department of Sociology
defnenazlibas@gmail.com

Chapter 15
Dr. İbrahim YÜKSEL
https://orcid.org/0000-0001-5173-4387
Ministry of National Education
iyuksel56@hotmail.com

Chapter 16
Dr. Murat ÇOBAN
https://orcid.org/0000-0003-2415-5747
Ağrı İbrahim Çeçen University
Department of Computer Education and Instructional Technology
mcoban2005@gmail.com

Chapter 17
Lecturer Mehmet Akif BAL
https://orcid.org/0000-0002-2368-8640
Süleyman Demirel University
Faculty of Sport Science
balakifmehmet@gmail.com

Chapter 18
Lecturer Muhammed Fatih AKSARAYLI
https://orcid.org/0000-0003-4883-4082
Süleyman Demirel University
Department of Business Management
mfatihaksarayli@gmail.com

Chapter 19
Prof. Dr. Andreas KAPLAN
https://orcid.org/0000-0002-5016-0841
ESCP Business School Berlin & Paris
kaplan@escp.eu

Chapter 20
Asst. Prof. Hema MANDAPANDA APPACHU
https://orcid.org/0000-0002-8932-3134
Bengaluru Central University
Indian Institute of Psychology and Research
hemamythili22@gmail.com
MA Student Biddappa MANDAPANDA APPACHU
https://orcid.org/0000-0003-1049-4561
St. Joseph's College
English Literature

Introduction

In today's world, which is called digitalization or technological era, our life practices, habits, and ways of doing business continue to change skin every passing day. Perhaps we are witnessing a period in which the concepts of time and space become difficult and complicated as the subject of a world that is running away. Topics such as artificial intelligence, Internet of Things, innovation, augmented reality, cloud computing, and cybersecurity have become part of our daily lives. Moreover, more than half of the 7.5 billion world population uses the internet, and two out of every three adults have mobile phones. People shop digitally, watch movies or TV shows in their native language, plan a vacation, and play games. People live and move with artificial intelligence, big data, learning machines, and new robotics.

Our lives, which differed after the Industrial Revolution, are progressing mostly through technological addictions and phenomena in today's world. In a way, a life connected to the digital has turned into a new social structure and culture. So to speak, humanity is shaped through social media trends, virtual agendas, and visuals, and virtual simulations determine our identity and personality. The Internet determines and transforms the individual, identity, and culture, especially through social media platforms. Today's humans' shuttle between the real and virtual lives and our definitions of social values, norms, privacy, communication, production, and consumption habits are radically changing.

In addition to the facilities provided by our digitalized social life in terms of access to data and access to information, it also carries concerns such as data pollution and insecurity. Concerns about robotization, Humanity 2.0, the surveillance society, and human enslavement are often voiced in future planning and lead to different debates about the nature of human–system interaction. Moreover, rather than where it is in the world, what is done and how it is made becomes important, and it is a reality to make the effects of digitalization, which has leaked into every area of our lives and become invisible by becoming ordinary, understandable on our social life.

From the way our digitalized social life processes information to our production processes, from socialization to our understanding of consumption, our diverse profiles stand in front of us in understanding the time we are in. Because the face-to-face relationships that exist in the traditional process are seeking their legitimacy in the virtual reality world, which is compared to reality and needs rational explanations. Concepts such as digital identity, culture, biotechnology, artificial intelligence, digital economy, and hypnotic society surround us and keep getting more and more evident.

This book in your hand deals with changes in the context of digitalization and the future of the digital society in the context of identity, culture, economy, surveillance, production, and consumption and includes examples of new life practices. In other words, the profiles of digital life are interpreted and discussed by different social scientists. I hope the study will contribute to the digital sociology literature and be useful, and I thank the academicians involved in the project.

Prof. Dr. Suat KOLUKIRIK

Gül AKTAŞ

A Period in Which Childhood Is Idled Away: Being a Child in the Digital World

1 Introduction

Technological developments changing at a striking speed, information-communication networks determining many parameters in all areas of human life have profoundly affected the meaning of world for children as well as adults. Today, the digital world has provided facilities that will save life, time and energy to make life easier in every segment of the society. In this way, individuals can easily access information wherever they are in the world, receive immediate news about the developments in the fields of economy and politics, and money flow becomes quick, thus providing a significant comfort area for individuals. But when we evaluate the comfort in question sociologically, it also deforms the life of the individual. While this subject is frequently mentioned in the field, psychologists, psychiatrists, and sociologists underline that the negative impacts of the digital world should not be ignored. According to experts, because social relations with the digital world have taken on a new quality, we are more often faced with alienation, withdrawal from real life and loneliness from time to time.

When we approach from this point, this study aims to make a sociological comment on the issue of the digital age and its implications creating important usefulness and opportunities as well as involving risk and threats in children's lives. It is an important process for children, who constitute an important part of the future of the society, to develop their creativity; to acquire skills focused on solving problems and crisis; and to provide the capacity to speak, which is one of the most essential requirements of being an individual, empathize, understand, and interpret. However, since digital technologies provide unlimited access, children move away from socialization and meet their play, friendship, and communication needs in a virtual environment.

In the premodern era, children practiced traditional learning methods—by living, experiencing and developing new learning strategies—in the process of understanding and interpreting the social, cultural, and economic world. Therefore, in times when digital tools had not penetrate our lives, children mostly learned by interacting with social groups and experiencing the necessities of being an individual. Traditional games helped children develop their roles and responsibilities, discover methods of reaching the truth by making mistakes, socialize with peer groups, and develop practical skills in problem-solving. Whereas today, digital tools, which provide a wide range of use, gradually eliminate the way children learn through trial and error. The digital world provides services to children through ready-made package programs, thus enabling them to be involved in the social space virtually without getting tired, with multiple mechanisms capable of instantly changing and correcting mistakes. In this respect, it is a preliminary matter to look at the areas where it creates problem without ignoring that digital tools have become an important part of our life and that they provide comfort and convenience. In this framework, it is desirable for children to try to recognize and understand social reality in social life. For this, children must have an awareness that can distinguish between the virtual reality created by technology and the profiles that the fictional world presents to them and the pure forms of social reality. As such, analyzing the perceptual world created by digitalization in a sociological context will also be important for us to see how technological changes affect children's different living spaces.

2 From Traditional to Modern Societies, Children Meeting the Digital World in a Nutshell

The age of information and communication is changing at a dizzying pace in many areas of human life. The globalizing world has changed people's lifestyles, class structures, and cultural and social environment from different perspectives. Twenty-first century is an era in which people's traditional habits, social values, customs, and traditions are deeply influenced. Traditional learning and application methods have changed; technological tools and equipment can do many things people take more time to do with less energy.

The age of information and communication has caused a multifaceted change in childhood, as in many stages of our lives. A few questions arise here. One of them is whether the change of the child and childhood corresponds to the same period with technological change. The other one is how the digital age affects which stages of children's life.

In order to answer these questions, it is useful to clarify the concept of childhood primarily on the historical axis. The concept of family is the basis of the concept of childhood and the views on childhood. How the family is structured and the family models change in the historical process are reflected in childhood, thus reshaping it. In the antiquity, when children served roles such as good citizens, safeguards in intergovernmental treaties, and being a part of kinship relations, childhood was a concept toward which enough attention was not paid (Uğur, 2018: 228). During this period, children were regarded as "small citizens," who should be educated within the laws and culture of society. In the Middle Ages, the foundation of the institution of the family was based on the ideology of slavery, and family referred to the slaves of a father and a community consisting of wife and children. In this community, the father had all kinds of rights over the child and the children had no special status and rights to protect themselves. Being economically nonfunctional and not contributing to reproduction were the main dynamics behind the low status of children in the Middle Ages. In these times, when fatal diseases and difficult living conditions were common and lack of information about childcare caused high infant mortality rates, children were left indifferent to their existence and the concept of childhood as a separate stage of life did not develop. Until the sixteenth and seventh centuries, children were considered slightly more than miniature versions of adults. The view that childhood is a limited version of adult life has lost its effect due to various factors such as the formation of various currents of thought, change of production style, and the emergence of the bourgeoisie.

With the Renaissance, which enabled the discovery of the value of the individual, the meanings attributed to childhood began to change. The view that a child is an "innocent being" expected to reach adult life was adopted, and childhood started to be seen as a life cycle separate from adulthood. The development of awareness that childhood has a special nature created special interest in raising children in accordance with their nature. Enlightenment thinkers and childhood advocates such as Locke and Rousseau emphasized that it is an investment in the society that children, who can make and implement decisions by themselves, have high self-esteem, can express themselves, can solve problems faced by them, can control their behavior, and should be raised peacefully and in harmony with their own nature. It can be stated that the new conditions that emerged due to these developments starting with the Renaissance, Reformation, and Enlightenment changed the state of children in the society and the family and removed the obstacles to the recognition of the nature of children.

These developments also changed the establishment of the relations between the child and the family. In this new order, where all children had a universal childhood nature, it was the primary task of parents to raise children in accordance with their own nature, protecting them from religious dogma and immorality in society. The perception that children are innocent and naturally weak creatures led families to treat their children with care and compassion and to protect their children from the harsh realities of adult life. In addition to separating the worlds of children and adults, the new understanding of childhood also protected individual values by separating siblings. In this period, with the opinion that the child was not an entity under absolute control of the father, the father's authority shifts to institutions such as medicine, pedagogy, and education, which indicates that childcare has gained a professional quality. The mother–child relationship gaining importance with the discovery of motherhood was the result of the steps taken in this period. In the modern period, a childhood phase different from the thoughts of adults was designed. The bourgeois class led the idea of childhood with individual value throughout the eighteenth and nineteenth centuries, and by the twentieth century, the view of childhood as a separate individual began to spread among the middle classes and lower classes. In this context, it can be stated that a special childhood idea, which needed to be protected with the developments that transformed social, economic, political, cultural, and social conditions, began to occur, and this transformation, which primarily affected the values of the bourgeoisie about the child, spread to all classes of the society over time and formed the roots of today's modern childhood image (Uğur, 2018: 228–229).

In today's world, childhood is shaped around different variables. A wide range of internet and technological networks, social activities, cultural events, the world of games and entertainment, urban dynamics, and diversity in educational materials have encompassed the social and personal worlds of children. Children embraced by an intense digital world have gradually started to lose the distinction between the real and the virtual world. Especially the world of game on the internet is sometimes defined as an achievement and a tension point for children and puts any other perception of reality in their lives aside. İdris (2017: 599) evaluates the phenomenon of the child in digital age as follows:

> *The reality of children, born into a digital world after 2000s, opened the door to new times, the effect of which has not diminished, and is also irreversible. Even though adults entered before this new temporal door opened, it was seen that the finger of the child started to quickly move around the keys. As perhaps the first example in history, a period began when fathers learned something from their children. The children were quick. The digital environment has set up armies to keep up with their biological, cognitive and spiritual pace. Numerous software developers, artists and writers began to produce and spread extremely attractive programs and games for children/youth. Global village discourse turned into real-time action through games and interests. The children who were called to the table from the game on the street before the evening, started to be called from the game on the computer screen. The child, who was previously criticized as 'not entering the house from the street,' is now being criticized as 'not going out of their room'. The voices expressing the boring life of education in the face of the game world in the internet cafes and the visual world with plenty of effects on the screen are now on the agenda with high impact. The children, who were fleeing and 'escaping' from the school with joy when the last lesson bell rings, are now slowly and absent-minded by playing games with the smartphones and tablets they carry in their pockets and bags—despite some limitations. They shoot and broadcast each other's videos, 'sharing' their classes and themselves in digital platforms. They take and share other shares with all kinds of cultures and cookies in and out of them.*

3 Children Lost in Digital Life

In modern societies, where family communication is weakened, fragmented family structures are relatively increasing nowadays, internet usage covers a significant part of our lives, and important breaking points have been observed in the lives of children in all age groups.

Children who participated in games mostly with peer groups used their free time by interacting with toys and small social groups in traditional social structures. Onur (2009: 177–180) conveys the thoughts of different scientists on this subject, as follows, in his work titled "Child in Turkish Modernization" ("Türk Modernleşmesinde Çocuk"):

The low numbers of toy expenditure per child in Turkey (in the EU annual expenses of 730 million, 7–9 million in Turkey) cannot only be explained by economic reasons. The main reason is the lack of interest. Lack of 'interest' does not necessarily mean a lack of 'love'. There is no doubt that mothers and fathers in this country love their children as they do everywhere. In addition, we can say that general love and tolerance towards children is an outstanding feature of this society. However, in the final analysis, we have a society that has not grown and matured enough. A society made up of people who, as in every field, could not rationalize their emotions and make them useful for the child! Because they did not think and know that it is not enough just to 'love' the child. They probably did not hear the word 'pedagogy '; unaware of what the word says or not, whether they are aware of it or not. So they can do many things for their child, for example can fight to death with their neighbor, etc.; if necessary, they can spend large sums of money for the child, but they can never do small, simple things that are useful to the child. This is a very common attitude: 'Kid, stop there.' To say 'stop there' and wait for them to stop. If they do not stop, they do not know a cure except for two slaps. The ability to think and understand the child as a child is never worked, and therefore not developed. The parent can give their life for their child, but they do not know how to spend time with them efficiently.

Technological life, which increases children's engagement and drives them away from the problems of real world but immerses them in another spiral of problems, comes to the rescue of children who grow up without interest and love. It is useful to present the findings of a striking research on this subject. The results of the EU Kids Online Project (2010), which was carried out in 25 European Union countries, including Turkey, with children aged 9–16 in 2010, shows that children started to use the internet at age 10. In the case study conducted by Berger, Hoveyda, and Sinha (2011) to demonstrate the use of digital technologies in Turkey, the age to start using the internet by children was observed to be 10. According to the results of the study conducted by TURKSTAT in 2013 entitled "Information and Communication Technologies Usage in Children and Media" (Çocuklarda Bilişim Teknolojileri Kullanımı ve Medya"), it has been determined that children start using computer at age 8 and internet at age 9. In the same study, it was reported that children started using personal mobile and smart phones at age 10 and 60.5 % of the children used computers, 50.8 % internet and 24.3 % mobile phones. According to the results of the "Children's Media Habits in Turkey Survey," carried out by the Radio and Television Supreme Council (2013), among the children in the 6–18 age group, 73.7 % have computer/tablet in the house and 63 % have internet connection. In the same study, the rate of children who have a computer/tablet in their room is 75.6 % while 58.6 % of them have internet connection in their room. Similar results have been achieved in European countries. In their study, Duerager and Livingstone (2012) report that it is seen that almost half of the children in Europe are online in their rooms and one third of them are online via mobile phones or handheld devices. In a study in which data from the Turkey operation of EU Kids Online project was reported, about 46 % of children stated that they started using the internet between the ages 7–10. In the same study, it is reported that children's internet usage rate in

Turkey was 13 % among 7-year-olds, and this rate was around 59 % in Europe. As can be seen, the rate of spread of computer and internet technologies to the population tends to increase, and the age of children getting to know these technologies tends to decrease. This situation exposes children to the internet risks brought by the opportunities of these technologies (from Yaman et al., 2016: 175–176). Idris (2017: 600) talks about a problem in this context:

> *The child/youth (or adult) is losing their center, their world. This situation both destroys the centralization in the sense of intense and endless digital flow attention by decentralizing the subject, and a center in terms of spatial belonging is now blurred. This does not include a result imposed only by adults and the system, limited to some values being broken down forever. The already atomized identity components of semantic clusters of different scales such as family, street, neighborhood, society, this time implies the fragmentation in the self and singular identity and the state of disidentification/not belonging to a place. If you don't care what social or individual identity integrity means, you may not care about this fragmentation. But this disregard is not enough to make the consequences of the fragmentation of both identities trivial. Together with sociologists, psychiatrists are bursting to explain the chaos and problems that this digital age has brought, which they have not encountered before, and to solve them—if possible. The digital interest that starts with the game for the child and continues with the 'siege' in other fields (culture, consumption, ideology, etc.) is undoubtedly nothing more than the current version of the hegemonic approach of an extension of the system-child relationship in the digital field. The general logic of the system/adult's relationship and communication with the child continues here as well. But what's new and perhaps unpredictable is something else (as in most children's engineering) when you think you are wrapping the child under your arms in the face of a digital siege in accordance with your own mind; a side effect, another unpredictable outcome appears: digital poisoning of childhood.*

4 The Effect of Digitalization on Children's Daily Life Practices

Technological advances and change in the content of the services provided have affected different areas of interest in the development of children's education, play, and social skills. Now, children are included in the education and training channels with smart boards and tablets in the education environment, and they weaken their social environment and family ties by meeting the limitless information provided to them by the virtual world through personal phones and computers. It can be suggested that while the digital age offers opportunities and conveniences to the lives of children through different information channels such as education, play, and health, it also reveals actions and visual materials that contain many threats and risk factors. Selwyn states that technology opens the doors of creativity and a new source of income for children but also emphasizes that it is a door to risk and threat. According to Selwyn, it is necessary to focus on endemic (social, political, and economic) conditions. This will also ensure that children's rights are protected and looked after (from Bardakçı, 2018: 234).

The media stands out as one of the important social forces that play a role in shaping children's emotions and thoughts. Children now learn to use media tools from an early age, spend most of their time on the internet and television, thus living a life with media technologies. In this world, qualifications such as "television kid" and "computer kid" have started to be used for the new child. This situation changed the way of communication for children by bringing about the consequences such as indirect communication and socialization instead of face-to-face communication, the loss of the culture of reading a book, and the popular culture elements based on visuals to come to the fore (Karaboğa, 2018: 2).

According to Postman, cultural change is also accelerates with the increase of the shaping effect of television on daily life. This situation, which Postman describes as "cultural collapse," has deeply influenced childhood and has caused the loss of childhood and book culture. It is now much easier to look at fast, colorful, and audible images. Reading books requires time, effort and attention. There are no such obligations for television. It can remain open at any time of the day, even if attention is not directed at it, and the subject can be captured instantly in a couple of seconds of glance. Colorful fairy-tale books have left the scene to cartoons, electronic books loaded with a single click, toys telling fairy tales. Bookshelves are now filling with the technological toys of modern times (from Aşar, 2014: 8).

Television and internet contents manipulate children about consumption and make them buy continuously and be a good consumer. Instead of producing an image of a questioning, knowledgeable child, a consumer child who continuously consumes, purchases, and is not satisfied with it has been created. Children's culture is built as part of consumer culture; from clothing to books and toys to playgrounds, everything is produced by the industrial system. While many media program contents play a role in the production of the consumer child and their culture, advertisements, TV series, cartoons, and computer games play an important role as well. Through these programs, the media produces popular and imaginary heroes and uses them in sales and marketing activities. The child's consumption habits and desire to have products can change, and the child's consumption expectations, addiction, and passion for toys can increase through cartoons, series, and virtual game characters. Children can be affected by the deceptive and misleading messages of these program contents, which can negatively affect their physical, intellectual, and emotional development (Karaboğa, 2018: 2).

Children of development age may not be able to take a stand against the services provided by the internet and develop a selective attitude because many websites of unknown sources have been opened to the use of children. And families sometimes may not be aware that these sites, which cause ethical and security problems, are accessible to children. While it is possible for the contents of different websites to be followed and encrypted by the families in the home environment, the control mechanism in the public sphere gradually weakens. Internet access especially from internet cafes and mobile phones makes control over children difficult. The main problems of digitalization are shares and games that propagate cyber bullying and mental violence, which may cause emotional disorders such as fear and depression, observed in recent years; lead to stealing personal information and sharing items that encourage substance and alcohol addiction; and establish friendships in a virtual environment where identities are not known.

It is an important responsibility for the families and educators to be informed about the risks of the internet and to convey this information to the children correctly. In the study titled "Internet Usage and Family" ("İnternet Kullanımı ve Aile"), conducted by the General Directorate of Family and Social Research in 2008 on this matter, the following study was included:

> *In the study conducted by Sotoyoma et al., to measure their knowledge about healthy computer use, it was determined that 60 % of both primary and high school students and 30 % of secondary school students were not given any information about computer ergonomics. 30 % of average primary and high school students and 60 % of secondary school students have received limited healthy computer use training. Overall, 10 % of schools actively integrate students' knowledge of healthy computer use in computer education. This shows that society does not have enough information about healthy computer use.*

Özkan and Hira (2017: 258–261) interviewed parents in their qualitative study, where they investigated the relationship between digital media and socialization. In this study, parents' observations and thoughts were included in order to analyze the interaction of children between the ages 6–12 with digital media. During the interviews, the parents stated that they complain about their children's excessive indulgence toward using the internet. Parents stated that although children use the internet for many purposes, they get carried away especially while playing games and postpone basic activities such as eating and going to toilet. According to the parents,

> *Children face the problem of addiction by playing or chatting with a computer or mobile phone for a long time. For this reason, their health is negatively affected because they meet their basic needs (such as eating and sleeping in front of TV or computers, or not being able to hold it) in the wrong places.*

Therefore, when the internet is not used correctly and effectively, it poses a number of problems for children who are sensitive and vulnerable. Among these, adverse effects on physical and mental health, age and content incompatibility, communication problems, online abuse, and security and information privacy violations can be shown as examples. It has been reported that obesity develops as a result of low physical activity and overeating behavior in children who live sedentary lives in front of the screen for a long time. In addition, a study conducted in our country showed that there is a relationship between problematic eating behavior and internet addiction among high school students. The relationship between using excessive digital technology in preschool period and increased mass body index has been shown. In a study, it was reported that there is an increase in depressive behavior in adolescent girls who use the internet more than 7 hours a day and excessive internet use causes psychological problems in both girls and boys. In another study, long-term use of social media in adolescents was associated with low sleep quality, low self-esteem, high anxiety, and depression frequency (Topçu, 2018: 28).

Finally, it can be said that internet and television prevent parents and educators, who are the main sources of information, and even assume the role of child-care. Yağbasan and Kurtbaş, who worked on this subject, state that children had two sources of information/influence in the past: parents and educators. Today, television has been added to these sources, on top of new communication technologies such as social media. The limitation in the diversity of the daily practices of children and the loss of the traditional family role have placed television as the basic device that builds and creates the child's scientific knowledge, besides assuming a "caregiver role." Şirin, who is working on this subject, states that a new child in this world is called the "television kid." Children, who used to ask the unknown to their parents, now prefer to learn the world from the internet and television. Accordingly, parent–child shares are also gradually decreasing (from Karaboğa, 2018: 5).

5 Digitalization and Children's Rights

In Article 1 of the *United Nations Convention on the Rights of the Child*, "a child means every human being below the age of eighteen years unless under the law applicable to the child, majority is attained earlier." The United Nations puts great responsibilities on parents. The following is highly stressed about parents in the mentioned convention:

> *(...) [c]onvinced that the family, as the fundamental group of society and the natural environment for the growth and well-being of all its members and particularly children, should be afforded the necessary protection and assistance so that it can fully assume its responsibilities within the community, [r]ecognizing that the child, for the full and harmonious development of his or her personality, should grow up in a family environment, in an atmosphere of happiness, love and understanding, (...).* (www.unicef.org)

The widespread use of the language of love and tolerance is important for children to be healthier and happier individuals in family and social life. This will reach the desired level with the contributions of both the family and the social, cultural, and political authority. When we approach from this framework, it is an important requirement to use the technology that encompasses the lives of children as well as adults, to be designed not to harm children's individual rights of life but to make them feel valuable, to eliminate inequalities in access to information, and to create an environment free from the spiral of violence and fear. Because when we look today, children who use digital tools more widely to access educational materials have fun and communicate with peer groups. Therefore, when they are online on the internet access network, it is important that they stay away from problems such as mental, spiritual, and physical risk factors.

Gasser, Director General of the Internet and Community Center, calls for more inclusion of children in digital rights in his statement titled "Taking Children Seriously" under the United Nations Convention. According to him,

> *Educators have to be trained so that they are not only effective teachers, but good mentors who engage creatively as well as responsibly with children, using the Web and other digital technologies. Adequate support and oversight structures need to ensure that the risks associated with the use of the Web in connected spaces of learning and engagement of children are properly addressed and the opportunities maximized. Serious work with children on the digital rights agenda also requires financial resources—for instance, to enable access and foster digital skill-building—and time commitments by caregivers, educators and, ultimately, policymakers. Above all, efforts require a strong collective commitment to take children and young people seriously as we seek to close digital gaps and overcome disparities, and to hear their views on matters that affect them, as the Convention demands. Whether in the context of the Convention on the Rights of the Child or the World Wide Web, concerted and global efforts must engage children more extensively in conversations about their future. If we fail to do so, we run the risk during the next 25 years of formulating treaties, laws and policies on digital rights that are disconnected from the realities of an increasingly digitally connected generation of young people, and won't succeed in making the world a better place, especially for the most vulnerable and marginalized children.* (www.unicef.org)

UNICEF evaluates children's following the digital world closely and the frequent use of technology in two ways. In this context, UNICEF emphasizes that the digital world facilitates children's access to opportunities and life chances with its work titled "The State of the World's Children 2017: Children in a Digital World" in 2017 and reveals what awaits children in the future. The point that draws attention in this study is that if digital technology is used correctly and is accessible to everyone, it can create radical changes especially in the lives of disadvantaged children. In this context, UNICEF said,

> *If leveraged in the right way and made universally accessible, digital technology can be a game changer for children being left behind—whether because of poverty, race, ethnicity, gender, disability, displacement or geographic isolation— connecting them to a world of opportunity and providing them with the skills they need to succeed in a digital world. But unless we expand access, digital technology may create new divides that prevent children from fulfilling their potential. And if we don't act now to keep pace with rapid change, online risks may make vulnerable children more susceptible to exploitation, abuse and even trafficking – as well as more subtle threats to their well-being. These risks can also create invisible threats that can harm children. Although attitudes can vary from culture to culture, children often refer to their peers first when faced with online risks or dangers. This makes it difficult for parents to protect children.* (www.unicef.org)

One of the important issues regarding children's rights in the digitalization process is that the measures to protect the mental, spiritual, and physical health of children who participate in the world of games and entertainment through digital media should be carefully considered. Especially children who go unwittingly to games that are not suitable for their age, gender, and development process can surrender to processes such as abuse, fear, and violence. Children who refrain from informing their families about this issue are also deprived of the control mechanism.

Today, children are considered as consumer assets to gain certain consumption habits, especially for companies, and the media is emerging as a powerful tool in the production of the consumer child. In recent years, the image of consuming children has become widespread, and companies have started to target children with advertising sales techniques. In this period when everything is tried to be integrated into consumption, children experience the feeling of dissatisfaction more.

Most of the new consumption tools, directly or indirectly, carry out marketing activities for children and young people. This target audience, which marketers and advertisers are keeping their eyes on, is in a vulnerable and unequipped state. In the hustle and bustle of everyday life, children who are most vulnerable to the culture of consumption, where we deliver our hearts, brains, dreams, and hopes, need protection (Karaboğa, 2018: 7).

Studies on the use of social media support that children who have access to these opportunities face more risks. Livinstone and Brake have said that social networking sites, which are rapidly adopted by children, especially adolescents and young people around the world, allow people to present themselves, learn new things, establish broad network relationships, and manage their own subjectivity and privacy. There are concerns that social networking sites increase the likelihood of new risks as well as opportunities. Some of the risks posed by the use of social media have been identified as loss of private space, exposure to cyberbullying, and damaging relationships. Cyber pitfalls such as deception, temptation, and use for their own interests require the ability to control, be aware of, and protect personal data and to use security settings actively, especially for children (from Karaduman, 2019: 694).

The American Academy of Pediatrics (APA) recommendation in the 2016 directive suggests that children under 18 months should avoid using screen media other than video chat. Parents of 18- to 24-months-old children who want to know digital media are recommended to choose high-quality programs and watch it with their children to help them understand what they see (from Topçu, 2018: 29). In order for such a process to take place, parents must also have a conscious media literacy and choose digital communication contents in a healthy way and present them to the interests of children.

Conclusion

The positive aspects of digitalization are an undeniable reality in terms of easy access to information, eliminating inequalities by providing ease of use to the educationally disadvantaged children, and helping children develop their skills and abilities. However, it is another reality that we see on the other side of the coin, that the digital world reveals personal risk and security problems as much as the opportunities it presents to children. It is observed that children face inappropriate content such as fear, violence, and harmful habits due to the fact that they use the internet more and more every day. It has been frequently mentioned by scientists in recent years that this segment is vulnerable and quickly persuadable and that they do not have the ability to cope with wrong choices in the digital game and entertainment world and with the problems that arise. According to experts, problems such as anxiety, attention deficit, insomnia, and communication disorders in children are a pile of important problems triggered by digitalization. Among the most important reasons for this is that the children who do not have enough information about how to eliminate the pressure elements carried out in the digital environment struggle with mental problems alone. Therefore, it can be said that the effects of the digital world that spread to all areas of the society have created important effects on the mental and physical lives of children.

Children who need to socialize with family, games, and peer groups are mostly trapped in closed spaces with the physical structure of the city and the diversity of people created by an increasingly cosmopolitan life. The problem of social trust has moved children away from the playgrounds outside. During the period when the game culture continued with traditional forms, children improved their communication skills in a healthier way in the relationships they established with their friends in the fields of common interaction. Children shaped their professions, friend preferences, and individual and social identities, which they would choose in traditional social structures, with the games they played with their peers within the dynamics of neighborhood. So, they learned life by interacting and experiencing small groups. The children of the present day do not find the games played in the axis of interaction, such as playing house, dodgeball, and marble interactive. Especially children are wrapped up with technological tools from the moment they are born. Families entertain children by giving them tablets and phones in order to chat with their friends or to continue their daily work and ultimately prepare the ground for the creation of a generation that is unaware, absentminded, and careless. Nowadays, children socialize by being alone in their single rooms and establishing virtual friendships. It can be said that with the rapid spread of internet and information communication tools with the transition to modern societies, the game and communication styles of children have undergone a radical change.

Today, the fact that children spend most of their time away from social dialogues on the internet also presents discussions about a different type of addiction called "internet addiction." In this regard, the World Health Organization (WHO) stated that the use of the internet, computers, smart phones, and other electronic devices has increased significantly in recent years, and this increase has caused not only enormous benefits to users but also negative health consequences. In an increasing number of countries, the problem also raised a major public health concern. The WHO's meeting on the effects of addictive behavior on public health with the participation of several countries was conducted in Istanbul, Turkey in 2017. In this meeting, discussions were held on behavioral disorders caused by game addiction on the internet and how to solve this with clinical methods. As a result of the discussions at this meeting, according to the WHO, video game addiction was defined as "mental health disorder" (www.who.int).

Children will always want to take advantage of the space and comfort provided by the digital world. Considering that children are vulnerable and easily persuadable individuals, it is a social responsibility to organize the content and presentations of games, advertisements, and other internet services for them, away from harmful effects. Otherwise, the harm done to children's spiritual, mental, and physical life will weaken the hope of a healthy next generation. In this regard, it is an important issue for media authorities, as well as political authorities and families, to take responsibility as control mechanisms.

Bibliography

Aşar Öğün, Ebru, (2014), "Keşfedilen Çocuk, İletişim Teknolojileriyle Yitip Giden Çocukluk ve Çocuk Kültürü", academia.edu/32366541/Keşfedilen_Çocuk_İletişim_Teknolojileriyle_Yitip_Giden_Çocukluk_ve_Çocuk_Kültürü (05.03.2020).

Bardakçı, Salih, (2018), " 'Dijital Yaşamda Çocuk Kitabı' Üzerine Bir İnceleme", *Eğitim Teknolojisi Kuram ve Uygulama Dergisi,* Vol. 8, N. 2, pp. 232–242.

İdris, Mevlana, (2017), "Çocukluğun Dijitalleşmesi Çocuğun Dijital Zehirlenmesi", *TRT Akademi Dergisi,* N. 4, pp. 598–602.

Karaboğa, Mehmet Tahir, (2018), "Medya Çağında Çocuk ve Çocuk Kültürü: Şiddet ve Tüketim Kültürünün Yansımaları", *Uluslararası Çocuk Edebiyatı ve Eğitim Araştırmaları Dergisi,* Vol. 2, N. 2, pp. 1–17.

Karaduman, Sibel, (2019), "Yeni Medya Okuryazarlığı: Yeni Beceriler/Olanaklar/Riskler", *Erciyes İletişim Dergisi,* Vol. 6, N. 1.

Onur, Bekir, (2009), *Türk Modernleşmesinde Çocuk* (First Edition), Ankara: İmge Kitabevi.

Özkan, Abdurrahman and Hira, İsmail, (2017), "Dijital Medya ve Sosyalleşme: 6–12 Yaş Çocukların Sosyalleşmesine Dair Ebeveyn Görüşleri (İstanbul Örneği)", *Kesit Akademi Dergisi,* Year: 3, N. 9, pp. 245–270.

T.C. Başbakanlık ve Aile ve Sosyal Araştırmalar Genel Müdürlüğü, (2008), *Internet Kullanımı ve Aile,* (Prepared for publication by Sadık Güneş), Ankara: T.C. Başbakanlık ve Aile ve Sosyal Araştırmalar Genel Müdürlüğü Yayınları.

Topçu, Selda, (2018), "Çocuk ve Sanal Ortam", *Kocatepe Tıp Dergisi,* N. 19, pp. 27–33.

Uğur, Sinem Burcu, (2018), "Geçmişten Günümüze Şekillenen Çocukluk Algısı ve Çocuk Yetiştirme Pratikleri", *Süleyman Demirel Üniversitesi Sosyal Bilimler Dergisi,* N. 45, pp. 227–247.

UNICEF, (1989), "Çocuk Haklarına Dair Sözleşme", https://www.unicef.org/turkey/%C3%A7ocuk-haklar%C4%B1na-dair-s%C3%B6zle%C5%9Fme. Accessed on 07 April 2020.

UNICEF, (2014), "Çocukları Ciddiye Almak: Dijital Haklara İlişkin Söylemde Çocukların Daha Fazla Yer Alması İçin Çağrı", https://www.unicef.org/turkey/media/2601/file/TURmedia_UNICEF-CRC_at_25_Gasser_Turkce.pdf.pdf. Accessed on 07 April 2020.

UNICEF, (2017), "The State of the World's Children 2017 Children in a Digital Word". www.unıcef.org/publications/index_101992.html. Accessed on 15 April 2020

WHO, (2018), "Public Health Implications of Excessive use of the Internet and Other Communication and Gaming Platforms", https://www.who.int/news-room/detail/13-09-2018-public-health-implications-of-excessive-use-of-the-internet-and-other-communication-and-gaming-platforms. Accessed on 15 April 2020

Yaman, Fatih, Dönmez, Omur, Kabakçı Yurdakul, Işıl, Odabaşı, Ferhan, (2016), "Dijital Dünyanın Ailesi İnternEtik Aile", (İşman, Aytekin, Odabaşı Hatice Ferhan, and Akkoyunlu, Buket, Eds.), *Eğitim Teknolojileri Okumaları 2016*, Ankara: TOJET.

Abdilcelil KOÇ

Capitalism 5.0: Digital Capitalism and Its Political Economy in the Context of the Evolution of Capitalism

1 Introduction

Capitalism, which has a history of more than 600 years, has undergone significant transformations and is evolving toward digital capitalism numbered as Capitalism 5.0. The political economy approach, which has an essential place in economic thought, focuses on understanding and explaining the capital accumulation mechanism in every stage of capitalism. The main purpose of this study is to consider the mechanisms of capital accumulation from a historical perspective. How this mechanism will evolve toward the near future is another matter of curiosity. In particular, the focus of the study is digital capitalism, expected to mark the near future, and its main components.

A two-way importance of this study can be mentioned. The first is to trace the past, present, and possible near expected continuity of capital accumulation. The second is to raise awareness through the literature by asking questions about the capital accumulation tools in the digital economy, which is the essential component of digital capitalism, new "opportunities" in this field, and possible problem areas.

The study consists of five main sections after the introduction. Each chapter examines the different forms of capitalism in its historical evolution. In these sections, the theoretical description method is used by making use of heterodox social scientists' studies. In the last chapter, the concepts and questions that stand out in the digital capitalism model are discussed. The topics are handled with a critical perspective and complemented by a general evaluation in the conclusion chapter.

2 An Overview of the Evolution of Capitalism

In this study, the trace of capital accumulation is followed. The radical changes in the capital accumulation mechanism show that capitalism is entering a new phase. In this context, when looking at the historical development of capitalism, it is seen that there are four different phases[1]. Each phase has its capital accumulation

1 Although there is no complete consensus on the periodization of capitalism, an almost consensus was achieved on commercial capitalism, industrial capitalism, and monopolistic capitalism (imperialism), which constitute the first three phases. However, there are different definitions of the fourth phase. See detailed readings (Tongyu & Weimin, 2012:17–24; Wright, 2009:211–16).

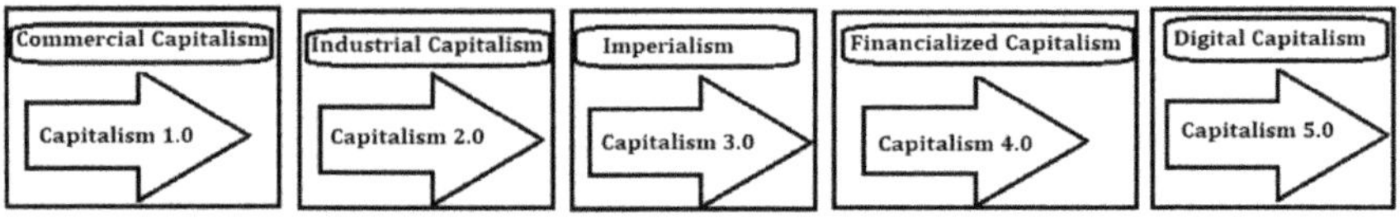

Fig. 1: The Evolutionary process of capitalism

mechanism and has a dialectical relationship with the past and next phase. In every capital accumulation phase, while the main mechanism continues to operate, the accumulation mechanisms of the previous periods continue to function until they disappear. In time, new capital accumulation channels are formed depending on the scientific-technological developments. Therefore, each capital accumulation phase of capitalism contains three different characteristic features or accumulation channels. In other words, past, current time (today), and future coexist. First of all, it is the main accumulation mechanism that marks the current period and represents the current time. Second, it is the accumulation channel from the previous period, representing the past. As time progresses, the relative importance of this channel continues to exist, although it gradually decreases. The third is the emerging capital accumulation channel, which will leave its mark on the next period. This third is very critical. Because this mechanism is becoming more and more widespread and getting stronger, conditions become mature and "when the time comes" dominates the system and creates a new capital accumulation mechanism. A multifaceted social transformation is necessary for the third structure to dominate the entire system. Undoubtedly, this process also requires a long time. In the evolutionary process of capitalism, this tripartite mechanism always seems to work.

The criterion taken into account in the numbering of capitalism's stages has been the main element that constitutes the accumulation of capital[2].

The evolution process of capitalism for about six centuries is shown collectively in Fig. 1. Each phase mentioned above will be briefly summarized in the following subheadings since the focus of the study is not the periodization of capitalism.

2 In the numbering of capitalism, some authors made different numbering based on different criteria. Among them, Rodrik (2012) named the period that Adam Smith called the "market miracle" in the early nineteenth century as Capitalism 1.0. Capitalism 2.0 is based on mixed economy and the Keynesian regulations of the 1930s, which lasted until the 1970s. The period we live in, Bretton Woods proposed Capitalism 3.0, which is based on updating the consensus, moderate globalization, and ensuring global communication.

2.1 Capitalism 1.0: Commercial Capitalism[3]

When we look at sixteenth, seventeenth, and eighteenth century Europe, it is seen that the main factor of the accumulation of capital is the commercial profits obtained by commercial capital over time, and this phenomenon gradually dominates the system. The Simple circulation of commodity: C(Commodity)=>M(Money) => C(Commodity) as in this formula, along with commercial capitalism, money is transformed into commercial capital and is shown as M=>C=>M'...M''. M' > M and M' – M= Profit. Here, the commercial capital (trader) buys the goods of the producer or provides the conditions for them to produce in manufactures. It sells these goods in domestic or foreign markets, fairs, and even overseas colonial countries and purchases new goods from there and converts them into money by selling them in other markets. The profits obtained from these transactions form capital accumulation, and the capital accumulation cycle continues continuously. Karl Marx calls this notation "The General Formula of Capital" (Marx, [1867] 2011: 152–159). It is how financial capital increases. M=>M'. Here, M'> M and M'–M = Financial Profit (includes speculative gains in the stock market that have just started operating in some countries, especially interest).

In this first phase of capitalism, although the primary source of capital accumulation is commercial profits from the commercial capital, accumulation based on land rent represents the past. In the period of commercial capitalism, the mechanisms representing the future, the third leg of capital accumulation, are the income from financial transactions and the newly emerging surplus value.

2.2 Capitalism 2.0: Industrial Capitalism

Significant developments took place in Western Europe between the 1760s and 1870s. The social conditions that emerged as a result of scientific and technological developments reaching a certain level revealed the industrial revolution in England. In this process, the entrepreneurial industrial bourgeoisie, which represented industrial capital, played an essential role in the emergence of factories with the state's support. Another characteristic of the industrial revolution is that the poor peasants, whose lands have been taken away, constitute the working class that will work in the factories that appear in the cities. Also, the industrial revolution was financed by looting raw materials and other resources in the colonies and transferring them to England.

In the first volume of *Capital*, Marx's masterpiece, he made the most serious systemic criticism of capitalism by scientifically revealing the "surplus-value" that emerged in the capitalist mode of production. The industrial bourgeoisie, while it spends some surplus-value-based profits obtained through labor exploitation to consumption, transfers the remaining portion to new investments to increase its

---

3 Also called mercantilist capitalism.

capital. Therefore, the accumulation of capital is provided in this way. The general formula of capital is M=>C=>Cm =>M'...M'' and (M'>M). Here, M is the first starting capital. That is to say, "Constant capital," such as factory building, machinery, tools-materials, and raw materials for production; he also invested in the workforce as a "variable capital" (Marx, [1867] 2011: 210). Here, Cm is the goods produced for the market. M' is the gross income obtained from selling the produced goods in the market. M'–M = Profit. This cycle must continue for the formation of capital accumulation. In the period of industrial capitalism, a particular part of the capital also provides capital accumulation in the financial sector by making money from money by reproducing their money with speculative gains (Interest, stock market, foreign exchange trade, etc.). With the formula of Marx, M'=>M...M''; M'>M and M'–M = Financial Profit.

In this period, when the working class came to the stage of history as a social class, many economic, social, and political gains made in the capitalist countries as a result of the organized class struggle marked the social development. However, accumulation based on land rent and commercial profits has transformed into a function representing previous periods and has taken its second place. Activities based on other service sectors, primarily financial activities, have served as the accumulation mechanism representing the future. This phase, called industrial capitalism, in this study is symbolized as Capitalism 2.0.

2.3 Capitalism 3.0: Imperialism

There was severe competition among the capitalists during the period of industrial capitalism. Every new product that entered the market had to compete with other companies' products to be included in the market. This situation may have caused the prices to decrease and the profits to decrease from time to time.

The "Capital Age" of Hobsbawm ended with the crisis that started with the collapse of the Vienna Stock Exchange in 1873. With the narrowing of the money supply, many banks went bankrupt. The period of 1873–1896 was called the "Great Depression." Profits and prices went upside down. Compared between 1850–1873 and 1873–1890, the annual average growth rate fell at different rates in different countries. In Germany, from 4.3 % to 2.9 %, the USA from 6.2 % to 4.7 % and in the UK from 3.0 % to 1.7 % (Faulkner, 2013: 224–225).

Industrial Capitalism had its first severe crisis, and Marx's "The tendency of the rate of profit to fall" and "Overproduction Crisis" took place. With the crisis in 1873, new trends emerged in industrial capitalism, which Marx noted. The first is "Centralization of Capital," while the other is "Concentration of Capital." In the process that develops with the crisis, new formations began to emerge. In the early twentieth century, social scientists, who analyzed some changes in the characteristic features of capitalism, new formations, and the first crisis, brought essential concepts to the political economy literature. Chief among these is the

Marxist economist Rudolf Hilferding. In his book "Das Finanzkapital[4]," where he examines the monopolization phenomenon of capital based on Germany, he drew attention to the dominant structure of the economy, both by a few big monopolists firms in the industry and the few Berlin Banks. He explained how big banks dominated industrial companies; they used the credit mechanism and the course of investments for their benefit, through the credit mechanism, or by taking over critical shares of joint-stock companies, and, thus, seizing a significant portion of the profits. He named the dominant power of this intertwined bank and industrial capital as "Finance Capital" (Hilferding, 1910).

Vladimir İ. Lenin, in his work titled *Imperialism: The Highest Stage of Capitalism*, compiled from the studies of the leading writers of that period such as Hobson (1902), Luxemburg ([1913] 1986), Hilferding (1910) and Bukharin ([1917] 2009), defined "Imperialism as the monopolistic stage of capitalism" (Lenin [1917] 1999:92).

2.4 Capitalism 4.0: Financialized Capitalism

The "Golden Age" came to an end in the mid-1970s. Until the 1980s, capitalism had an intermediate period with a severe crisis. To get out of this grave crisis, it entered a new era with the 1980s and the beginning of this turn is that the "New-Conservative" parties, who will implement "Neo-Liberal" policies, came to power. These governments initiated intense attacks to rapidly destroy the working class's gains and implemented structural transformation programs in favor of capital. Thus, this phase in which capitalism enters is called as "Financialization process" in this study. In this new period, the focus of capital accumulation shifted from industrial profits to speculative gains in financial markets. Unfortunately, there is no consensus on defining financialization. Here, the definitions of financialization that stand out in the heterodox social science literature are briefly summarized[5].

Epstein (2005: 3) describes financialization as follows: "Increasing the role of financial motives, financial markets, financial actors and financial institutions in the functioning of the local and international economy." Krippner's (2005: 181) defines financialization as "The accumulation model of profit obtained through increasing financial channels rather than profits from commodity production and trade." According to Foster (2010), financialization is the long-term shift of the center of gravity of the capitalist economy from production to finance.

In financialized capitalism, financial markets are at the center of capital accumulation. These markets are the places where the capital of that country operates

---

4 For a broad introduction, criticism, and short life story of the author of Hilferding's "Das Finanzkapital" (1910), see Öztürk (2011).

5 For a broader literature review on financialization, see Koç (2013), Krippner (2011: 1–57), Lapavitsas (2011), Orhangazi (2008: 3–93; 2016: 248–263).

within the borders of a nation-state and global markets in which entrances and exits are incredibly free. Very high octane financial transactions are carried out in these markets, and trillions of dollars change hands every day. Therefore, severe financial profits occur.

After all these explanations, financialization can be defined as becoming an essential element of capital accumulation of increasingly speculative revenues in financial markets in the process of sharing the positive externalities caused by surplus-value created as a result of real production.

2.5 Was Global Crisis Starting in the USA in 2007 and Spreading All Over the World the End of Financialized Capitalism?

The crisis that erupted in the USA's housing sector toward the end of 2007 spread to the whole economy in a short time and turned into a global crisis in 2009.

Arrighi emphasized that in the late 1970s, with the Reagan, the US Systemic Cycles of Accumulation had entered the period of fiscal expansion. Also, this financialization process today – this crisis which started in the subprime[6] the housing market in the USA and influenced the whole world in 2009 – evaluates it as an indication of the final stage of US hegemony and its final collapse (Arrighi, 2010). The evolution of capitalism toward a new stage is mostly after a troubled period after a serious global crisis. The process we are in now is similar to the mid-1970s to the mid-1980s. That is, it remains unclear where this process will evolve.

3 Where Does the Capital Accumulation Model Emerging After the Global Crisis[7] Evolve?

Despite the vast monetary expansion of the US Central Bank (FED) and other large central banks, funding the markets, the state's wing to the markets, the banks' saving their debts by various methods, and the support of the banks and financial capital packages could not be eliminated in 10 years. The capitalist system is trying to find a way out and is going through a process to regain capital accumulation. What will be the inner dynamics of this process?

An anti-capitalist alternative to this form of capitalism whose foundations have been shaken thoroughly does not seem to be in question. However, it is now too early to assess the most significant economic crisis to date, perhaps caused by the breakdown of supply chains, production losses, and the collapse of financial

---

6 It is the name given to mortgage loans given a relatively higher interest rate to consumers who have a low score according to the individual credit assessment criteria: the risk of not being able to pay the loan.

7 For the relationship between the global crisis and financialization, see Akçay and Güngen (2014: 19–106).

markets due to the dangerous spread of Covid-19 virus to all countries. Therefore, the sharpening of capital and the conflicts of interest between countries, the course of global power relations in the world, and, most importantly, the severity, scope, and impact of the struggle that the working masses will develop in line with concrete demands and targets on a global scale will be decisive. However, it is a fact that the system to be established after 2021 will bring radical changes in economic, financial, social, and political terms.

Here is the question to be asked right at this time: "Where is the capital accumulation model evolving after the global pandemic and economic crisis?" Increasing technological innovations since the 1980s shows that there is a radical transformation in every area of life, from production processes, finance, management systems to security, entertainment, culture and arts, media to education, and health. Therefore, in the next period, the concepts that will mark the definition of political systems globally and at the level of countries are most likely "digital capitalism, digital socialism, or digital democracy." Alternatively, the worst will be "digital dictatorship and digital fascism." In this study, the capital accumulation mechanism of digital capitalism is discussed.

4 Capitalism 5.0: Is Digital Capitalism on the Agenda?

Capital accumulation is provided as a result of the mass production of some essential technological products in every period of capitalism. Machinery, weaving looms, trains, and steamboats powered by coal and steam energy formed Industry 1.0. With Industry 2.0, generators, railways, submarines, automobiles, airplanes, and large tonnage ships were produced with electrical energy. Mass production technology started to be applied. Heavy industries based on iron and steel developed. In Industry 3.0, electronic, computer, and communication technologies provided automation. The introduction of the Internet and mobile phones into our lives has deeply affected social life. Therefore, capitalism provided capital accumulation from the technology, which made its mark on the age and the surplus-value obtained from the products produced based on this technology.

Today, technological innovations continue to change at a dizzying pace. It is called the fourth industrial revolution, or Industry 4.0, which expresses digitalization at all levels and marks the time we live now. The innovations in Industry 4.0 are artificial intelligence, robotics, Internet of Things, autonomous vehicles, 3D printing, nanotechnology, biotechnology, materials science, energy storage, and quantum computing. All these new technological breakthroughs are surprisingly intertwined in various fields and are expected to bring radical changes in economic, social, political, and cultural structures (Schwab, 2016: 9).

Considering the capital accumulation mechanism that will occur with these technological innovations, which may be likely to imprint in the coming period, and the social-class relations that will form around it, this stage of capitalism can be called "digital capitalism." The concept of digital capitalism was first used by Schiller (2000). The concept of "digital capitalism" in the neo-Marxist approach, to name the last period of capitalism's evolution, namely, the "new" form of capitalism, has been developed. Various theories have been developed in this context. The most common of these theories are Digital Capitalism Theory, Autonomist Theory, Cognitive Capitalism Theory, Transnational Informational Capitalism Theory, and Cognitive Materialism Theory. According to Schiller's theory of digital capitalism, digital capitalism represents a "phase change" within the history of nearly 500 years of capitalism. Schiller argued that as a result of the capitalist system entering a structural crisis after the 1970s, digital capitalism was designed as an essential part of the global "capitalist restructuring" strategy to exit this crisis (Akçoraoğlu, 2019: 530–531).

Moulier-Boutang (2011: 47–48) claims that "Cognitive capitalism," which developed based on the accumulation of immaterial capital, the spread of knowledge, and the knowledge economy, emerged from the last quarter of the twentieth century (as cited in Akçoraoğlu, 2019: 536). Jeon (2010: 90) explains that "according to the theory of cognitive capitalism, advanced capitalist centers of the world have transformed into a new form of capitalism as a result of the profound developments in the last 30 years" (as cited in Akçoraoğlu, 2019: 536). However, the period that marked the last 40 years of capitalism is when capitalism restructured under the guidance of neo-liberal ideology and financialized and reached a global dimension. Of course, the technologies that led to digitalization applications, which are becoming increasingly common today, were sown in the 1970s and 1980s. However, this does not indicate that "digital capitalism" started in the 1970s or 1980s. Technological innovations such as Industry 4.0[8] and digital finance (Fintech), which will be at the center of the accumulation of capital, and new generation products, institutions, and production relations produced with these technologies will form digital capitalism. Therefore, this is what distinguishes this study from other "digital capitalism" theories. In this study, the author suggests that the period of "digital capitalism" will begin after the 2020 global crisis.

Fuchs (2018) explains digital capitalism in the context of the transnational digital media companies making high profits as a result of the international division of labor. In this division of labor; labor forms are paid labor, unpaid labor, reproductive labor, and users' digital labor. Transnational digital media companies try to maximize capital accumulation by minimizing labor.

How will the functioning of digital capitalism, working conditions, and social-class structure, property, and relations of production, new exploitation relations created by the formation of surplus-value and new capital accumulation mechanisms? Can the conversion to "digital socialism" or "digital democracy" occur due to the dialectical relations of social dynamics? What could be their conditions? Or what is the danger of "digital dictatorship" or "digital fascism" as the worst possibility?

---

8 "Industry 4.0," a term coined at the Hannover Fair in 2011 in Germany. For more detailed information, see Schwab (2016).

All similar problem areas of the new social life will be the discussion topics of social scientists in the coming years.

4.1 Capital Accumulation Channels of Digital Capitalism

In this section, the possible capital accumulation channels of digital capitalism that are emerging in this process that we are going through will be tried to be determined, and awareness will be raised for the discussion areas of the future by asking questions about the problems that may arise in these fields.

4.2 Traditional Sectors

As can be seen in the previous forms of capitalism, traditional capital accumulation channels of previous periods will continue to live even if they lose their relative importance. It will be possible to see them in all areas of the economy. The traditional capital accumulation model will continue to operate in agriculture and many sub-sectors of the industry and all service sectors.

4.3 Digital Production (Industry 4.0)

Digital production, called Industry 4.0, covers a vast area and the topics included are: smart factories, Internet of Things (IoT), big data and artificial intelligence, intelligent robots or advanced robotics, self-driving cars(smart cars) or driverless cars, digitally driven organizations, autonomous vehicles, 3D printing, new materials, and smart cities. With the digitalization of production, changes can be expected in production relations. Here are the questions and problem areas related to the possible developments expected in these areas:

How will the previously produced goods and their production methods be affected by these changes, as new generation products enter the market quickly? How will the new generation marketing models match the personal preferences and customer purchasing conditions be in harmony with the old classic models? How much longer can the older generation production and marketing businesses continue to live?

With blockchain technology, can essential variables such as production, supply chain, demand, consumption, sales, inventory control, etc., which are at the center of the economy, become more transparent? Can planning and crisis management become more productive? Since this digital transformation in production processes will also cause a change in cost structure, what effect will this have on the prices of goods and services?

What impact will this digital transformation on manufacturing processes have on the following variables: new investment and new company models, employment structure, labor-capital relations, property, ownership of means of production, new contradictions, and conflicts of interest in intra-capital relations, profit formation, and sharing?

It is expected that the market share of smart cars and vehicles, which will work with new generation fuels, which can be used on land, air, and water, is increasing. As a result, where will the economic, social, and political contradiction and conflicts with the old generation industrialists, the representatives of new sectors, have increased in importance due to the decrease in production and profit, to evolve?

With which main criteria will smart freeways and smart cities using these vehicles be created? What new advantages will these provide, or what new problems will they create? The decrease in the price, production, and profit will result from the gradual decrease of the relative importance of oil, which is the traditional energy source, in the new economy. How will it affect reliable oil companies and countries? Where will the economic, social, and political contradiction and conflicts between these companies and capital groups investing in new generation energy resources evolve?

These examples can be considered for any industry where digital production will become increasingly widespread.

4.4 Digital Finance

Digital capitalism has made its fastest development in the financial sector to date. Rapidly spreading credit cards after the 1990s, Internet banking, the fact that all exchanges are made electronically, and e-commerce applications were essential steps in the transition to digital capitalism. However, the leading critical development is that Bitcoin operates on blockchain technology and cannot be controlled by any central authority (States, central banks, etc.) and which emerged in the 2009 global crisis and has survived to the present day successfully. The rapid development of Bitcoin[9] in this process increased its recognition and increased the demand for this digital asset with limited supply. In the cryptocurrency exchanges formed in this process, excessive volatilities in Bitcoin price started. The market's high price volatility has been accepted as a new financial instrument by the speculator, as it is very convenient for speculative gains. The number of people who trade Bitcoin in these exchanges, especially the younger generations, has increased gradually.

The fact that its supply was limited, its infrastructure was based on a robust technology such as blockchain, and the fact that its launch coincided with the period when most central banks, especially the American Central Bank (FED), offered excessive liquidity to the market after the global crisis caused severe increases in its price.

9 For more detailed information, see Çarkacıoğlu (2016).

	29.03.13	3.01.14	2.01.15	1.01.16	2.01.17	17.12.17	1.01.18	3.01.19	10.01.19	2.01.20	14.02.20	16.03.20
Bitcoin Price(USD)	142	825	314	430	1018	19842	13277	3893	13012	7150	10248	4882
Market Cap. Million USD	1581	10061	4305	6466	16374	332315	222740	67981	231758	129680	186691	89213
Daily Volume (Million USD)	0	52,6	6,9	42,1	208,8	14784	10169	4900	27300	19338	43699	43872

Tab. 1: Price and market value of Bitcoin (2013–2020). Source: The data are taken from Coinmarketcap (2020)

4.5 Cryptocurrency Market Data

According to Coinmarketcap.com data on 22.04.2020, there are 5392 different cryptocurrencies in the world, traded on 21590 different exchanges. The total daily trading volume in the crypto money market is 121.6 billion USD, and the total market value of all cryptocurrencies is approximately 206 billion USD. Bitcoin's market value, which ranks first in terms of market value, is 130.7 billion USD and corresponds to 63.4 % of the entire crypto currency market. Daily trading volume is approximately 33 billion USD. Ethereum is in second place with a market value of 20.3 billion USD and a daily trading volume of 17.8 billion USD. The third-ranked XRP has a market value of 8.3 billion USD and a daily trading volume of about 1.8 billion USD. As can be seen, Bitcoin alone constitutes 63.4 % of this market, while all other Altcoins only have 36.6 %.

If the traditional money and capital market instruments (stock, bond, and exchange rates), which have hundreds of trillion market volumes, are compared to the cryptocurrency markets, although the cryptocurrency markets have reached a high number in terms of diversity, it seems to be quite undersized in terms of market value and daily trading volume.

Tab. 1 shows the price of Bitcoin in USD between 2013–2020 and its total market value and daily trading volume. In these eight years, it is observed that it has volatility in terms of all three variables. While 1 Bitcoin was 142 USD in 2013, it reached 1018 USD in early 2017. It showed a very rapid rise in 2017 and approached 20,000 USD at the end of the year. It declined in 2018 and fell to 3900 USD at the beginning of 2019. After rising again to 10248 USD in the first two months of 2020, it decreased to 4,000 USD levels again with the effect of the Covid-19 pandemic. As of the fourth quarter of 2020, when this study is about to be published, according to Coinmarketcap.com data, there are 7970 different cryptocurrencies in the world, and they are traded on 34213 different exchanges. The total daily trading volume in the cryptocurrency market is 101.3 billion USD, and the total market value of all cryptocurrencies is approximately 562.3 billion USD. The total market value of Bitcoin, which ranks first in terms of market value, is 355.3 billion USD, corresponding to 63.4% of the entire cryptocurrency market. Ethereum is in second place with a market value of 66.7 billion USD and a daily transaction volume of 9.6 billion USD. Bitcoin price also increased rapidly and reached around nineteen thousand USD. Despite this volatility in bitcoin prices, there is still considerable interest, especially among the younger generation. It is also seen that the interest of institutional investors is increasing as of the second half of 2020.

4.6 Is There a Difference Between the Central Bank's Digital Moneys and Cryptocurrencies?

In recent years, the central bank officials of many countries, including the International Monetary Fund, have made a statement that they have been working on issuing digital money. Even the Covid-19 pandemic, which emerged in China in the last months of 2019 and spread worldwide quickly, can speed this process up. It is predicted that the digital money to be issued by the central banks will not significantly change the capitalist system's money-created mechanism; it will reduce the use of only physical (cash) money with the current estimates and gradually destroy it. Thanks to the monopoly power of the central banks arising from the power to print money, the seigniorage revenues, and the power to implement the monetary policy continue.

However, cryptocurrencies working on blockchain technology threaten this feature of the system. Cryptocurrencies, whose money supply cannot be controlled by a particular center, such as Bitcoin, which has limited supply, and the amount of which to be supplied to the market depends on a particular algorithm, are not desirable by governments and central banks. In times of financial and economic crisis, governments want to prevent the collapse of the financial system by allowing the economy to expand by printing unrequited money. This monetary expansion practice (quantitative easing) was widely used in the period of financialized capitalism.

Likewise, the emerging concept of decentralized finance (Defi) is a candidate to be a serious alternative with the convenience it offers to the banking system's applications such as money transfer, borrowing, deposit creation, and the use of credit and the solutions it brings to high costs.

In the light of the practices so far, it is not known what the attitude of nation-states, especially the USA, will be against Bitcoin and other cryptocurrencies and decentralized finance applications. In the early years of Bitcoin, Bitcoin bans existed in some countries due to some illegal use in the underworld (such as money laundering). However, despite this, the increasing interest has increased its trade like a commercial asset. Therefore, instead of introducing a ban in the future, some legal regulations regarding its use are expected.

Therefore, considering this feature, it seems quite difficult for an Altcoin that offers Bitcoin or similar decentralized finance applications to be "A new money" that can be used in daily life on a global scale. However, thanks to the volatility and increasing interest in the coming years, it is possible that digital capitalism will function as an alternative capital accumulation tool. However, the condition of performing this function should not cause any of the following problems. These are lack of trust in the system, a deficit of security in blockchain technology, the collapse of its technological infrastructure, the double-spending problem, and the new technological innovations that will enable it to be controlled by a particular center.

If cryptocurrencies continue to live, can it be the dissident or democratic face of digital capitalism? What kind of problems and questions are on the agenda in digital finance, which is likely to mark the coming period? Here are some ofthem:

Will legal regulations be made within the scope of the principles and standards to be determined on an international scale in the following subjects? These are the production of cryptocurrencies, storing and insuring crypto, or digital money, placing them on the market, operating principles of institutions such as stock exchanges, brokerage houses, new generation banks. Will new institutional structures be created on an international scale to take responsibility in this area? How will the relationship between these global institutions and nation-states be? How will the duties and responsibilities, working conditions, and wages of employees working in institutions in the cryptocurrency sector be determined? How will the taxation of all earnings in the crypto money sector be made? How will the inheritance process be realized with digital assets and cryptocurrencies?

4.7 Digital Law

How to eliminate the shortage of police, prosecutors, judges, and lawyers who are experts in combating illegal activities such as crypto money theft, digital fraud, terrorism, and the use of crypto money in drug trafficking? How will the standards of digital law, including digital economy, be determined? How will the international criteria for digital crimes and penalties be formed? Is it known to public authorities that delays in this matter will cause serious problems? What kind of preparations are there in this regard?

4.8 Health

The health system is a candidate to be the most critical capital accumulation channel of digital capitalism. The new health system is primarily based on personal health data; many other developments such as drugs and vaccines developed against global virus outbreaks, new methods of diagnosis and treatment of artificial intelligence robots, patient-specific drug production, and organ production with three-dimensional printers will increase the rapid commercialization in the field of health and will be an essential area of capital accumulation.

Another problem is to move away from social state practices. In other words, the health system is developing in a direction dominated by market and profit-oriented private companies. In this case, class inequalities will increase, and the poor and those not included in the system for various reasons will have difficulty accessing health; therefore, their right to life will be under threat.

What advantages and disadvantages can organ production create with three-dimensional printers, which will play a significant role in the digitalization of production? Will it increase the commercialization process that starts in health? How much is the risk of monopolization of production in this field? Will the emergence of emerging virus outbreaks (Corona Virus, etc.) create monopoly power and large profit areas for biotechnology companies requiring a large amount of R & D investment in diagnosis, protection, and treatment? Do not genetic changes made primarily on cancer and human embryo through new generation technologies cause more health commodification? Moreover, future unethical/illegal interventions cause severe problems for the future of the human generation? How much can society benefit from these treatment methods?

4.9 Environment

What will be the impact of the new digital production on the environment? Will new profit areas arise as a result of transforming carbon emissions into global goods due to global warming? Will the commodification of fresh air, water, soil, and natural food resources accelerate?

4.10 Security

Will war equipment and digital security systems produced by leading institutions/ companies of digital technology constitute a monopoly power for these companies? How will these global companies, which produce strategic equipment and weapons, have relations with the states? What kind of problems will the use of new generation robots working with artificial intelligence in terrorism and wars cause?

One of the most sensitive issues is the threats to be created by recording all critical information about individuals from birth to death. That is, it is the risk of critical information being used against people. While this information functions in one's favor in some cases, it also involves risks such as blackmail and being discriminated against and is considered a potential criminal.

4.11 Media and Entertainment Sector

Where will digital media tools, digital culture-art products, digital entertainment industry evolve? With the emergence of block chain-based new media tools, will it be an alternative that solves the problems such as controlling (monopolizing) traditional media from a center, censoring, and capital control? Can art and culture be more creative and freer, thanks to blockchain technology? In this way, will they obtain the rights they deserve from the works created by all artists by eliminating the exploitation of the intermediary?

4.12 Education

Are there any training plans for training the workforce, such as managers, experts, intermediaries, and educators, who will have a great need in all fields, especially digital finance and cryptocurrencies? How will the new vocational education meet the rapidly changing needs? Will there be no "school-desk-teacher-blackboard," that is, the traditional school system? Will digital media schools be formed? How will rapid transformation in education be achieved to learn new digital professions as a result of the rapid disappearance of many professions? Will commodification in the education sector accelerate as in health? In case of moving away from social state practices, as the education sector gradually becomes more privatized by decreasing free weight, just like in health care, the replacement of compulsory and free public education by expensive and private education institutions will cause severe inequalities in the society. Children of families who cannot enjoy the right to a good education have the risk of turning into slaves like in the past.

All these problems will be joint discussion, problem, and solution areas of many disciplines, especially political economy in the coming period.

Conclusion

In this study, the continuous feature of capitalism, which has a history of about six centuries, is that it always progresses by creating capital accumulation as a result of production relations based on private ownership of the means of production. However, it has come to this day by adapting itself to many changes in this process. Of course, it is known how important the first industrial revolution was in the emergence of capitalism. Likewise, the subsequent technological revolutions have also enabled it to survive by creating new products, new profit areas, and new class and production relations, changing its form with new infrastructure and superstructure institutions.

In this study, various questions were asked in light of current discussions about the new capital accumulation channels of digital capitalism and the problems in these areas. Since these listed questions are still in the maturation stage, it is not easy to answer them yet. It is already out of the scope of this study.

After the last two global crises that took place 12 years apart, capitalism is rapidly evolving toward "digital capitalism," which we call 5.0 by changing form. In this process, it will be essential to follow the developments in the USA and China carefully, which are two global powers because the size of the hegemonic power struggle between the USA and China in the coming period will have an essential place in the evolution of this process.

Probably, as in their previous forms, it will be possible to see different versions of digital capitalism in different countries in the coming years. Some countries will adapt to digital capitalism quickly, while others will take a long time to adapt. Some will go into more extreme applications, while others will prefer more hybrid applications. These differences will be determined by the socioeconomic, demographic, and cultural characteristics; historical experiences; and political preferences of the countries. All humanity will experience this process carefully.

Bibliography

Akçay, Ümit and Ali Rıza Güngen, (2014), *Finansallaşma, Borç Krizi ve Çöküş: Küresel Kapitalizmin Geleceği*, İstanbul: NotaBene.

Akçoraoğlu, Alparslan, (2019), "Yeni Kapitalizm Teorileri, Dijital Devrim ve Türkiye Kapitalizmi", *Mülkiye Dergisi*, Vol. 43, N. 3, pp. 525–575.

Arrighi, Giovanni, (2010), *The Long Twentieth Century: Money, Power, and the Origins of Our Times*, London: Verso.

Blockchain Türkiye, (2019), *Blokzincir Teknolojisi Terminoloji Çalışması*, İstanbul: Türkiye Bilişim Vakfı.

Bukharin, Nikolai, (1917), *Imperialism and World Economy*, Retrieved from https://www.marxists.org/archive/bukharin/works/1917/imperial/index.htm(10.04.2020)

Coinmarketcap, (2020), https://coinmarketcap.com/currencies/bitcoin/ (15.04.2020)

Çarkacıoğlu, Abdurrahman, (2016), *Kripto Para Bitcoin*, Sermaye Piyasası Kurulu Araştırma Dairesi Araştırma Raporu. Ankara: Türkiye

Epstein, Gerald A., (2005), "Introduction: Financialization and the World economy", (Gerald A. Epstein, Ed.), *Financialization and the World Economy*, Cheltenham: Edward Elgar, pp. 3–17.

Faulkner, Neil, (2013), *A Marxist History of the World: From Neanderthals to Neoliberals (Counterfire)*, London: Pluto Press.

Foster, John B., (2010), "Financialization of Accumulation", *Monthly Rewiev: An Independent Socialist Magazine*, Oct 01, 2010. https://monthlyreview.org/2010/10/01/the-financialization-of-accumulation/ (10.02.2020)

Foster, John B., (2007), "The Financialization of Capitalism", *Monthly Review: An Independent Socialist Magazine*, Vol. 58, N. 11, https://monthlyreview.org/2007/04/01/the-financialization-of-capitalism/ (10.02.2020)

Fuchs, Christian, (2018), "Capitalism, Patriarchy, Slavery, and Racism in the Age of Digital Capitalism and Digital Labour", *Critical Sociology*, Vol. 44, N. 4–5, pp. 677–702.

Hilferding, Rudolf, (1910), *Finance Capital- A Study of the Latest Phase of Capitalist Development*, https://www.marxists.org/archive/hilferding/1910/finkap/ (10.04.2020).

Hobson, John A., (1902), *Imperialism: A Study*, New York: James Pott & Co.

Jeon, Heesang, (2010), "Cognitive Capitalism or Cognition in Capitalism? A Critique of Cognitive Capitalism Theory", *Spectrum: Journal of Global Studies*, Vol. 2, N. 3, Summer 2010, pp. 90–117.

Koç, Abdilcelil, (2013), *Dünyada Finansallaşma Süreci ve Türkiye Deneyimi*, (Unpublished Doctoral Dissertation), Çukurova University, Institute of Social Sciences, Adana: Türkiye.

Krippner, Greta R., (2005), "The financialization of the American Economy", *Socio-Economic Review*, Vol. 3, N. 2, pp. 173–208, DOI: https://doi.org/10.1093/SER/mwi008 (15.02.2019).

Krippner, Greta R., (2011), *Capitalizing on Crisis: The Political Origins of the Rise of Finance*, Cambridge, MA and London, UK: Harvard University.

Lapavitsas, Costas, (2011), "Theorising Financialization", *Work, Employment and Society*, Vol. 25, N. 4, pp. 611–626.

Lenin, Viladimir İ., ([1917] 1999), *Imperialism, the Highest Stage of Capitalism*, Australia: Resistance Books. Retrieved from https://www.readingfromtheleft.com/Books/Classics/LeninImperialism.pdf (15.02.2019).

Luxemburg, Rosa, (1913), *The Accumulation of Capital*, Retrieved from https://www.marxists.org/archive/luxemburg/1913/accumulation-capital/index.htm. (156.02.2019).

Marx, Karl, ([1867] 2011), *Kapital: Ekonomi Politiğin Eleştirisi*, 1.Cilt, (Çev./Trans.) Selik, Mehmet and Satlıgan, Nail, İstanbul: Yordam.

Moulier-Boutang Y., (2011), *Cognitive Capitalism*. (Emery, E., Trans.), English Edition, Cambridge, UK: Polity Press.

Orhangazi, Özgür, (2008), *Financialization of the US Economy*, Cheltenham, UK and Northampton, MA: Edward Elgar.

Orhangazi, Özgür, (2016), "Contradictions of capital accumulation in the age of financialization", (Turan Subasat, Ed.), *The Great Financial Meltdown: Systemic, Conjunctural or Policy Created?* Cheltenham, UK and Northampton, MA: Edward Elgar, pp. 248–265.

Öztürk, Özgür, (2011), "100. Yılında Das Finanzkapital", *Devrimci Marksizm: Üç aylık Teorik / Politik Dergi*, Sayı:12, pp. 133–142.

Rodrik, Dani, (2012), *The Globalization Paradox: Why Global Markets, States, and Democracy Can't Coexist*, Oxford, UK: Oxford University Press.

Schiller, Dan, (2000), *Digital Capitalism*, Cambridge, MA: MIT Press.

Schwab, Klaus, (2016), *The Fourth Industrial Revolution*, World Economic Forum, Switzerland.

Tongyu, Zhang and Weimin, Ding, (2012), *Sermayenin Tarihsel Diyalektiği ve Marx'ın Toplumsal Sermaye Teorisi*, (C. Kızılçeç, Der.), (Çev./Trans.) Üçoluk, E., ve Muhaddisoğlu, A., İstanbul: Kalkedon.

Usta, Ahmet and Doğantekin, Serkan, (2016), *Blockchain 101*, İstanbul: BKM, Bankalararaarası Kart Merkezi.

Wright, Erik Olin, (2009), "Marksist Birikim ve Bunalım Teorisinde Almaşık Perspektifler", (Nail Satlıgan ve Sungur Savran, Der.), *Dünya Kapitalizminin Krizi*, 2. Baskı, (Çev./Trans.) Çakır, N., Satlıgan, N., Kayaalp, Y., Arın, T., İstanbul: Belge.

Subhasree Basu ROY and Daniel W. NGUYEN

Digitization in Health Care: A Global Perspective

1 Introduction

Telemedicine, also commonly referred to as telehealth, is defined by the World Health Organization (WHO) as:

> *The delivery of health care services, where distance is a critical factor, by all health care professionals using information and communication technologies for the exchange of valid information for diagnosis, treatment and prevention of diseases and injuries, research and evaluation, and for the continuing education of health care providers, all in the interests of advancing the health of individuals and their communities (World Health Organization [WHO], 2010).*

The rise of telemedicine has corresponded with the digitization of the world economy as more nations become comfortable with the sharing and storing of data digitally. Health care professionals have taken advantage of the widespread digitalization by making use of new technology to store patient data and to perform remote consultations. This use of telemedicine has reduced traditional barriers to health care access such as time and distance. Many patients now have the option to consult with a medical professional on demand any time day or night, and hospitals are better equipped to track patient data through online storage systems.

The changes in global health care have resulted in previously insurmountable barriers being broken down by new technology. Barriers based upon gender as well as disparities in health care access between nations have long been challenges to health care systems around the world. In this chapter, we explore how gender barriers to health care access have been overcome by telemedicine, and we discuss the role of international cooperation in the adoption of telemedicine around the world. Next, we reflect on the role of telemedicine during the recent COVID-19 (coronavirus) pandemic. Finally, we look to the future of telemedicine adoption and the challenges that remain.

2 The Role of Gender in Telemedicine

Gender roles can have a significant impact on the delivery of health care in communities undergoing the effects of globalization. In 2005, the Commission on Science and Technology for Sustainable Development in the South (COMSATS) launched a telemedicine project in northern Pakistan in order to increase access to health care services and education in rural areas (Khan & Hayee, 2009). The program aimed to increase the standard of living in the Baltistan region of Pakistan, and it relied heavily on an understanding of the cultural norms and practices of the area.

Women in rural Pakistan typically face limitations in social interaction, which hinders their ability to seek out care. In the absence of telemedicine, women must travel long distances to receive consultation and are dependent upon the availability of their male family members to accompany them (Khan et al., 2009). This kind of limitation can discourage women from receiving care by limiting their ability to travel to a health care facility and eliminating their privacy when there. In order to overcome this cultural barrier to health care services, it was imperative that COMSATS partner with a regional player, the Baltistan Health and Education Foundation.

Prior to the introduction of COMSTATS into northern Pakistan, the Baltistan Health and Education Foundation was dedicated to improving health care services for women in the rural Baltistan regions of Pakistan. The Baltistan Health and Education Foundation overcame gender barriers by working alongside community members and introducing technology that could address their concerns. While community members had access to landlines, they were often of poor quality and expensive for villagers to use. COMSTATS addressed this by establishing an internet service provider to connect rural areas to medical specialists in Islamabad (Khan et al., 2009).

The results had the desired effect of increasing access to care for women in the villages by eliminating travel barriers. Women accounted for 61 % of those utilizing the telemedicine to speak with specialists in Islamabad, and the program had a 15–20 % rate of follow-up cases, which indicated trust in the service. Within the first four months of the program, a total of 361 patients had received specialist consultation via telehealth. This was accomplished without a promotional campaign and instead relied on the program being spread by word of mouth (Khan et al., 2009).

The COMSTATS program is an example of how telehealth can address the gender specific barriers to health care consultation. Similar barriers exist around the world in various cultures and economies. In Saudi Arabia, women are forbidden by law from being admitted into a government hospital without permission from a male, and in some areas of the country a male relative may prevent a woman from consulting with a male medical professional even during an emergency (Mobaraki & Soderfeldt, 2010). This myriad of gender-specific laws prevent equitable access to health care and can exacerbate the greater travel distances that are existent in rural areas.

Access to health care also depends upon the availability of certain services at health care facilities, and populations with gender-specific health care needs may need to travel long distances to receive care. A relevant example is found with women veterans in the rural United States. At least 25 % of women veterans live in rural communities, and many of them are reliant upon a Veterans Administration (VA) clinic to receive health care. This reliance on the VA coupled with the fact that

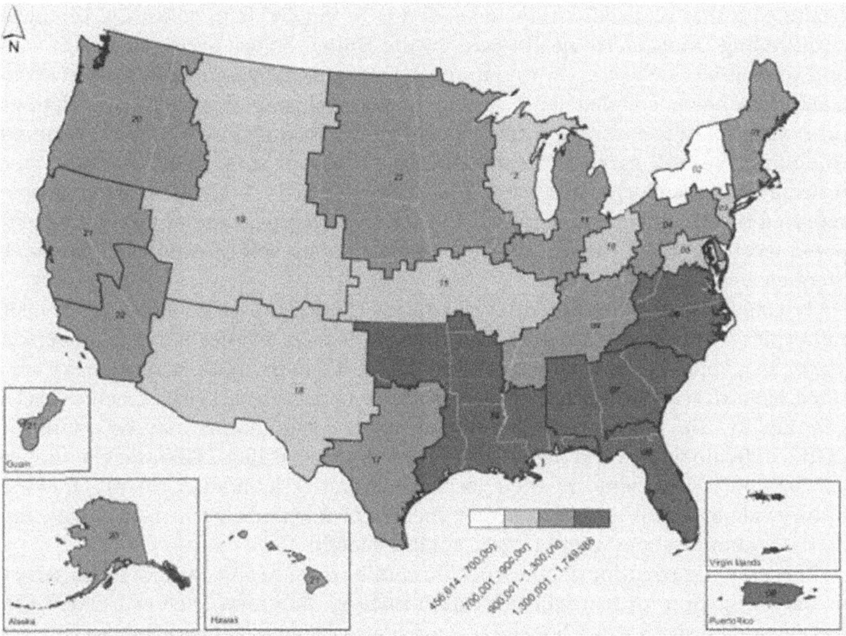

Fig. 1: Veteran Population by Veterans Integrated Service Network (VISN): Fiscal Year 2015. Source: (National Academies of Sciences, Engineering, and Medicine, 2018)

most US veterans are concentrated in rural, southern states can present unique challenges for demographics that have unique health care needs. Refer to Fig. 1.

Although the number of women veterans continues to grow, the United States military is still a majority male. As a result, the density of women at any given VA clinic can be small relative to the men (Goldstein, Giersch, & Zullig, 2017). This could have impacts on the ability of women veterans to receive timely and cost-effective care since the VA is still serving a largely male population and is only beginning to experience an influx of women veterans. While the VA has addressed this issue by expanding its ability to offer gynecology and maternity care to women veterans, longer travel times still have the potential to interfere with a patient's ability to receive care. Initiatives such as the Virtual Integrated Multisite Patient Aligned Care Team seek to address this by providing remote consultation to women veterans at VA clinics that lack women's health providers (Hayes, 2019). As the patient population of VA clinics becomes more diverse, telemedicine can be used to bridge the gap in care as clinics adjust to a new list of patient needs.

The growing, diverse list of patient needs in rural clinics also has an impact on the transgender population. Transgender individuals often face difficulty receiving health care that is specific to their needs due to the divide in cultural acceptance surrounding transgender health care. In the United States, access to health care and regulations regarding hormone replacement therapy can vary from state to state. This being coupled with certain doctors refusing treatment on religious or moral grounds leads many transgender individuals to travel farther distances to clinics that will agree to treat them. In a study of rural LGBT (lesbian, gay, bisexual, transgender) health care, it was found that 14 % of transgender people reported traveling more than an hour to see their primary care physician in compared to only 5 % for the rest of the cisgender population (Whitehead, Shaver, & Stephenson, 2016).

Telemedicine has the potential to address the longer travel times for gender minorities by providing remote video consultations as an alternative to in-person visits. This could be especially important for LGBT individuals in rural areas who often have to travel to urban centers in order to access an LGBT-specific health care facility. The ability to utilize remote video consultations may be useful for LGBT individuals in rural areas since discussion regarding LGBT-specific issues can be considered taboo in their local communities (Landman, 2018). Drawing upon evidence regarding the current success of telemedicine in rural areas, the LGBT community may stand to gain similar benefits.

The growing need for gender-specific care in rural health care facilities serves as the foundation for many health care initiatives. Programs such as COMSTATS has shown the success of telemedicine in remote communities with gender barriers to health care. Similar issues are found outside of Pakistan in economies of various development and cultures with different norms. The growing number of women in the US military and the growing number of transgender individuals seeking specialized care are populations that could stand to benefit from telemedicine in the same way that women in rural Pakistan benefited. In the age of digitalization, similar gender barriers across the globe can find similar solutions with shared technology.

3 International Cooperation

In December 1988, a major earthquake occurred in the former Soviet Union that devastated the medical infrastructure of Armenia (Johnson, 2015). It was a unique time in world affairs as the United States and the Soviet Union were involved in a tense Cold War that had almost ended in nuclear calamity in 1963. At the time, grassroots organizations were attempting to address misconceptions and promote cooperation between the two governments. "Soviets, Meet Middle America" was a program established by the Center for Citizens Initiative where 400 citizens of the Soviet Union visited hundreds of cities in the United States. The goal of the program was to create "cultural ambassadors" out of Soviet citizens who would then return to the Soviet Union and debunk myths about the American way of life (Koshkin, 2018). This informal, behind-the-scenes diplomacy was a way for two populations to cooperate with one another before their governments were prepared to do so.

By the time the earthquake took out Armenia's medical infrastructure, an independent agency of the United States known as the National Aeronautics and Space Administration (NASA) had already developed a Communications Technology Satellite (Dunbar, 2020). In response to the destruction, NASA coordinated their efforts with Armenia to link medical centers in the United States, Yerevan, Armenia, and Moscow together through a telecommunications connection called Spacebridge. Spacebridge was used to provide consultation to those affected by the earthquake, and the system was later used to treat children that were injured in a train explosion in Ufa, Russia (Johnson, 2015).

This moment in history represented a stark departure from the near disastrous events of the Cuban Missile Crisis and the Berlin Airlift that preceded it. By way of new technology in an emerging digital age, two countries that were in an undeclared Cold War were capable of working together and saving lives in a time of natural disaster. The use of telemedicine in Soviet–American relations was an example of how two countries that were on opposing sides of world events could come together for the purpose of improving health care.

That same cause has motivated European nations to unite with their former African colonies to deliver health care via telemedicine to isolated populations. In the former French colony of Mali, Geneva University Hospitals launched a telemedicine program known as RAFT (Reseau en Africue Francophone pour la Telemedicine) to webcast interactive courses to medical professionals in the region (WHO, 2011). The program serves as an echo of intertwined histories, as the courses are broadcasted to French-speaking African nations whose use of the French language is itself a lasting detail of their former status as a French colony.

The use of telemedicine in the region arises out of a chronic need of medical expertise and lack of health care infrastructure. Africa has been experiencing a "brain drain" of medical professionals to Europe and America, which has put a strain on its capacity to deliver health care to rural populations (WHO, 2011). In addition, health care workers in sub-Saharan Africa tend to be clustered in urban centers, which exacerbates the already existing problem of access in rural areas by creating a sense of "professional isolation" for rural medical professionals. Without access to the Internet and suitable roads for travel, many of these doctors remained isolated from their peers in urban centers (Bediang et al., 2014).

RAFT attempts to alleviate this issue by providing teleconsulting services between medical professionals in Europe and medical professionals in rural sub-Saharan Africa. RAFT has designed its program to run on unstable and low-bandwidth connections and has made use of text messages as a reliable form of communication. In addition, the video lectures that RAFT produces to provide education for local doctors is converted to HTML5 so that rural practitioners can access the videos regardless of web browser (Bediang et al., 2014).

This use of telemedicine has addressed "professional isolation" by giving doctors in sub-Saharan Africa access to training, research, and development with health

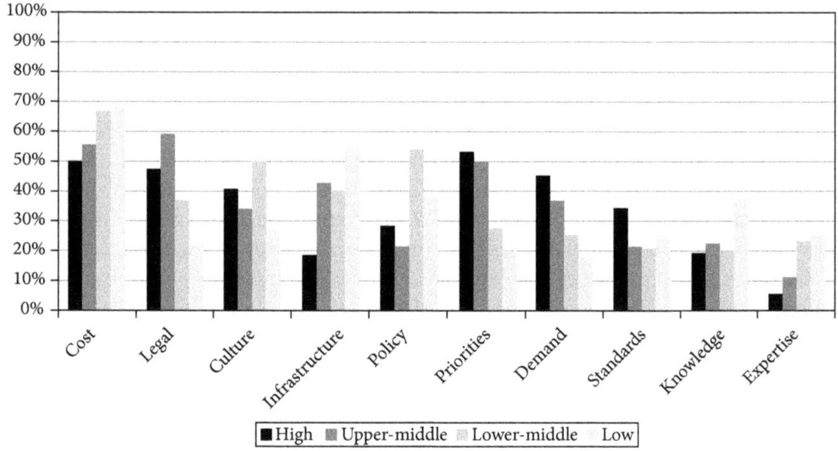

Fig. 2: Barriers by World Bank income group. Source: (World Health Organization – Opportunities and developments in Member States, 2010)

care professionals around the world. Each year, doctors from Geneva University Hospitals visit the area to perform surgeries, train local health care workers, and develop the telemedicine program (WHO, 2011). This approach of both personal and digital consultations creates a network of medical professionals that can better treat patients in areas that are slowly losing health care workers. By creating a global community dedicated to patient care, vulnerable regions are better positioned to adapt to issues of "brain drain" and "professional isolation."

Other nations have built upon the success of RAFT by developing their own telemedicine capabilities. In Burkina Faso, an eHealth division has been created within the Ministry of Health and is led by the local RAFT coordinator. Similarly, in Niger, Mauritania, and the Republic of Congo, governments are using RAFT as a model to implement their own telemedicine program (Bediang et al., 2014).

Challenges still exist for the nations that wish to create a robust and effective program for telemedicine. Cost remains the most prevalent barrier to the adoption of telemedicine by low and lower-middle income countries according to survey data by the WHO (2010, p. 72). Refer to Fig. 2.

Significant cost barriers may cause nations to seek foreign aid in order to build the necessary infrastructure, and the issue of profitability for private entities is a concern in sparsely populated areas. The absence of economies of scale is a problem with the expansion of telemedicine and one that may be solved with greater government involvement or "national clusters" that aggregate patients to justify the costs (Bediang et al., 2014).

Another example of international cooperation between two nations with intertwined histories is the United States and Cambodia. The Vietnam War brought millions of Americans into Southeast Asia and entangled the United States in a conflict that many dubbed a "quagmire." While the war in South Vietnam garnered the most media attention, the United States was also involved in stopping the rise of Communism in the neighboring country of Cambodia. Under President Richard Nixon, the US government aided Cambodian president Lon Nol in his fight against the Khmer Rouge and sent combat troops into Cambodia in 1970 to dislodge the North Vietnamese Communists (Karnow, 1984). The effort ultimately failed on both counts, as the Khmer Rouge took over Phnom Penh in 1975 and the North Vietnamese captured Saigon in the same year.

The reign of the Khmer Rouge in Cambodia brought about the death of millions and left behind a decimated health care system that was ill equipped to handle the host of issues following one of the worst genocides of the twentieth century. In 1979, it was estimated that Cambodia had less than 50 doctors left in the country after the mass killing of intellectuals during Pol Pot's reign (Heinzelmann, Chau, Liu, & Kvedar, 2009). This left the health care system overwhelmed with infectious diseases, malnutrition, and psychological disorders, which left many Cambodians relying on traditional medicinal practices as a result.

Fast forward a few decades, Cambodia is facing many of the same issues that sub-Saharan Africa faces. Most health care workers are clustered in urban centers despite 85 % of Cambodians living in rural communities. While foreign assistance in the 1990s expanded internet access in Cambodia, only 2 out of every 1,000 citizens became Internet users by 2002 (Heinzelmann, Chau, Liu, & Kvedar, 2009).

In 2001, Operation Village Health was launched with the goal of increasing the self-sufficiency of the Cambodian health care system. The initiative was a project of the Center for Connected Health of the Partners HealthCare system in Boston and the Sihanouk Hospital Centre of HOPE in Phnom Penh and incorporated the work of foreign and local health care professionals (Heinzelmann, Chau, Liu, & Kvedar, 2009). With the history of Cold War conflict in the background, the goal of the American and Cambodian doctors was to bridge the digital divide in a war-torn nation.

The program ran into the initial problem of health care literacy in rural Cambodia. The misuse of prescription drugs was a widespread issue, and local health practices often proved dangerous and counterproductive. Most residents of the rural Rovieng District had never met a Western-trained physician and had relied on traditional herbal remedies and animistic healing practices to cure ailments. The program incorporated local health facilities with centers in Phnom Penh and Boston by delivering information via email for consultation from outside physicians. This process strengthened the ability of local health care professionals to care for their communities and allowed them to provide input on how the program was being received (Heinzelmann, Chau, Liu, & Kvedar, 2009). This was especially helpful considering many patients were unfamiliar with Western medicine.

A review of Operation Village Health in the Rovieng District from January 2005 to May 2006 found that two-thirds of all patients sought repeat telemedicine consultation and that the majority of those seeking help were female. Health care workers from Phnom Penh traveled to the village for in-person consultation and submitted results from physical and diagnostic testing to clinics in Phnom Penh and Boston for recommendations (Heinzelmann, Chau, Liu, & Kvedar, 2009). These recommendations strengthened the self-sufficiency of the Cambodian health care system by providing an educational opportunity for physicians hoping to become independent from foreign assistance.

The data also indicated that Operation Village had success in treating chronic conditions. Each year of the program saw the proportion of patients referred to offsite facilities for care drop by 51 %, which indicated that chronic conditions were being addressed and local facilities were successful in treating patients (Heinzelmann, Chau, Liu, & Kvedar, 2009). This result also displayed the long-lasting impact of the program, as the health care system in Cambodia began to regain its footing and increase its capacity to respond to public health concerns.

The success of telemedicine in Africa and Cambodia has led to the expansion of existing programs to new regions hoping to bridge the digital divide. RAFT has expanded beyond Africa to Bolivia and Nepal and is connecting health care professionals from all over the world. This worldwide community of professionals globalizes the delivery of health care and provides the framework for future cooperation in the digital age. In Cambodia, Operation Village Health has expanded to include districts outside of Rovieng that are even more isolated from Western medicine (Heinzelmann, Chau, Liu, & Kvedar, 2009). The success of telemedicine in rural Cambodia can give guidance to future policymakers on how to best administer telemedicine in an age of globalization. By relying on local expertise and fostering self-reliance, Cambodia serves as a model for expanding the benefits of digitization to new communities.

4 Role of Telemedicine During the Coronavirus Pandemic

A global pandemic adversely affecting the world population occurs about once a generation. This unprecedented time is difficult for our society; it is something that people often never experience in their lifetime. The present coronavirus-associated outbreak of acute respiratory disease called COVID-19 is the third documented spreading of an animal coronavirus to humans that is causing a major pandemic in the last two decades. Such pandemics pose serious challenges to the health care delivery system.

During this pandemic, the main fear is that hospitals will overflow with patients and there will not be enough health care resources to treat all ill patients that walk through the door. As health care professionals try and take appropriate measures to ensure that individuals with low-risk diseases and conditions do not take up the already limited resources that the hospitals have. Hospitals also want to make sure that those who are seriously ill are receiving the appropriate treatment and care to combat this illness (Portnoy et al., 2020). This is where telemedicine can be helpful. Telemedicine has the potential to help by permitting the mildly ill patients to get the supportive care that they need. Patients who do not need to be hospitalized but are still sick can receive this alternative. Patients who take this route can get the care they need without being exposed to other individuals who may have underlying health conditions.

The United States with more than 1.5 million infected cases and over 100,000 deaths has been the worst hit by the coronavirus pandemic. To prevent the choking of the country's health care system and restrict the spread of the virus through contact, telemedicine has been strongly promoted in America during the COVID-19 pandemic. Nearly all health care insurance plans in the US offer some form of telemedicine coverage. These different health care insurance plans are a part of the three-part stimulus package[1] passed by the Congress. In response to the COVID-19 pandemic, the Centers for Medicare and Medicaid Service and commercial health plans have largely waived co-pays for telemedicine visits to incentivize and encourage its use. For some people under their circumstances, it can be riskier to leave their homes. Telemedicine offers ways to get the services a person may need in the comfort of their own home.

According to a recent Sykes survey conducted among 2,000 Americans, two-thirds of respondents say that COVID-19 has increased their willingness to try telehealth in the future. Fully one-quarter of respondents had not considered this as an option before. More than 15 % of respondents still report that COVID-19 had not increased their use of telehealth. Most of that group fell within the 35–45 age group and were predominately in the southern United States. Only a small

1 The first legislation change is called "The Coronavirus Preparedness and Response Supplemental Appropriations Act, 2020." The first stimulus bill authorized the Centers for Medicare and Medicaid services to temporarily lift the existing geographic originating site restrictions to Medicare beneficiaries. Prior to this, Medicare beneficiaries residing in nonrural areas could not receive access to telemedicine services or receive care in their own homes. However, under the new legislation, all Medicare beneficiaries will be able to access telemedicine services during a national emergency. This is true regardless of geographic region, site of care, and beneficiaries' home. This stimulus bill also applies to non-COVID-19-related issues that have been approved on the Medicare fee-for-service list. The second legislation is called "The Families First Coronavirus Response Act." This legislation made clarifying changes to the first package to ensure that all Medicare beneficiaries have access to health services they may need. The third legislation is called "The Coronavirus Aid, Relief, and Economic Security (CARES) Act." The most recent legislation added additional telehealth flexibility. The law temporarily removed the requirement that a beneficiary must have a preexisting relationship with a provider to be eligible for services. The law also allows Health and Human Services to waive the requirement that providers give out telemedicine services through virtual platforms. It allows beneficiaries to access providers off their devices.

percentage of respondents knew someone who had used telehealth to be screened for COVID-19, primarily in the 25–34 and 18–24 age groups.

While there are benefits of it, there are some barriers. In general, patients may be unaware that they have telemedicine as an option and may not know how to even access it. It would seem as if a lot of people have either heard of telemedicine or have a general understanding of what it is but may not know the benefits and barriers that can come from it (Portnoy et al., 2020).

5 Future Barriers to Telemedicine

One effect of globalization has been the mass migration of people between nations. This transfer of people and ideas across borders has the potential to change the health care landscape of several nations. The United States, a long-time hub for international migration, has a culturally diverse population within its own borders. In the field of health care, this can lead to several cultural barriers that policymakers must contend with.

Several immigrant groups may be hesitant to seek care due to limitations imposed by culture or lack of proficiency in English. For example, Asian Americans in the United States experience a similar rate of diagnosable mental illness when compared with their white counterparts but are less likely to seek medical attention. This can be due to multiple reasons, but research has pointed to cultural norms and language barriers as reasons why this occurs. Many Asian cultures see mental illness as being caused by a lack of discipline, and many Asian Americans experience shame for undergoing these illnesses (Ye et al., 2012).

One strategy to address the disparity in health care utilization is the use of telepsychiatry. A form of telemedicine, telepsychiatry provides interactive videoconferencing for those suffering from psychiatric issues. Telepsychiatry is especially useful for those who may face stigma for walking into a mental health facility, and it provides immigrant populations with access to doctors across the nation who share their language and culture. A study of Korean immigrants found a high level of acceptance of telepsychiatry and the cultural sensitivity that it provided (Ye et al., 2012). Patients had access to a service that they may otherwise feel stigmatized for using, and the remote nature of telepsychiatry meant that it could be utilized by Korean populations in several different areas.

Despite the success that telemedicine has had with certain subgroups, other demographics have been reluctant to adopt it. In focus groups with African Americans and Latino Americans, both groups of people were asked to express their thoughts on the use of telemedicine. While both groups seemed receptive to reduced wait times for care and increased access to medical information, African Americans expressed more concerns related to privacy and the loss of an in-person interaction with a specialist. This disparity may represent a greater distrust of the health care system among African Americans due to a history of past abuses in American society. In comparison, Latino Americans are more likely to be recent immigrants, which may eliminate the concerns related to past abuses (George, Hamilton, & Baker, 2012).

Privacy concerns in health care have led to regulations and operating procedures to protect consumer information in the United States. The Health Insurance Portability and Accountability Act protects the privacy of personal health information, and Fair Information Practice Principles provide guidance on how to properly collect health information (Hall & McGraw, 2014). Despite the existence of privacy protections, many Americans are still concerned about the possibility of personal health information being shared with third parties. This concern has led to the opposition of electronic databases for health information.

One such type of database, Prescription Drug Monitoring Programs (PDMPs), tracks controlled substance prescriptions across a given state (CDC). The database can be accessed by pharmacists and can reduce the number of unnecessary prescriptions. Some states have even gone as far as to require pharmacists to check PDMPs before dispensing to a patient (CDC).

The use of PDMPs have been shown to have a significant impact on the opioid crisis in the United States. The rate of increase for opioid-related poisonings in states without PDMPs is shown to be higher than those with PDMPs. Pharmacists benefit from access to the PDMPs because it gives them a clearer patient profile, and state health agencies benefit from it by gathering a clearer understanding of widespread opioid abuse and strategies to curb it. In addition, states with a more robust reporting requirement are the ones that experienced the lowest rates of increases for opioid-related poisonings (Pauly, Slavova, Delcher, Freeman, & Talbert, 2018).

While the use of PDMPs had been existent in 49 out of the 50 US states and touted by the Centers for Disease Control, one state went years without taking advantage of a PDMP. The Missouri Senate did not pass legislation to establish a PDMP until March 12, 2020 and did so against political concerns that had held it back for years prior (Driscoll, 2020). Concerns over the role of government reflects America's cultural distrust of government and their values of privacy and independence. Opponents in Missouri expressed their concerns of privacy and the unintended consequences that could come along with a statewide PDMP. Concerns included the belief that a PDMP was in violation of an amendment to the US Constitution as well as fears that a PDMP could be used to prevent firearm sales (Rosen, 2020).

Privacy concerns are not unique to the United States, as many societies have their own privacy concerns related to telemedicine. In Vietnam, privacy was one of the chief barriers to the adoption of SMS communication for health care consultations. The Hanoi University of Public Health attempted to increase access to health care by using SMS to communicate information remotely to patients. The program provided a free counseling hotline for female migrant workers as well as information on sexual and reproductive health services (Lam et al., 2018). This direct messaging of health information served as an important means of communication for topics that may have been considered taboo, and it provided a way for health care professionals to dispel myths associated with taboo topics.

The continued success of telemedicine in Vietnam depends upon the willingness of households to adopt it. Phone sharing is common in many Vietnamese households, which hinders the ability of specialists to reach specific individuals as well as prevent many Vietnamese from sending confidential information over a shared device (Lam et al., 2018). This lack of confidentiality may lead many Vietnamese to abandon the telemedicine in favor of an in-person consultation or may lead them to forgo a consultation altogether. While this concern is mainly one of privacy, it reflects a broader barrier in the Vietnamese economy, where the lack of access to individual telecommunication devices may hamper the future implementation of telemedicine.

The broader barrier to the adoption of telemedicine programs in countries of various stages of economic development continues to be the cost and availability of infrastructure to support its expansion. While many telemedicine programs in the United States are aimed at rural communities, there remains a "digital divide" between urban and rural areas in terms of broadband access. Refer to Fig. 3.

A smaller percentage of the population in rural counties has access to Internet download speeds greater than 25 Mb/s, the speed necessary for video-based telehealth visits, than populations in urban counties. In 2016, a report by the US Federal Communications Commission found that 39 % of rural Americans lacked access to this broadband speed (American Lung Association, 2018). Many of these communities lacked proper access to primary care physicians and psychiatrists, yet these same communities were the ones that reported the lowest access to adequate broadband speeds. The paradox of need versus access illustrates that the communities that stand to gain the most from telemedicine programs are the same communities that lack the infrastructure for these programs. A similar issue exists in Europe where infrastructure barriers remain highly concentrated in Eastern Europe. Refer to Fig. 4.

The United States has attempted to address this infrastructure concern by appropriating $29 million for each fiscal year of 2021 through 2025 for a telemedicine grant program in rural areas of the country (Moss et al., 2020). Private sector actors have also attempted to fill the gaps in access by donating money for the purpose of developing telemedicine for psychiatry and other forms of specialty care (Prina, 2018). In Europe, similar initiatives have taken place, such as RuNe, a €250 million project for increasing Internet speeds in rural areas that is funded by both public and private actors (Shaping Europe's digital future, 2020).

Future action to close the gaps in health care access will depend upon the cooperation of both public and private entities. Providing broadband access is less profitable in low-density population areas, which can be a barrier to expanding Internet access to vulnerable communities. There is considerable risk in expanding operations into rural areas, and profitability can be limited in communities with low population density (Reardon, 2018). Therefore, the infrastructure barrier in the expansion of telemedicine will be closely linked to the ability of nations to overcome the initial barriers to entry for broadband providers. Public and private

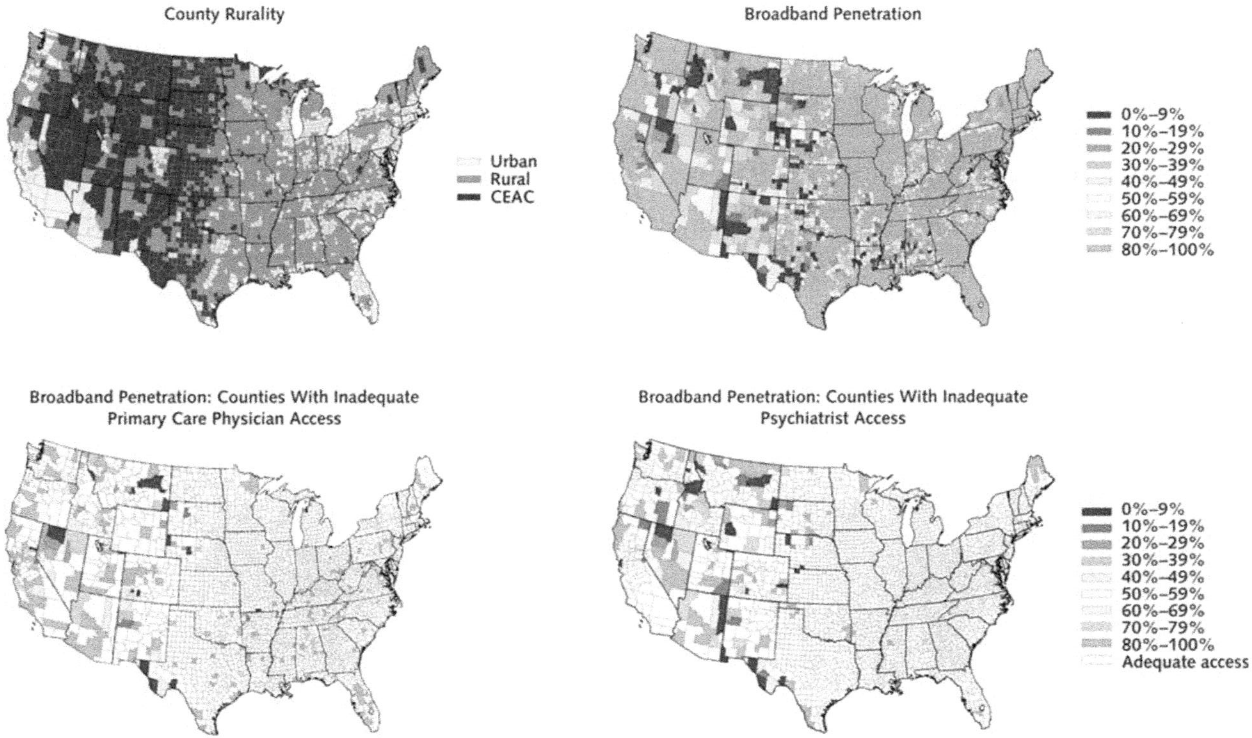

Fig. 3: County rurality, broadband penetration, and inadequate access to primary care physicians and psychiatrists. Source: (Annals of Internal Medicine – The Limitations of Poor Broadband Internet Access for Telemedicine Use in Rural America: An Observational Study, 2019)

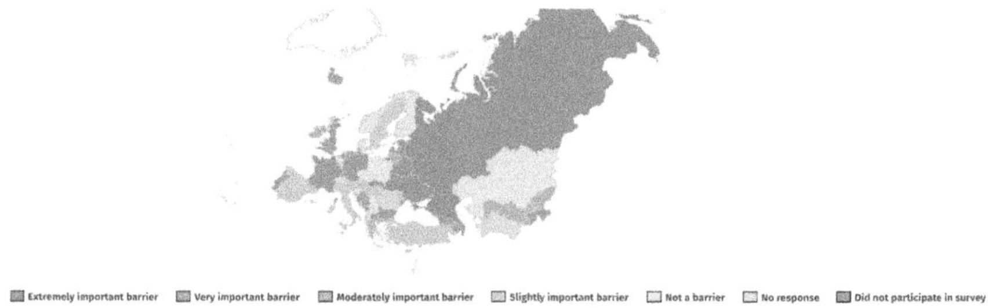

Fig. 4: The importance of infrastructure (lack of equipment and/or connectivity as a barrier to telehealth supporting universal health coverage. Source: (World Health Organization – The importance of infrastructure as a barrier to telehealth supporting universal health coverage, 2015)

grants as well as industry innovations will be needed to introduce broadband into new areas, and rural telemedicine will need to be capable of adapting to a broadband landscape with limited resources.

5 Conclusion

The use of telemedicine in health care is a topic of growing relevance in an age of global digitalization. National governments, universities, private enterprises, and individual patients have all had an impact on how telemedicine has been perceived and adopted in communities across the globe. Examples such as RAFT in Africa or COMSTATS in Pakistan illustrate successful telemedicine adoption and provide a framework for future programs in new areas. Bringing in both foreign and domestic actors can break down cultural barriers and engender greater cooperation between health care professionals around the world. As the case studies showed, building trust with local populations and encouraging self-sufficiency were important factors in the success of a telemedicine program.

Despite the success of several telemedicine programs, challenges still remain for proper access to health care. Cultural backlash to telemedicine due to concerns with privacy or a historical distrust of the health care system continue to limit the adoption of telemedicine. The greater interconnectedness of society runs counter to many communities' cultural norms, and past abuses by governments may cause patients to distrust health care initiatives enacted from above.

In addition to these concerns, the traditional barrier of distance still exists for patients in rural communities. The lack of broadband and the unwillingness for private actors to expand coverage into vulnerable areas further exacerbates the problem and widens the digital divide between rural and urban communities. The expansion of telemedicine will depend upon the expansion of broadband access into areas that lack proper infrastructure. Doing so will require greater international cooperation and incentives for private industry to expand Internet access into remote areas.

Further research may be done in order to address the remaining challenges in telemedicine and to track the future of digitization in health care. Telemedicine is still a new subject in many areas of the world, but it will continue to expand as globalization continues to grow. As economies become more interconnected, so will their capacity to provide health care to others.

Bibliography

Bediang, G., Perrin, C., Ruiz de Castañeda, R., Kamga, Y., Sawadogo, A., Bagayoko, C. O., & Geissbuhler, A. (2014). The RAFT telemedicine network: lessons learnt and perspectives from a decade of educational and clinical services in low-and middle-incomes countries. *Frontiers in Public Health, 2,* 180.

Drake, C., Zhang, Y., Chaiyachati, K. H., & Polsky, D. (2019). The limitations of poor broadband internet access for telemedicine use in rural America: an observational study. *Annals of Internal Medicine, 171*(5), 382–384.

Driscoll, J. (2020, March 12). Prescription Drug Monitoring Likely Coming To Missouri, The Last State Without It. Retrieved from https://news.stlpublicradio.org/politics-issues/2020-03-12/prescription-drug-monitoring-likely-coming-to-missouri-the-last-state-without-it

Dunbar, B. (2015, April 16). NASA Glenn's Historical Timeline. Retrieved from https://www.nasa.gov/centers/glenn/about/history/timeline.html.

George, S., Hamilton, A., & Baker, R. S. (2012). How do low-income urban African Americans and Latinos feel about telemedicine? A diffusion of innovation analysis. *International Journal of Telemedicine and Applications.* Volume 2012, Article ID 715194, 9 pages doi:10.1155/2012/715194

Goldstein, K. M., Gierisch, J. M., Zullig, L. L., Alishahi, A., Brearly, T., Dedert, E. A. ... & McDuffie, J. (2017). Telehealth services designed for women: an evidence map. VA ESP Project #09-010; 2017.

Hall, J. L., & McGraw, D. (2014). For telehealth to succeed, privacy and security risks must be identified and addressed. *Health Affairs, 33*(2), 216–221.

Hayes, P. (2019). *Statement of Dr. Patricia Hayes Chief Consultant For Women's Health Services Veterans Health Administration (VHA) Department of Veterans Affairs (VA) Before The House Veterans Affairs Committee Subcommittee on Health* House Committee on Veterans Affairs.

Heinzelmann, P., Chau, R., Liu, D., & Kvedar, J. (2009). Cross cultural telemedicine via email: experience in Cambodia and the USA. In *Telehealth in the Developing World.* London: IDRC, pp. 135–148.

Johnson, C. (2014). Crossroads: how the intersection of technology, medicine, and the law, impact the administration of healthcare in Florida and Puerto Rico. *University of Miami Inter-American Law Review, 46,* 209.

Karnow, S. (1984). *Vietnam: A History. 1983.* New York: Penguin Books.

Khan, H. A., & Hayee, I. (2009). Telehealth for mountainous and remote areas of northern Pakistan. *Telehealth in the Developing World,* 242–251.

Koshkin, P. (2018, June 18). Five lessons on U.S.-Russian relations thirty years after the Reagan-Gorbachev Summit. Retrieved from https://www.wilsoncenter.org/blog-post/five-lessons-us-russian-relations-thirty-years-after-the-reagan-gorbachev-summit

Lam, J. A., Dang, L. T., Phan, N. T., Trinh, H. T., Vu, N. C., & Nguyen, C. K. (2018). Mobile health initiatives in Vietnam: scoping study. *JMIR mHealth and uHealth, 6*(4), e106.

Landman, K. (2018, May 5). Telemedicine takes transgender care beyond the city. Retrieved from https://www.npr.org/sections/health-shots/2018/05/05/608472248/telemedicine-takes-transgender-care-beyond-the-city.

Mobaraki, A. E. H. (2010). Gender inequity in Saudi Arabia and its role in public health. *Eastern Mediterranean Health Journal, 16*(1), 115.

Moss, K., Wexler, A., Dawson, L., Long, M., Kates, J., Cubanski, J., Musumeci, M., Meredith, F., Ramaswamy, A., Ranji, U., Pollitz, K. (2020, April 10). The Coronavirus Aid, Relief, and Economic Security Act: summary of key health provisions. Retrieved from https://www.kff.org/global-health-policy/issue-brief/the-coronavirus-aid-relief-and-economic-security-act-summary-of-key-health-provisions/.

National Academies of Sciences, Engineering, and Medicine. (2018). *Evaluation of the Department of Veterans Affairs mental health services.* National Academies Press.

Pauly, N. J., Slavova, S., Delcher, C., Freeman, P. R., & Talbert, J. (2018). Features of prescription drug monitoring programs associated with reduced rates of prescription opioid-related poisonings. *Drug and Alcohol Dependence, 184,* 26–32.

Portnoy, J., Waller, M., & Elliott, T. (2020). Telemedicine in the Era of COVID-19. *The Journal of Allergy and Clinical Immunology: In Practice, 8*(5), 1489–1491.

Prina, L. L. (2018, December 1). Foundation funding for telehealth. Retrieved from https://www.healthaffairs.org/doi/full/10.1377/hlthaff.2018.05150 (7.5.2020).

Reardon, M. (2018, October 23). Rural America needs better internet, here's why it doesn't have it. Retrieved from https://www.cnet.com/news/why-rural-areas-cant-catch-a-break-on-speedy-broadband/ (7.5.2020).

Rosen, C. (2020, February 25). Will Missouri remain the only state without a prescription drug monitoring program? Retrieved from https://www.kansascity.com/news/politics-government/article240460171.html (25.3.2020).

Shaping Europe's Digital Future. (2020, February 14). The Connecting European Broadband Fund drives investment in rural and remote areas. Retrieved from https://ec.europa.eu/digital-single-market/en/news/connecting-european-broadband-fund-drives-investment-rural-and-remote-areas (7.5.2020).

Telehealth as a Vehicle to Support Tobacco Cessation. (2018). American Lung Association, p. 8.

What States Need to Know about PDMPs. (2017, October 3). Retrieved from https://www.cdc.gov/drugoverdose/pdmp/states.html (10.5.2020)..

WHO European health information at your fingertips. (n.d.). Retrieved from https://gateway.euro.who.int/en/indicators/ehealth_survey_46-infrastructure-as-a-barrier-to-telehealth/visualizations/#id=31724 (5.5.2020).

World Health Organization. (2010). Telemedicine: opportunities and developments in member states. *Global Observatory for eHealth Series, 2,* 1–88.

World Health Organization. Geneva University Hospitals – RAFT Network. (2011, July 23). Retrieved from https://www.who.int/workforcealliance/members_partners/member_list/hugraft/en/ (25.3.2020)..

Ye, J., Shim, R., Lukaszewski, T., Yun, K., Kim, S. H., & Rust, G. (2012). Telepsychiatry services for Korean immigrants. *Telemedicine and e-Health, 18*(10), 797–802.

Bahar MURATOĞLU PEHLİVAN

Deindividuation, Disinhibition, and Anonymity on Social Media: A Field Study on Online Behaviors of Turkish Users

1 Introduction

Social networks sites are an important part of digital era. Even though they have lots of advantages, create new opportunities, and in many ways make an individual's life easier, they also often promote aggressive and violent behavior toward others, lynch culture, and bullying.

Theories of social psychology have long studied how and why individuals behave differently, more aggressively, atypical etc. or, in other words, with decreased inner restraints, when they are part of a group or a crowd and/or they are anonymous. These theories might give insight on the kind of online behaviors stated above. When applied to behaviors on social network sites, deindividuation, crowd psychology, disinhibition, and anonymity can help explain online crowd behavior. Before discussing online behavior, however, first we should revisit these concepts stated above.

In his work on crowd psychology, Gustave Le Bon (1997: 19–26) argues that masses are more than the sum of individuals and can have different characteristics than people who form the crowd. That means, individuals can behave differently in crowds than they would when they are alone, and do not curb their behaviors, albeit temporarily. Le Bon also points out that these crowds are not necessarily formed by individuals that are physically together; people who take action and gather together with a common emotion triggered by a specific event also form the same crowd psychology.

Convergence theory makes a similar point as it suggests that individuals join groups due to their shared needs and characteristics, so that they can release their otherwise controlled behaviors. Thus, it is possible to predict the behavior of the mass considering these repressed motives (Johnson and Johnson, 2003: 270).

There are two main common points of these conceptualizations stated above. One is the change in individuals' acts when they are in a crowd. The other is while individuals control their behaviors when they are by themselves, they tend to release some of this control when they are part of a group. The mechanisms underneath these changes are also a question of debate on deindividuation theory.

Starting from a parallel perspective, deindividuation theory argues that it can be easily observed that same people act differently when in groups versus individually. According to deindividuation theory, like other theories mentioned above, the base of this difference is the release of inner restraints and disinhibition individuals experience when they are part of a group. Thus, they commit acts they would not commit when they are alone. This means when in crowds, people feel they have more freedom and they allow themselves to release some of the constraints or rules they set for themselves. In other words, if people are not identifiable in a crowd, they would loosen their boundaries and behave in ways they would not in individuated conditions (Festinger et al., 1952: 382). Here, deindividuation theory explains the difference in individuals' behavior by their identifiability. In other words, if individuals are identifiable, there are inner restraints; but if they are not, the restraints are loosened.

On the other hand, Philip Zimbardo (as cited in Johnson and Johnson, 2003: 270–271) states three components of deindividuation: *conditions of deindividuation*, *state of deindividuation*, and *deindividuated behavior*. *Conditions of deindividuation* refers to anonymity, decreased responsibility, being in a group, and arousal. *State of deindividuation* means that due to decreased self-awareness and altered experience, individuals' internal restraints are lowered. And finally, *deindividuated behaviors* occur, which are polarized, atypical and extreme.

One of the notable points in Zimbardo's theory is lowered responsibility in individuals while in deindividuating conditions. If individuals do not feel as responsible when they are anonymous in a group, decrease in inner restraints can be considered as a predictable outcome. On the other hand, decreased self-awareness is another remarkable point, as it suggests individuals may not be fully conscious of their behavior, and thus not be able to evaluate them accurately.

Diener (1976: 497) also argues that under deindividuated condition, self-awareness and self-evaluation decrease. What's more, an individual is less worried about the evaluation of others. This situation further diminishes the internal restraints a person has. Being a member of a group plays an important role here, as individuals can lose their personal identities when they are part of a group. Diener (1979: 1161–1168) adds that deindividuating conditions make people closer to the group, thus their attention is more on the group and less on themselves or other things. This creates a decrease in self-awareness and results in disinhibited behavior. Diener's study found a "correlation between disinhibited behavior and lack of self-consciousness." Another point Diener (1976: 498) indicates is, one important concept of deindividuation theory is anonymity, because when anonymous, people are not regarded as individuals.

In other words, according to deindividuation theory, while in crowds individuals become anonymous and this anonymity reduces individual responsibility; thus, their inner constraints are weakened. This process also results in decreased self-awareness and freedom from normally internalized norms. This loss of self-identity, for that matter, promotes aggressiveness (Mann et al., 1982: 261). Thus, when in a crowd, people are able to satisfy their needs they cannot satisfy in an individuated state because of their internal constraints. This makes groups who allow deindividuation more attractive (Festinger et al., 1952: 382–387).

Similarly, one view of deindividuation argues that aggressive or deviant behaviors are under control if a person is identifiable. However, when an individual is not discriminable, this control will become loose and nonnormative behaviors will be released (Dipboye, 1977: 1057–1059). Research also showed that when individuals are not identifiable, they tend to commit antisocial behavior more (Silke, 2003: 496).

In conjunction with this view, according to a study, deindividuation has different effects on individuals. If an individual is self-differentiated, she/he keeps internal standards and restraints, even if there is anonymity. Self-differentiated individuals have a strong sense of distinction between self and nonself, independent of the changes in situations and social environment. Thus, they conform group pressures less. Undifferentiated individuals, however, continue to rely on their inner standards when identifiable, but guided by other group members' actions more when anonymous. Thus, they seem to be influenced more from environment, and their distinction between self and nonself is weakened in case of anonymity (Nadler et al., 1982: 1129–1134). Another study's findings also supported that anonymity leads to increased aggression. Results suggest that self-awareness functions as a mediator of aggression, and deindividuation state decreases self-consciousness (Mann et al., 1982: 271).

On the other hand, contrary to classical deindividuation theory, the social identity model of deindividuation (SIDE) suggests that groups do not destroy self and they do not cause individuals to lose their sense of self. It is rather when individuals are part of a group, their identification changes from personal to social level. Social identity model also argues that even though individuals' criteria for their behavior changes to social level when they are a group member, this does not mean they necessarily lose control over their actions. According to SIDE, if individuals identify themselves with the group, this means their perceptions of group members and other individuals outside of the group are defined by social categorical level. Therefore, group members internalize the group's understandings, and chose their actions accordingly (Reicher et al., 1995: 177).

In sum, according to deindividuation theory, anonymity in a group leads to decreased responsibility, loosened inner restraints, lowered self-awareness, uncontrolled behavior, and atypical acts. In today's world, this raises the question of what happens if individuals are anonymous not in a physical group but anonymous online. Internet allows people to become part of an online group; what's more, it also allows people to become anonymous individually. These points inspired research on how mechanisms of deindividuation work on digital platforms. The following section takes a look on this research.

2 Online Deindividuation

In the digital era, masses are not necessarily groups that are physically together but groups that get together online. These masses are free from constraints of time and space, and their messages can instantly be sent anywhere in the world. Online masses are similar to real-life masses in the sense that they can have similar mechanisms for collective action. They are, however, different in terms of speed, size, and scope (Russ, 2007).

Social network sites allow individuals to come together and form a mass without being physically in the same space. What's more, social media gives the opportunity to connect anonymously with others. It can be argued that this anonymity, along with online deindividuation, might cause disinhibition. In other words, individuals' inner restraints might be weakened while communicating online. This might also cause violent, aggressive, and destructive behavior, which often comes up in social media as cyber bullying, lynch, and violent behaviors alike.

A research on this issue, for example, found that the anonymity social network sites provide, along with deindividuation, encourages cyberbullying (Lowry et al.., 2016). Barlett's (2015: 77) findings were similar as they suggest anonymity, along with the absence of power differential, is linked to cyberbullying. Another study on online game players also found that, online anonymity cause disinhibition and more antisocial acts (Chen et al., 2009: 343). These findings indicate that if experienced online – much like when experienced in real life masses – anonymity can cause more aggressive and uncontrolled behavior.

Similarly, a study on anonymous versus non-anonymous comments on online forums of newspapers found that anonymous comments showed more uncivil statements and tend to have more xenophobia, hate, and bigotry (Santana, 2014). Likewise, a study on a website on technology news stories also found that anonymous comments had more swearing and less positive emotion. Conversely, nonanonymous comments were more relevant to the news story (Omernick and Sood, 2013: 533). Another study on online anonymity showed that for bloggers, anonymity affects the amount of self-disclosure (Qian and Scott, 2007: 1442). These findings, like other research mentioned above, suggest that being anonymous online has a significant effect on individuals' inner restraints and therefore provides insight on online deindividuation.

On the other hand, a study showed that the effect of deindividuation in computer mediated communication increases polarization and group identification (Lee, 2007). Another study examined the differences between face-to-face communication versus computer-mediated communication in terms of deindividuation. Results showed that in computer mediated communication, participants are more immersed and perceived their team members as individuals less. However, this did not result in more negative reactions, which means aggressiveness cannot be attributed to deindividuation solely (Coleman et al., 1999). These results show that deindividuation mechanisms such as group identification, immersion and polarization are in works in online conditions, likewise in real-life conditions. Therefore, it can be argued that online deindividuating situations create similar outcomes as real-life deindividuating situations.

On the other hand, a research studied Le Bon's view of crowd psychology in terms of online instant communications and found that Le Bon's concept proved to be productive while analyzing online behavior (Stage, 2013). This means Le Bon's conceptualization of masses is also applicable to online crowds and can be used to understand online behavior. Therefore, from this point, it could be argued that when in an online community, people would behave differently and control their behaviors less than usual.

These deindividuation effects that are discussed above are also related to the concept of online disinhibition, as when deindividuated behavior occurs, individuals experience disinhibition and let go of their inner restraints to some degree. Therefore, the following section discusses online disinhibition effect.

3 Online Disinhibition Effect

According to Joinson (2007: 75), individuals often behave differently online than they would otherwise. Joinson argues that this difference occurs because individuals worry less about their self-representations and what others think of themselves when they are online. Thus, their inner restraints are not present (Joinson 1998, as cited in Joinson 2007: 75). This situation is similar with deindividuation effects and suggests a state of disinhibition online.

Suler (2004: 321) defines online disinhibition effect as individuals' less restrained behavior they exhibit online. He also states that there are two types of online disinhibition effect. One is called *benign disinhibition* and indicates situations where individuals weaken their restraints and reveal things about themselves they would not reveal or keep secret otherwise. This kind of disinhibition also might make individuals more kind or generous. *Toxic disinhibition*, on the other hand, indicates that individuals behave more rude, aggressive, and harsh when they release their inner constraints. However, Suler also states that these two effects may always not be distinguishable easily and the differences between them may be ambiguous in some situations.

Suler (2004) also introduces six factors related to online disinhibition, stating that they too might intersect and compose a complex effect. These factors are *invisibility, dissociative anonymity, asynchronicity, solipsistic introjection, dissociative imagination*, and *minimization of status and authority*. According to Suler's conceptualization, *invisibility* means that, especially in a text-driven environment, people do not see each other; they even may not know if the other person is present or not in a specific moment. These environments also give the opportunity to be physically invisible and not to worry about how they look or sound, etc. They also do not see the immediate expression of the other person while communicating, and these factors create disinhibition. *Dissociative anonymity* is a similar factor and refers that when online, it is possible for individuals to reveal only the amount of information they would like about themselves, and this gives them the opportunity to not be identifiable. So they can separate their online selves from their real life selves and have different online/offline identities. *Asynchronicity* means when people communicate online, there is no obligation for it to be in real time, and when individuals do not have to face the other individual's reaction immediately, this creates disinhibition. *Solipsistic introjection* refers to the situation when a people read a text online, they might experience it as a voice in their head and feel it as if the other's psychological presence. This process makes the other person a combination of inner representational system and how other person represents herself/himself. *Dissociative imagination* means that when individuals create imaginary characters online, they feel like what happens in the digital world does not apply to real world, as if they happen in a different or alternative dimension. Lastly, *minimization of status and authority* refers that while online, authority figures cannot represent themselves with body language, environmental arrangements, etc. Also, everyone has the same opportunity to speak their mind. Thus, status and authority diminish while communicating online.

Following Suler's conceptualization, a study on online disinhibition effect found that both benign and toxic disinhibition effect is related to cyberbullying (Udris, 2014: 259). Similarly, a different research found that when predicting cyberbullying, online disinhibition is explanatory (Lai and Tsai, 2016). Another research on young people showed that online disinhibition is linked to cyberbullying and cyberhate (Wachs et al., 2019). One study found that online disinhibition, false self, and cyber aggressiveness are related to each other (Kurek et al., 2019). And another research showed that lack of eye contact – which allows for invisibility – is an important factor on online disinhibition resulting in negative behavior (Lapidot-Lefler and Barak, 2012: 440). The findings of these studies are related to what Suler called toxic disinhibition. They show a relation between aggressive online behavior and online disinhibition. Therefore, the result of these studies may help us understand the mechanisms underneath the hate speech, bullying, and violent behaviors alike on social network sites.

On the other hand, online disinhibition is also found to be linked to the amount of self-disclosure on the Internet, as found in a study (Schouten et al., 2007). Similarly, another study also showed that online disinhibition positively predicts self-disclosure (Machackova and Dedkova, 2014). A meta-analysis also indicated that there is a connection between anonymity – which is an important factor for both online disinhibition and deindividuation – and self-disclosure online (Clark-Gordon et al., 2019). These findings represent what Suler calls benign disinhibition. They do not necessarily show aggressive or harsh behavior, but they also indicate loosened inner restraints as increased self-disclosure.

In conclusion, online disinhibition effect, much like online deindividuation, cause people to act differently than their real-life situations. The mechanisms at work here allow people to loosen their restraints and behave more freely online, whether it is anonymity, invisibility, or diminished sense of authority. This may cause nonviolent behavior such as increased self-disclosure or acting less inhibited, but it can also cause aggressive behaviors that can damage other individuals. Therefore, this study is designed to take another step forward to understand online disinhibition effect on Turkish social media users. The method adopted for the study and the findings are discussed below.

4 Method

This study was designed to have an insight on the effects of online deindividuation and online disinhibition on Turkish social media users. In other words, the purpose of the study was to understand how Turkish social media users' behavior change when they are online. To measure the disinhibition effect and change in online behavior, an online questionnaire was applied to social media users.

In the beginning of the questionnaire, participants were told that their names and personal information would not be asked and the information they provided would only be used for research purposes. They were also told that there were no right or wrong answers and they should choose the answer that represents them the most. Then the first part of the questionnaire asked participants their demographics, such as age, sex, and education. After that, the following question was presented: "Do you have an account which does not include your formal name and surname on any kind of online (dictionaries, digital games, forums, blogs, social media network sites such as Twitter, Facebook, Instagram, Reddit, etc.) platform?" This question was followed by Turkish translations of 11 items with 5 point Likert Scale (1 – Strongly Disagree, 5 – Strongly Agree) from the Online Disinhibition Scale, developed by Udris (2014). This scale has two factors, one being benign disinhibition and the other being toxic disinhibition, and uses Suler's (2004) disinhibition concepts of *invisibility, dissociative anonymity, asynchronicity, solipsistic introjection, dissociative imagination*, and *minimization of status and authority*. After that, 6 items, inspired by the study of Messinger et al. (2008), were presented. The original study measured the differences between real-life and virtual life behavior (represented by behaviors in the online game Second Life). Researchers asked participants to rate how shy versus outgoing, introvert versus extravert, risk averse versus risk taking, reserved versus loud, and superficial versus thoughtful in real life and when playing Second Life. This study, on the other hand, to measure how different individuals behave online versus real life, developed 6 items with 5 point Likert Scale (1 – Strongly Disagree, 5 – Strongly Agree) asking:

I am more outgoing on internet than real life
I am more extravert on internet than real life
I am more risk taking on internet than real life
I am more loud on internet than real life
I am more more superficial on internet than real life
I am more more rude on internet than real life

The questionnaire was shared from various social media platforms and also shared with university students from different universities via their professors. In total, 178 participants (109 female [61.9 %], 66 male [37.5 %], 3 unspecified [0.6 %]) filled the questionnaire. Age distribution of participants are as follows: 18–24 (44.4 %), 25–34 (20.8 %), 35–44 (18.5 %), 45–54 (8.4 %), 55–64 (4.5 %), and 65 and above (3.4 %). Educational levels of participants are as follows: primary and secondary schools (0.6 %), high school (8.4 %), university (79.2 %), and postgraduate (11.8 %).

5 Results

In total, 93 participants (52.2 %) stated that they had an anonymous account (an account that does not include their formal name and surname) on an online platform, while 85 (47.8 %) stated that they did not have an anonymous account.

Reliability analysis showed that six questions asking participants to rate how differently they behave online versus real life had Cronbach's Alpha of 0.727. As for online disinhibition scale, benign disinhibition factor had Cronbach's Alpha of 0.673 and toxic inhibition had 0.628.

Independent samples t-test was applied to measure if there was a difference between participants who had anonymous accounts versus participants who did not have anonymous accounts. Results showed that participants who owned anonymous accounts had scored higher in all three sets of questions (benign inhibition and toxic inhibition factors of online disinhibition effect and six questions on how different individuals behave online versus real life), and the results were statistically significant ($p < 0.05$). Tab. 1 shows mean scores and standard deviations for groups.

It should be noted that, although statistically significant, toxic disinhibition effect's alpha value was found to be very close to the 0.05 threshold (0.046). Also, the reliability of this factor was relatively low (Cronbach's alpha of 0.628). On the other hand, benign inhibition effect and online versus real-life questions showed much stronger significance ($p=0.00$). These results indicate support for existing literature that anonymous individuals behave more disinhibited.

Another result for toxic disinhibition effect was men scored higher (M=1.6364, STD=0.70326) than women (M=1.3601, STD=0.65390). The result was statistically significant ($p < 0.05$). Although both groups scored relatively low. Other notable difference between male and female participants was their score on the question "I am more superficial on internet than real life". Men scored significantly ($p < 0.05$) higher (M=1.70, STD=0.960) than women (M=1.25, STD=0.683). Again, both groups scored relatively low.

Tab. 1: Mean Scores and Standard Deviations for Groups

	Participants who own an anonymous account	**Participants who do not own an anonymous account**
Benign inhibition	M=2.9432, STD=0.82661	M=2.5070, STD=0.73064
Toxic inhibition	M=1.5538, STD=0.7685	M=1.35, STD=0.55528
Online versus real life questions	M=2.2061, STD=0.86791	M=1.7620, STD=0.59744

Tab. 2: Mean Scores and Standard Deviations for All Participants

	Mean	**Standard deviation**
Benign inhibition	2.7349	0.81008
Toxic inhibition	1.4565	0.68106
Online versus real life questions	1.9940	0.78129

On the other hand, in general, participants scored highest in benign disinhibition effect and lowest in toxic disinhibition effect. Tab. 2 shows mean scores and standard deviations for all participants.

For benign disinhibition effect, the highest scored question was the one asking participants if it was easier to communicate online because they could reply anytime they wanted (M=3.47, STD=1.446), while the lowest scored question was the one asking whether the participants felt like a different person online (M=1.87, STD=1.234). For toxic disinhibition effect, the scores were generally low, with the lowest one being the item stating insulting online is not bullying (M=1.13, STD=0.586). On the other hand, the highest scored item for online versus real life questions was: "I am more more superficial on internet than real life." (M=2.72, STD=1.421), and the lowest scored one was: "I am more more rude on internet than real life." (M=1.43, STD=0.829).

Conclusion

This study was designed to measure the effects of online deindividuation, online disinhibition, and online anonymity on individuals' behaviors. In other words, the purpose of this study was to understand how and why individuals behave differently when they are online, in relation to deindividuation theory and online disinhibition effect. One important finding of the study was individuals who had accounts that did not include their formal name and surname, engaged more disinhibited behavior, and acted more differently when they were online. In other words, if individuals had anonymous accounts, they were more disinhibited. Both deindividuation theory and a perspective on online disinhibition argue that anonymity causes individuals to loosen their inner restraints. Therefore, the findings of the study support existing literature.

On the other hand, in general, participants scored low in toxic inhibition effect (M=1.4565) than benign inhibition effect (M=2.7349). This result indicates that even though individuals have a tendency to loosen their restraints online, they do not necessarily show aggressive behavior. However, Suler (2004) stated that these two factors are not always distinguishable and can be ambiguous. In other words, they are not independent but related factors. On the other hand, this finding relies on individuals' statements of themselves and people have a tendency to show themselves more positively and favorably while answering (Funham, 1986: 385). Thus, there is a possibility that they might have understated their aggressive behaviors.

Another finding showed the difference between male and female participants. Although both of them scored low, men scored significantly higher (M=1.6364) than women (M=1.3601) for toxic disinhibition effect. Men also scored significantly higher (M=1.70) for the question "I am more superficial on internet than real life" than women (M=1.25). These results show that men engage in toxic disinhibition more and act more superficial on the Internet. This might be related to gender roles and expectations of the patriarchal order of society, but further discussion is out of the scope of this study.

On the other hand, in general, the highest scored question was the one asking participants if it was easier to communicate online because they could reply anytime they wanted (M=3.47). According to Suler's (2004) conceptualization, this question is related to the concept of *asynchronicity*, that is, individuals experience disinhibition when they are not communicating in real time, because they do not have to face the other's reaction immediately. While the highest scored item for online versus real life questions was: "I am more more superficial on internet than real life." (M=2.72). This suggests that individuals do not behave as sensitive as real life when they are online and can be evaluated as related to the Suler's (2004) concept of *dissociative imagination*. The concept means that individuals feel what happens in the digital world does not apply to real world. Therefore, they are able to act without their inner restraints.

In conclusion, this study showed support for existing literature, which suggests that anonymity results in less restrained and more disinhibited behavior online. This result has notable implications, as in the digital world online behavior is an important part of social life. Thus, how behaviors change online can have significant effects on both individuals and society. On the one hand, as discussed above, online disinhibition may cause more aggressive and violent behavior and can harm other members of society; on the other hand, it may cause individuals to reveal more about themselves, or share secrets online, which may have serious consequences for them. Therefore, online disinhibition and deindividuation are important concepts for digital society and understanding online behavior.

Bibliography

Barlett, Christopher P., (2015), "Anonymously Hurting Others Online: The Effect of Anonymity on Cyberbullying Frequency", *Psychology of Popular Media Culture*, Vol. 4, N. 2, pp. 70–79.

Chen, Vivian H., Duh, Henry B., and Ng, Chiew W., (2009), "Players Who Play to Make Others Cry: The Influence of Anonymity and Immersion", *Proceedings of the International Conference on Advances in Computer Entertainment Technology*, New York: Association for Computing Machinery, pp. 341–344.

Coleman, L. H., Paternite, C. E., and Sherman, R. C., (1999), "A Reexamination of Deindividuation in Synchronous Computer-Mediated Communication", *Computers in Human Behavior*, Vol. 15, N. 1, pp. 51–65.

Clark-Gordon, Cathlin V., Bowman, Nicholas D., Goodboy, Alan K., and Wright, Alyssa, (2019), "Anonymity and Online Self-Disclosure: A Meta-Analysis", *Communication Reports*, Vol. 32, N. 2, pp. 98–111.

Diener, Edward, (1976), "Effects of Prior Destructive Behavior, Anonymity, and Group Presence on Deindividuation and Aggression", *Journal of Personality and Social Psychology*, Vol. 33, N. 5, pp. 497–507.

Diener, Edward, (1979), "Deindividuation, Self-awareness, and Disinhibition", *Journal of Personality and Social Psychology*, Vol. 37, N. 7, pp. 1160–1171.

Dipboye, Robert L., (1977), "Alternative Approaches to Deindividuation", *Psychological Bulletin*, Vol. 84, N. 6, pp. 1057–1075.

Festinger, Leon, Pepitone, Albert, and Newcomb, Theodore, (1952), "Some Consequences of De-individuation in A Group", *The Journal of Abnormal and Social Psychology*, Vol .47, N. 2S, pp. 382–389.

Furnham, Adrian, (1986), "Response Bias, Social Desirability and Dissimulation", *Personality and Individual Differences*, Vol. 7, N. 3, pp. 385–400.

Johnson, David W., and Johnson, Frank P., (2003), *Joining Together: Group Theory and Group Skills*, Boston: Pearson – Allyn andBacon.

Joinson, Adam N., (2007), "Disinhibition and the Internet", (Jayne Gackenbach Ed.), *Psychology and the Internet: Intrapersonal, Interpersonal, and Transpersonal Implications*, Massachusetts: Academic Press, pp. 75–92.

Kurek, Anna, Jose, Paul E., and Stuart, Jaimee, (2019), " 'I Did It for the LULZ': How the Dark Personality Predicts Online Disinhibition and Aggressive Online Behavior in Adolescence", *Computers in Human Behavior*, Vol. 98, pp. 31–40.

Lai, Cheng-Yu, and Tsai, Chia-Hua, (2016), "Cyberbullying in the Social Networking Sites: An Online Disinhibition Effect Perspective", *Proceedings of the 3rd Multidisciplinary International Social Networks Conference on SocialInformatics 2016*, New York: Association for Computing Machinery, pp. 1–6.

Lapidot-Lefler, Noam, and Barak, Azy, (2012), "Effects of Anonymity, Invisibility, and Lack of Eye-Contact on Toxic Online Disinhibition", *Computers in Human Behavior*, Vol. 28, N. 2, pp. 434–443.

Le Bon, Gustave, (1997), *Kitleler Psikolojisi*, (Y. Ender, Trans.), Istanbul: Hayat Publications.

Lee, Eun-Ju, (2007), "Deindividuation Effects on Group Polarization in Computer-Mediated Communication: The Role of Group Identification,

Public-Self-Awareness, and Perceived Argument Quality", *Journal of Communication*, Vol. 57, N. 2, pp. 385–403.

Lowry, Paul B., Zhang, Jung, Wang, Chuang, and Siponen, Mikko, (2016), "Why Do Adults Engage in Cyberbullying on Social Media? An Integration of Online Disinhibition and Deindividuation Effects with the Social Structure and Social Learning Model", *Information Systems Research*, Vol. 27, N. 4, pp. 962–986.

Machackova, H., and Dedkova, L., (2013), "Self-Disclosure in Online Communities", *16th European Conference on Developmental Psychology*, Bologna: MEDIMOND International Proceedings, pp. 145–146.

Mann, Leon, Newton, James W., and Innes, J. M., (1982), "A Test Between Deindividuation and Emergent Norm Theories of Crowd Aggression", *Journal of Personality and Social Psychology*, Vol. 42, N. 2, pp. 260–272.

Messinger, Paul R., Ge, Xin, Stroulia, Eleni, Lyons, Kelly, Smirnov, Kristen, and Bone, Michael, (2008), "On the Relationship Between My Avatar and Myself", *Journal For Virtual Worlds Research*, Vol. 1, N. 2, pp. 1–17.

Nadler, Arie, Goldberg, Marta, and Jaffe, Yoram, (1982), "Effect of Self-Differentiation and Anonymity in Group on Deindividuation", *Journal of Personality and Social Psychology*, Vol. 42, N. 6, pp. 1127–1136.

Omernick, Eli, and Sood, Sara O., (2013), "The Impact of Anonymity in Online Communities", *2013 International Conference on Social Computing*, Massachusetts: IEEE Computing Society, pp. 526–535.

Qian, Hua, and Scott, Craig R, (2007), "Anonymity and Self-Disclosure on Weblogs", *Journal of Computer-Mediated Communication*, Vol. 12, N. 4, pp. 1428–1451.

Reicher, Stephen D., Spears, Russell, and Postmes, Tom (1995), "A Social Identity Model of Deindividuation Phenomena", *European Review of Social Psychology*, Vol. 6, N. 1, pp. 161–198.

Russ, Chriss, (2007), "Online Crowds-Extraordinary Mass Behavior on the Internet", (Klaus Tochtermann, Werner Haas, Frank Kappe, Arno Scharl, Tassilo Pellegrini, and Sebastian Schaffert, Eds), *Proceedings of i-Media, 7*, Graz: Verlag der Technischen Universität Graz, pp. 65–76.

Santana, Arthur D., (2014), "Virtuous or Vitriolic: The Effect of Anonymity on Civility in Online Newspaper Reader Comment Boards", *Journalism Practice*, Vol. 8, N. 1, pp. 18–33.

Schouten, Alexander P., Valkenburg, Patti M., and Peter, Jochen, (2007), "Precursors and Underlying Processes of Adolescents' Online Self-Disclosure: Developing and Testing an "Internet-attribute-perception" Model", *Media Psychology*, Vol. 10, N. 2, pp. 292–315.

Silke, Andrew, (2003), "Deindividuation, Anonymity, and Violence: Findings from Northern Ireland", *The Journal of Social Psychology*, Vol. 143, N. 4, pp. 493–499.

Stage, Carsten, (2013), "The Online Crowd: A Contradiction in Terms? On the Potentials of Gustave Le Bon's Crowd Psychology in an Analysis of Affective Blogging", *Distinktion: Scandinavian Journal of Social Theory*, Vol. 14, N. 2, pp. 211–226.

Suler, John, (2004), "The Online Disinhibition Effect", *Cyberpsychology & Behavior*, Vol. 7, N. 3, pp. 321–326.

Udris, Reinis, (2014), "Cyberbullying Among High School Students in Japan: Development and Validation of the Online Disinhibition Scale", *Computers in Human Behavior*, Vol. 41, pp. 253–261.

Wachs, Sebastian, Wright, Michelle F., and Vazsonyi, Alexander T., (2019), "Understanding the Overlap Between Cyberbullying and Cyberhate Perpetration: Moderating Effects of Toxic Online Disinhibition", *Criminal Behaviour and Mental Health*, Vol. 29, N. 3, pp. 179–188.

Sevilay Ece GÜMÜŞ ÖZUYAR

Evaluation of Taxation of Cryptocurrencies in the Context of Schumpeter's Creative Destruction

1 Introduction

Tax can be expressed as compulsory payments taken depending on the financial situation of individuals and demanded by the state due to its sovereign right. Generally, tax revenues are given greater importance than other public revenues all over the world since they are the most important source of income of the state. Also, they give greater contribution to meet the increasing public financing needs depending on social and economic developments so far. Besides, tax revenues are significant in terms of their income distribution function.

Yet, in order for a tax system to show the desired success or for providing the health of the taxation activities, it is of great importance that the basic rules regarding tax practices are understood, terminology is clearly stated and the public have positive tax perception since interviews, surveys and experiments in this field show that tax behaviors of individuals are directly related to willingness of payments, audits and fines (Puviani, 1903; Scmölders, 1976; Fischer et al, 1992; Blackwell, 2007). Tax perception, on the other hand, is shaped by many different items that can be evaluated as internal, including personal characteristics, individuals' knowledge of tax laws, their tax morals as well as tax consciousness (Kirchler, 2007). In fact, tax psychology affecting the perceptions of individuals is the most important factor that determines the limits of taxation (Schmölders, 1959). In this context, the intrinsic values of individuals, such as perceptions or impulses about tax practices, give an idea about how new money systems as a tax payment instrument will be affected. Now therefore, individuals' perceptions and their financial psychologies must be taken into consideration very carefully in the taxation of cryptocurrencies still in development and formed as part of the blockchain system.

Cryptocurrencies, which can be defined as an innovation from the Schumpeterian perspective, in terms of new information creation, production by entrepreneurs in the system and having a widespread effect by creating new markets, operate differently than price money due to the dynamic structure of the blockchain mechanism. In this mechanism, any state wishing to tax cryptocurrencies will have to struggle to make itself an actor in the system. In this context, the state that aims to collect taxes becomes an entrepreneur in the system, and if it chooses to fight to achieve the desire to succeed in collecting taxes, it also captures the "new spirit" that Schumpeter deems necessary for the destruction of the old. If the state makes the taxation process adaptive to the continuous renewal of the mechanism and the system itself, the entire taxation process will become dynamic and economic balances can be reorganized by the state for its own favor according to each new situation.

This study is based on the examination of where the taxation process of cryptocurrencies can stand in Schumpeter's concept of creative destruction and whether the change of taxation structure can create or be a Schumpeterian creative destruction. For this purpose, this study is structured in three sections. In first section, assumptions, entrepreneur-innovation relationship and creative destruction theory of Schumpeter are analyzed. The second section introduces the cryptocurrencies in the framework of Schumpeterian theories. Finally, the last section discusses new taxation scenarios of cryptocurrencies in the world and whether it can define or accept as a force which creates radical changes in taxation procedures theoretically.

2 Assumptions of Schumpeterian Theory and Creative Destruction

In neoclassical economics, the relationship between input and output is defined by the production function, and due to the static equilibrium in its structure, technological development is not an internal variable, on the contrary, an external variable that can only be defined by the shift in the production function upwards. Due to the definitions imposed by this static structure, The neoclassical economics approach is also quite lacking in explaining innovations and technologies (Coricelli & Dosi, 1988: 15). The difference of Schumpeter solely is that the factors previously evaluated externally in the analysis are included in the analysis (Schumpeter, 1954: 964). In this sense, instead of static analysis, which does not take into account the accumulation of capital, Schumpeter dynamically analyzed the theory of creative demolition in an evolving cumulative process by constantly evolving (Schumpeter, 2010).

According to Schumpeter, evolution is a sum of radical change which has five different stages and types: philosophy, Marxist evolution, historian evolution, intellectual evolution and biological evolution, and no matter what stage it is in the economic sense, progress can be observed (Schumpeter, 1954:435–445). This development cannot be achieved through small steps from old to new, but through the processing of qualitative and large processes (Schumpeter, 2005: 113–115). These processes must be achieved, otherwise entrepreneurs and incompetent governments that have lost their dynamism and openness to innovation will end capitalism (as cited in- Dolgun, 1006: 176).

Schumpeter's explanation of the aforementioned innovations and the effects of these innovations across the economy have been through multiple studies. Each study almost completes and explains the previous one. The systematization of the theory has been due to the concretization of the effect of the innovations on the workflow in the process of cyclical fluctuations as well as defining the destructive effect created by these innovations.

The main purpose in this study is to investigate the effect of creative destruction in the taxation process of cryptocurrency, which is an acceptable innovation; Schumpeter's creative destruction process and the economic impacts created by the innovations and their cyclical processes in the workflow are collectively evaluated, and the general assumptions of the theory are aggregated.[1] According to the works of Schumpeter (1927, 1934, 1939, 1950, 1954, 1959, 2005), the assumptions can be listed as follows:

- Individuals are hedonistic in the aim of maximizing the utility levels.
- Variables can be classified internally and externally. While internal factors can be listed as individual characteristics, technology and innovations, the external factors can be collated as extraordinary situations such as war and earthquake and personal and social habits.
- The past and the expected values of the variables are also included in the analysis. In this context, a situation assessment is made that is far from static, where time is also included in the analysis and is beyond quantity and price relationship.
- The understanding that the motive that directs and manages firms is explicit rational calculation is rejected.
- Entrepreneurs are the most fundamental but destabilizing forces.
- Every static analysis should also be supported by a historical observation.
- Invention is a result of intellectual accumulation while innovation is an economic phenomenon.
- Innovations are constituents that change the routine and cause it to evolve into a different phenomenon.
- Evolved innovations occur suddenly and radically in chaotic situations.
- The economy is constantly being renewed by dint of innovations, yet it does not show a regular development in this dynamism.
- Development is not an acceptable phenomenon until it can be observed in a cyclical flow or a certain equilibrium trend.
- Companies that innovate and adapt to innovations remain in the system, otherwise they disappear. So innovations constitute the driving force in capitalism.
- The wheels of the capitalist production process constantly revolve due to innovations and new entrepreneurs.
- Technological developments are internal factors in the creative destruction process.
- Economic developments occur as a result of technological developments. However, these economic developments are not internal; on the contrary, they are spontaneous and constantly interchangeable.

- Technological developments create innovation but inventions are at the core of innovations.
- Innovations are internal elements that constantly create imbalances in the economic system.
- There are basically two types of innovations: Mark I and Mark II. Innovations created by small firms working in a competitive market structure (Mark I) and innovations created by large firms working in an oligopoly market structure (Mark II).
- The leading actor that caused the emergence of innovations is entrepreneurs. In this sense, the commercial firm makes innovation; entrepreneurs are the ones who create these innovations.
- At the end of each innovation, there is an unpredictable result or progress.
- Since companies follow each other's innovations, innovations tend to create a cluster in a particular sector or industry.
- The economy adapts itself to the constantly developing and renewed process. However, a 50–60-year period is required for innovations and the work flows created by these innovations to achieve adaptation by providing a complete cycle.
- Cycles consist of four phases: prosperity, recession, depression and revival.
- When the old structure has been destroyed as a result of innovations during the adaptation process, some of the entrepreneurs can react creatively by making brand new innovations within the radical innovations that emerge.
- A Walrasian general balance approach has been adopted in which the entrepreneur plays an active role in the nonstatic structure. "*In real life markets can never balance.*"
- In this whole process, by reducing the costs of newly produced products, thereby opening the scissors between the determined prices of the manufactured goods and the prices of new goods, the capitalist wheels are also rotated with the residual value created by innovations.

Constitutively, Schumpeter's economic development theory is based on a triple trivet. These trivets are innovations, loans that provide financing for these innovations and entrepreneurs who implement these innovations (Schumpeter, 1934: 73–74). The trigger of the whole development process is technology.

Schumpeter introduced the concept of innovation by associating it with the production function. While the production factors can be changed to the extent of the marginal technical substitution rate, the production function does not change, Schumpeter stated that innovation has emerged by changing the shape of the function instead of the amount of factors in the production function.

Schumpeterian innovation concept can be defined under five headings (Schumpeter, 2012:66):

i. Creation or promotion of a new product that is not yet acknowledged since it does not exist in the market or it is a newer version of an existing product

ii. Creation or promotion of a brand new production technique that does not yet exist or is not used before
iii. The opening of a brand-new market that a certain sector of the country has not entered before
iv. Creation and use of new supply sources for raw materials or semifinished products required for production
v. Creation of new organizational sectors or markets

According to Schumpeter, a new product goes through three different stages. The first one is the new knowledge required for the emergence of innovation; in other words, invention. After new information is revealed, it participates in production through the entrepreneur; in this second phase, defined as innovation, new ideas are included in the market as marketable products and processes. In the last stage, innovation is included in the economy as a new product or process and it is used in new markets, which is expressed as the dissemination phase. Therefore, the entrepreneur appears as an actor that ensures the continuity of the dynamic structure of the capitalist system. Schumpeter considered his entrepreneurial activities as the third production factor after labor and capital, where the meaning of production is new combinations: innovation (Schumpeter, 2012: 74).

In addition, according to Schumpeter, the entrepreneur's main motivation for achieving innovation is the "new spirit" that is full of war and heroism rather than high profit motive similar to Weber (Schumpeter, 2012: 93–94). When the financing needs of the entrepreneurs who create innovations, both to create them and to ensure their sustainability, are met by drawing credit, a recovery in the economy will occur as a result of a series of economic activities depending on the demand created by this additional credit supply (Schumpeter, 1939: 107–155). This revival will also create a revival effect in the relevant sector or industry and will cause more entrepreneurs to enter the system with the motive to make a profit. This will cause the innovation to become ordinary as much as it is proliferated, and profit will be obtained less than expected. This period of recession will create a crisis in the area related to excessive uncertainties, high risk and the cause of myopia. However, since this process is actually an adaptation process, the system will renew itself after a while, and it will be structured more strongly by introducing new innovations. This is the process of creative destruction in which a new system is replaced instead of the old system.

The Schumpeterian concept of creative destruction has two important parts: "creation" and "destruction." Schumpeter defines new combinations: innovation as the elimination of the old, which defines creation (of the new one) and destruction (of the old one) (Schumpeter, 2012: 67). The destruction actually means the withdrawal of old-dated and inefficient companies that do not follow technological developments from the market. In any event, creative destruction is already the disruption of the cyclical economic cycle. In this regard, there is no innovation without inventions, an entrepreneur without innovation and a capitalist cycle without the activities of the entrepreneur (Wingham, 2004: 31). The ongoing innovation activities of the entrepreneur in uncertainties and imbalances will bring along a renewed equilibrium state in which evolution and economic development are achieved (Andersen, 2004: 4).

3 Blockchain and Cryptocurrencies

In the last two decades, there has been a significant increase in innovative activities with the effect of rapid movements in information technology. The path of electronic commerce sites in the process has ended the actions of many institutions that carry out intermediary activities in this context and directly affected financial services and transactions. Financial instruments also had a share in this movement and development process, and creative innovations caused diversification in financial instruments. The biggest evolution seen in financial instruments is the emergence of digital money. Electronic money can be evaluated in two categories: payment instruments created by transferring banknotes under the control of national central banks to digital media and payment instruments produced through private institutions or mechanisms that are outside the control of national central banks yet still carry all three money motives (Ozturk and Koç, 2006: 209). In this study, the second type of electronic money is referred to as digital or cryptocurrencies and its effects on taxation processes are evaluated.

For the first time, in the study titled "Bitcoin: A Peer-to-Peer Electronic Cash System" by Satoshi Nakamoto (2008: 1–2), the mechanism in electronic commerce was purified from the cumbersome and challenging nature of the central authority, and the mechanism shaped by the trust and transactions among users was proposed. This blockchain system, which was proposed after many different digital currency attempts in a cumulative process, has been created differently from similar applications that have an accounting-score system that is easily accessible to users or participants, based on a peer-to-peer network system, accessible to all participants or users simultaneously. It is a reconciliation system that does not allow to make transactions with digital money twice and, in this context, protects people and created account with more secure, encrypted special algorithms (Krause et al., 2016: 6–8).

Two types of chaining have been created in the blockchain system. The first is the chains that users or participants can enter without any permission, without a password, and the second is the private chains that can only be accessed with a permission. While the entry of the participant to the private chain is with a special invitation, the usage permission is also pursuant thereto defined. In this context, secret protocols can be made, and data security can be ensured at the highest level through workflows (Jenner, 2017: 1087–1088).

Following the common acceptance of mining in the blockchain system (consensus), users have their own proof of shares, and these provable shares become a motivation or incentive for users. As a rule, the first transaction in a block belongs to the creator of that block which is a special mechanism that initiates a new cryptocurrency. As there is no central control center for the mentioned currencies earned in return with all this share, the system creates incentives for its users to stay in the system for a longer period and continues producing digital money (Nakamoto, 2008: 3–4). Bitcoin is the name of customized cryptocurrencies that are available in the system, such as coins.

However, Bitcoin is not the only form of cryptocurrency created in the blockchain mechanism. Ethereum, HyperLedger, Ripple, Cardano, Litecoin, Stellar, Neo, Monero, Dash, Iota and Tron are the other major blockchain platforms and cryptocurrencies created in listed platforms. These real-time and mostly international money platforms are alternatives where the system is obstructed for Bitcoin using blockchain-based technology (Usta and Doğantekin, 2017: 79–91).

4 Cryptocurrencies and Schumpeterian Creative Destruction

In the blockchain mechanism, which is a technology that supports Bitcoin, all protocols are executed over the Internet through a consensus. This situation has destroyed the way money has been used for the last few hundreds, giving it a brand new form of electronic printing and usage. It is definitely an innovation in this respect. In addition, the answer to the question of whether this innovation called "cryptocurrency" is constantly improving and renewing itself is yes; since the first cryptocurrency, Bitcoin, a lot of new types of cryptocurrencies have emerged, and the process of creating and using these coins is in a continuous improvement movement.

Besides, the dominant actor that led to the emergence of this innovation is the entrepreneurs themselves. In this regard, it is a technological playground of the entrepreneurs. The entrepreneur removes the old but disrupts the current economic balance and creates a new balance, but this new situation is repeated and renewed due to innovations. The result is an imbalance with constant change (Praag, 1999: 320). Also, the system, which is based on technology, innovations in technology and personal characteristics of entrepreneurs and decision motives, is also affected by the habits and/or behaviors of the environment (in this case, the society of the consensus). This situation simply shows that not only the internal variables but also the externals can affect the system. While each entrepreneur wants to maximize his profits, it can be only done so to the extent permitted by the consensus. A situation approved by all of those in the system is shaped on the basis of past and future profitability, as none of those in the system will want to make losses compared to their previous positions. So much so that in this area, the only thing that whips entrepreneurs is not profit, there are also different sources of motivation, such as staying active in a system to ensure continuity and feeling part of the consensus.

One of the leading usages of the blockchain mechanism is to provide value transfer between users. In this manner, Schumpeter's concept of the effects of innovation and new technologies both in the industry and the economy is evaluated in a dynamic competitive framework, since the continuous variable valuations and productions of cryptocurrencies make capital accumulation dynamic due to the structure of the blockchain mechanism.

Both national and international markets and economies are making efforts to adapt themselves to this new developing process. Some countries internalize as this process is in Canada, when trying to catch up the process by including these currencies to the process of the usage of public services; some other countries, such as Germany and Turkey, even if they do not accept such crypto coins as legal currencies, by following the development process of this money carefully they are striving to adopt their national monetary authorities as well as their system. In this context, cryptocurrencies create radical changes in the whole world and managements. However, it is clear that decades will be needed to shape this process. The movements to be experienced will be dynamic, considering that they will bring new developments rather than static. So, it would not be wrong to say that cryptocurrencies provide all assumptions of the Schumpeter's creative destruction theory since it is quite revolutionary in this context.

5 Taxation Process of Cryptocurrencies in Turkey and Across the World

In the blockchain system where cryptocurrencies are created, the cost of money creation is around 0 to 10 Euro cents, and its use increases rapidly due to the high security, ease of access and its solid infrastructure (Boucher, 2017: 6–7). According to Jenner (2017: 1089), as cryptocurrency usage increases in the society, new applications are required to be created and implemented to adapt tax laws to the system. Especially the excessive return offered by the fluctuating structure of cryptocurrencies and their use in the black market causes actors and the state in this field to pay attention to policy development by closely following the technological developments (Boucher, 2017: 7, 9).

The taxation and financial identification methods of some of the countries selected according to the common use of crypto money are summarized in Tab. 1.

When the financial instruments included in the blockchain in the Turkish financial structure are evaluated within the scope of electronic commerce, it is seen that various problems are encountered in more than one scenario. In order to understand the taxation in the sales process of goods and services, the circumstances defined by the Republic of Turkey Tax Procedural Law (TPL) No. 213 Articles 229, 230 and 231, Turkish Value Added Tax (VAT) Law No. 3065 Article 10 and Special Consumption Tax (SCT) Law No. 4760 Article 3 needed to be well understood.

According to the TPL, the invoice, which is a commercial document, must have the name, surname, address and account number of the issuer as well as some relevant information of customer. According to TPL 231/6, if the person who issues the invoice requests, the customers have to declare their identity. According to VUK 231/5, the voucher or invoice must be issued within 7 days from the date

Tab. 1: Recognition and taxation of cryptocurrency in selected countries

Country	Acceptance as legal payment instrument	Financial identification	Taxation
Argentina	Not accepted	Not accepted as a financial instrument, such as fiat currency or foreign currency	Income Tax Corporate Tax *Sales of the digital currency is taxed rather than ownership.
Australia	Not accepted	Not accepted as a financial instrument, such as fiat currency or foreign currency *Considered as a financial asset (property) *These currencies treated as money in terms of GST to avoid double taxation	Income Tax Corporate Tax Goods and Services Tax (GST) Fringe Benefits Tax *Taxation is made on the capital gain from the sale of such money held for more than 1 year and on the increase in value gained from the sale of money held for less than a year. *If employee earns it as salary, then it can be taxed as fringe benefits or income due to the circumstances.
Brazil	Not accepted	Not accepted as a financial instrument, such as fiat currency or foreign currency	Income Tax *There is almost no taxation yet.
Canada	Accepted	Considered as a financial asset similar to money	Income Tax Corporate Tax *Active taxation takes place.
China	Not accepted	Not accepted as a financial instrument, such as fiat or foreign currency *Accepted as an asset that can be considered a virtual commodity	Usage Prohibited

(*continued*)

Tab. 1: Continued

Country	Acceptance as legal payment instrument	Financial identification	Taxation
Estonia	Not accepted	Not accepted as a financial instrument, such as electronic money or foreign currency	Income Tax Value-Added Tax Corporate Tax
France	Not accepted	Not accepted as a financial instrument, as fiat currency or foreign currency *Considered a portable asset (moveable property)	Income Tax Corporate Tax Gift Tax *It is taxed when sold (noncommercial or commercial profit due to the activity). *No taxation on barter transactions. *Free transfers can be a subject of gift tax. *Taxation process has not been clarified completely.
Finland	Not accepted	Not accepted as a financial instrument, such as money or foreign currency	Income Tax Value-Added Tax Corporate Tax
Germany	Not accepted	Not accepted as a financial instrument, such as fiat money or foreign currency *Kept in a special status, in the same legal status as local currencies	Income Tax Value-Added Tax *The income generated by disposing of digital money held for less than 1 year is taxed. *No taxation for transactions longer than 1 year. *There is almost no taxation yet.
Holland	Not accepted	Not accepted as a financial instrument, such as electronic money or foreign currency	Income Tax Value-Added Tax Corporate Tax

Tab. 1: Continued

Country	Acceptance as legal payment instrument	Financial identification	Taxation
Japan	Accepted	Not accepted as money or foreign currency *Yet, legal method of payment *Accepted as property value *Still considered as financial asset	Income Tax Corporate Tax *Taxation is made by considering earnings as if the foreign currency's change profit. *Taxed as other income and earnings.
Norway	Not accepted	Not accepted as a financial instrument, such as money or foreign currency	Income Tax Value-Added Tax * There is almost no taxation yet.
Russia	Not accepted	Not accepted as a financial instrument, such as money	Income Tax *Taxation process has not been clarified compared to other taxes.
Singapore	Not accepted	Not accepted as a financial instrument, such as money or foreign currency *Accepted as a service	Goods and Service Tax *Even if they are included in the barter trade when they are used to purchase products in digital media, they are not taxed.
Turkey	Not accepted	Considered as a financial instrument, such as electronic money	Income Tax Value-Added Tax Corporate Tax *No taxation has been made yet.
UK	Not accepted	Not accepted as money or foreign currency *Considered as financial asset (working capital) *Accepted as investment only for small-scale business	Income tax *Taxed as foreign currency earnings *No taxation below £12,000 or up to 45 %

(*continued*)

Tab. 1: Continued

Country	Acceptance as legal payment instrument	Financial identification	Taxation
USA	Not accepted	Considered as commodity asset or property	Income Tax Value-Added Tax Corporate Tax *The tax-generating event takes place through the sale, exchange and use. *The income generated by disposing of digital money held for less than 1 year is taxed. * This income is taxed as capital gain. *Taxation process has not been clarified compared to other taxes.

Source: This table is constructed by referencing Payment Service Act of Japan (2009), Banco Central de Brasil (2014), Farell (2015), Raiborn and Sivitanides (2015), Senate Economic References Committe, (2015), Laudonia (2018), Zhang (2018), CH (2019), IRAS (2019), Solodan (2019), Canada Revenue Agency (2020), and Legifrance (2020).

of realization on behalf of the person purchasing or intermediating this good or service. Whereas, cryptocurrency users and participants of the system in the blockchain system can choose to keep their identities secret once again when they are given the right to use these coins in electronic commerce. As this secrecy is already a systematic feature, the tax system must adapt itself to these creative destructions of digital currencies. Considering that the sale is a consignment sale, the bill should be issued to the side of the sales person, as stated in Public Ruling No. 395. In the same scenario, in the context of the security of personal information, the taxation process will not be carried out again if the identity of the person performing the sale is hidden, not the intermediary or the seller.

When it is evaluated in terms of VAT with the prior knowledge that there is a situation that does not issue invoices, evaluations should be made on consignment sales within the scope of subclause d of Article 10, in case of intermediate sales rather than direct ones. In consignment sales, the delivery of the goods from one intermediary to another is not considered within the scope of the tax-generating event. However, during the delivery of the purchased good to the receiving person by the last intermediary, the tax is generated simultaneously to the intermediaries and those who sell the goods. Considering an environment where these financial instruments are exchanged, the owner of the product must arrange this voucher or invoice on behalf of the person who sells or mediates this sale. However, in this kind of environment where both the seller and the buyer are anonymous, there is a legal gap regarding how to perform taxation.

According to Ruling No. 1457 of the Ministry of Finance and Treasury, the persons who mediate the sale of coupon-based services on the Internet receives commission income and the institution's earnings should be calculated as of the periods when these revenues are accrued. It is also stated in the aforementioned Ruling that the transfer payments to the accounts of those who engage in customer acquisition activities will be evaluated as an incidental commercial gain, but if these activities are permanent rather than temporal, then taxation should be made in accordance with the provisions of commercial gain. It is clearly seen from this Ruling that cryptocurrencies are encountered with more than one problem when it is evaluated by buying or selling on the block chain or in another medium. In cases where the determination of the persons cannot be made, taxation will not be possible; if the persons are not determined and the number cannot be determined or the relevant amount of the cryptocurrencies is not deposited in the Turkish financial institutions, the taxation will not be made again or will be incomplete.

Law No. 6493 is about Payment and Securities Settlement Systems, Payment Services and Electronic Money Institutions. Hereunder this Law, electronic-currencies are the ones that are "*exported in exchange for a fund accepted by electronic-currency issuer, used to perform the payment transactions defined in this Law and the monetary values that are accepted as payment instrument by real and legal persons other than the electronic money issuer.*" However, Law No. 6493 also emphasizes that the institutions that will issue and define these funds should perform this transaction through the banks specified in Banking Law No. 5411 (Unsal, 2006:12). However, the operating logic of the blockchain mechanism is quite outside this descriptive framework. Therefore, there is an uncertainty about controlling the economic consequences of using an activity not carried out through banks.

Despite the fact that the creators of the cryptocurrencies created within the Blockchain and those who mediate the purchase of these currencies by others create residual values and earn profits, the transfer of these moneys to block accounts cannot be tax in terms of VAT since these transfers are not counted either as delivery of goods or service performance according to the current taxation laws of Turkey. Also, since invoicing is not required in such systems, they cannot be considered as income. The taxation process can be operated by ignoring these rules only if the owner of the cryptocurrency orders to pay to banks recognized by Law No. 5411.

If the buyer and the seller hold digital coins to take advantage of the exchange arbitrage difference, then this activity should be considered as a capital gain (entering the area of increment value tax); yet, if this trading is being carried out continuously, it should be evaluated as a commercial gain according to Turkish Income Tax Law No. 193 (Kukrer, 2016: 594). However, it is essential both to know these buyers and sellers identity information and to make them bring the revenues of the crypto money they have obtained in the blockchain system to the banks specified in Law No. 5411. In any event, the definition of cryptocurrency as a security will makes it only subject to Income Tax by exempting it from VAT (Gunay & Kargi, 2018: 69).

According to Dogan et al. (2018: 30), if the income to be obtained from cryptocurrencies is evaluated as a security return or a capital market instrument, then it should be treated as an increase in value and evaluated in terms of Income Tax procedures. If cryptocurrencies are to be treated as a type of financial instrument, such as a foreign currency, then the foreign exchange should be evaluated by the market in accordance with Article 280 of the TPL. Kukrer (2016: 593–594) has suggested the idea that cryptocurrencies can be taxed by evaluating them as "*virtual stock market*" due to their structure dependent on supply and demand as exchange-earning factors. In this regard, in case of selling by cryptocurrency, the VAT base should be computed by determining the price based on the comparative value at the date of the invoice. Since there is neither such a stock market in Turkey nor an exchange rate determined by the Ministry of Finance and the Treasury or a declared period-end value by the Central Bank, according to Article 289 of the TPL, transactions must be made either on comparative or on equivalent values.

Moreover, in case of the cryptocurrencies are kept by taxpayers within the framework of a business activity or keeping these coins in the system without converting them to fiat currency, it should be taxed on comparative value when it is registered.

6 Schumpeterian Inferences in Terms of Taxation in Cryptocurrencies

With the advent of cryptocurrencies, as discussed under "*Taxation Process of Crypto-Currencies in Turkey and Across the World,*" ideas about the taxation of this type of money, which did not exist before but emerged due to the development of technology and that it is now obvious that it will continue to exist in the life of human beings, have begun to be developed and applications have been formed. When these practices are evaluated, it is understood that Schumpeter's innovation concept is directly within the scope of creating a new product or a new version of this product in the context of the introduction of new taxes and/or the emergence of new tax issues. Also, considering that there is a new production technique and/or new applications in taxation, it is clear that it fulfills this condition. In this regard, the taxation of cryptocurrencies is an absolute innovation, as cryptocurrencies will create a new tax system requirement for countries both theoretically and practically.

As mentioned earlier, the emergence and spread of innovation and its use in new markets is part of Schumpeter's creative demolition theory, where entrepreneurs who aim to realize this innovation will be financed mostly through loans, and this credit supply will create development and recovery in the economy. With more people entering the system with the profit motive, innovation will become ordinary, as the general accepted acceptance in the economic literature, the profit will decrease in the process and a period of recession will occur. This period, when uncertainties and risks are high, will also cause depression when combined with myopia. Here, in fact, this is an adaptation process.

When this process is formulated for cryptocurrencies, its inclusion in a wide range of transactions, from voting to electronic commerce, from stock market transactions to public offering, from the moment of its emergence shows that it creates new markets as an innovation and takes an active role in the spread of its own innovation. While those who want to enter the system provide their personal financing, these entries, on the other hand, increase the number of participants and users in the system with high profit probability. In this development process of cryptocurrency, the state will become an agent once it discovers the opportunity for taxation and from this second on it will try to adapt its system.

If the government adapts the taxation process to the constant renewal of this digital system itself, the entire taxation process will also become dynamic and economic balances can be reorganized by the state in favor of each new situation. In fact, taxation will not take place for a while or there will be significant disruptions due to systematic deficiencies such as the financial definitions or in which tax subject these currencies have not yet been clarified in taxation process. In consequence of this situation, the state will produce temporary adaptive solutions such as arbitrage gain-increment value gain. With the acceleration of cryptocurrencies' development and increasing usage, states will have to put cryptocurrencies into financial instrument status. With the realization of the financial definition and elimination of the uncertainties regarding the tax process, the state will collect more tax revenue from the creation and use of such money in the first place compared to the previous periods; however, such as the escapes from national currencies to such currencies, transactions made without including the national banking system will cause a decrease in tax income. In addition, clarifying the procedures and rules of taxation, which were not made before or made at a low rate or amount, will increase the loss of earnings for entrepreneurs and escapes from the system on an entrepreneur basis will increase as well. In this case, the state will have to make a new effort to adapt itself to this process, and the tax system will be transformed again. It is seen that this radical evolutionary process corresponds to the theoretical phenomenon that Schumpeter called creative destruction.

However, even if this situation is envisaged in terms of process, it is necessary to test whether there is a match with the assumptions dimension or not. Tab. 2 is created to investigate whether Schumpeter's creative demolition effect assumptions can shed light on the transformation in taxation time. Tab. 2 is an aggregated list of Schumpeter's assumptions in creative destruction theory. In this list, the third section on why cryptocurrencies are considered to have a creative destruction effect is visualized in the items examined verbally under the heading "*Cryptocurrencies and Schumpeterian Creative Destruction Theory.*" In this way, it is also desired to display inferences about why the taxation of these currencies is considered as a radical change factor.

As can be seen from the table, cryptocurrencies provide all the assumptions of the creative destruction theory, while the taxation process of these currencies is consistent with almost all assumptions (as previously detailed). The individual in theory can be described as the state when it comes to the taxation process since the state is the main actor along with the citizens themselves. In this case, one of their motivations in the taxation activity in their state is utility maximization. By courtesy of this maximization, it becomes possible to maximize the ultimate goal of collective utility.

Besides, it is shown in the first two paragraphs of this section that the taxation of cryptocurrencies is also a theoretical innovation. In this regard, with the prediction that innovation itself will become a new system in the long term as a tool for innovation, a projection parallel to the theory can be made in terms of the taxation process in terms of both internal and external variables. Considering that both the entrepreneurs as cryptocurrency users in the blockchain mechanism and the state that designs and manages the taxation process build all the calculations according to past and future valuations, as in fiat money, this assumption of the theory also seems to be accepted. In addition, the taxation process will be shaped over time, in the adaptation process of the system, and will even transform the taxation into a new process that is different from the current phases. Yet, as with all evolutionary processes, this process is predicted to take place in the long term. Evidence of this prediction is hidden in the process of paper money and electronic money being included in the economy. All the stages of these money types have created important innovations in the economy over time and have radically affected and changed the taxation compared to the previous situation. Concordantly, when all processes from the creation of the digital currency to the present are analyzed, it is expected that cryptocurrency will have closely similar effects created by others. Therefore, there is no reason to not say that it fulfills the fifth, sixth and fourteenth assumptions.

Also, profit is one of the reasons entrepreneurs use crypto money or join the system. However, profit is not the only motivation, as described in previous chapters. Similarly, although the government's basic motivation to tax is the acquisition of income, it is undoubtedly not the only motivation. Taxation can serve more than one notion, such as financing public spending to provide public goods and services to citizens, bringing markets closer to economic stability, ensuring the continuity of the social security system and forming a basis for borrowing. In this context, adding a new tax or making additional changes in the subject and scope of a tax does not only create radical changes in the system itself but also causes a radical change in its motives. Whereas a tax on deposits reduces demand for money with prudence, expanding the scope of the tax on stock exchange transactions will adversely affect the speculative motive. Considering that dozens of different cryptocurrencies have been created and included in the economic system in the same logic in less than a decade and considering some of these currencies specialize

Tab. 2: Evaluations of cryptocurrencies and their taxation process in terms of creative destruction theory's assumptions

Assumptions	**Cryptocurrencies**	**Taxation of cryptocurrencies**
Individuals seek to maximize their utilities.	✓	✓
Internal variables: personal characteristics, technology and innovations.	✓	✓
External variables: extraordinary situations such as war, social habits and environmental factors.	✓	✓
Profits are determined by including the past value and the expected value in the calculations.	✓	✓
Time is an important dynamic dimension that affects individuals and changes the system.	✓	✓
Every dynamic analysis requires a historical observation.	✓	✓
It is not only clear rational accounts that motivate companies or individuals, but also there are some other motivational factors.	✓	✓
Innovation is the change in routine, is a radical evolution toward being a different factor.	✓	✓
The economy develops in a continuous motion due to innovations.	✓	✓
The leading actor that caused the emergence of innovations are entrepreneurs.	✓	✓
Innovations are clustered in an industry or industry, as firms follow each other's innovations.	✓	✓
The economy adapts itself to the constantly developing and renewed process.	✓	✓

(*continued*)

Tab. 2: Continued

Assumptions	**Cryptocurrencies**	**Taxation of cryptocurrencies**
It takes a period of 50–60 years to achieve adaptation and complete evolution by providing a complete cycle.	✓	✓
A cycle consists of four phases: prosperity, recession, depression and revival.	✓	✓
The old structure is destroyed during the adaptation process and collapses completely as a result of innovations.	✓	✓
The balance resulting from adaptation is dynamic, rather than static. Even it constantly changes, it never occurs again at the same equilibrium point.	✓	✓
In this process, it decreases the costs of newly produced products, thereby opening the scissors with the determined prices of the manufactured goods, and revolving the capitalist wheels with the residual value created by innovations.	✓	✓

Source: Author

for certain business and transactions (i.e., Ripple stands out in commercial activities), it is obviously seen that the economy is still in the process of adaptation, and for this reason, the need for a taxation procedure to be created by the states to be adaptive in such a way that it is constantly renewed emerges automatically. In this case, the validity of the seventh, eighth and ninth assumptions, which are shown in Tab. 2, becomes visible.

Moreover, it is clear that the most important leading factor in the creation and use of cryptocurrencies is entrepreneurs. The entrepreneur can be come into prominence in the taxation process since in order to realize the taxation, it is a necessity to have some sort of earnings. On the other hand, it should be underlined that the state also plays a dominant role as much as the entrepreneur. Since the state is the only management factor that has the power of collecting taxes, unlike all the administrative elements. In this regard, how use and spread of cryptocurrencies cannot be mentioned without being an entrepreneur, no taxation can be mentioned without the state. So, the state replaces the entrepreneur as a creative factor in taxation. In addition, the state can become an actor in this market through some of its functions such as interbank and Takasbank, and can be both participatory and normative. In this case, it becomes a main actor of the process from a different perspective again. Consequently, another assumption is accepted in the taxation process of cryptocurrencies.

The inclusion of entrepreneurs who have been pursuing profit maximization since the day cryptocurrencies emerged has brought together a cluster in the consensus structure, both as a user and a participant. This natural process is as usual as the cryptocurrency creation process and the entry into the blockchain mechanism for other individuals who want to maximize profits. When it comes to the taxation of these currencies, as can be seen – under "Title IV" – each country tries to establish the legal basis for the taxation of these money within the framework of its own laws and/or aims to be included in the system. While this is the case, it is obvious that there is and will be a cluster in terms of similar practices since the leader country's successful experiences are used. Therefore, another assumption becomes valid.

Besides all these, it is foreseen that the economy will look for a number of ways, take precautions and carry out practices regarding the use and prevalence of these currencies both in parallel with the decisions taken by the country administration and in the light of the reactions of the national and international markets. This prediction becomes reality based on the fact that some countries already use this kind of money in the definition of money and use it in public applications. The economy, which tries to adapt itself to cryptocurrencies and the blockchain system, will also give evolutionary reactions to the decisions regarding the taxation of these coins. Because the government's discovery of the opportunity for taxation and decisions made with the desire to generate more tax revenues, on the one hand, the moves that strengthen the hand of the public, on the other hand, may be factors that discourage the entrepreneur's determination and enthusiasm. Forbye, the weakness of governments' enforcement, the failure of the national monetary mechanism and the existence of mostly informal transactions are also a challenge for national governments. As a result, this process will have a long swing effect in parallel with the determination of legal limits, old habits will be abandoned and will eventually come to balance. However, only time will determine whether it will create a general equilibrium or partial equilibrium conditions. In this context, the fact that the process creates new effects in the long run and reaches a general or partial balance is exactly what the creative destruction theory points to.

7 Results

Cryptocurrencies and the blockchain mechanism have rapidly changed economic balances since cryptocurrencies were theoretically drawn and their applications started. These changes have become so radical that even the institutions that are the main actors of the national money markets have begun to transform their systems into harmonization in order to be included in this mechanism. Whether these radical changes have a structure that will transform the economy in a better way by shaking the old structure from the foundation has been investigated by many theorists. The evaluation of the effects of cryptocurrencies from the perspective of Schumpeter (1927), who mentioned this process as a creative demolition process by talking about the fact that technology is a variable in the process of evolving from the old to the new and that the innovative elements will create radical changes in the economy, has begun to attract new attention of scientists as an emerging topic. However, researches on the taxation of these currencies have focused only on the dimension of how taxation can be done, and its theoretical transformation has not been addressed before. So, here the purpose of this study is to complete this missing piece of the puzzle and to determine whether changes in the taxation process of cryptocurrencies can be evaluated within the scope of Schumpeter's theory of creative destruction.

When the assumptions of the theory and the stages of the relevant taxation process are compared, it is determined that this process, which is experienced and will continue to be experienced, fits to the assumptions. This result is a prediction that a strong analogue of the effect that the paper money has created against alternative valuation instruments in the period when it first appeared will be made to fiat money by cryptocurrencies, at least in the long term. In this regard, Schumpeter (1927)'s idea that "*completing the cycle of operation in each innovation requires a long process*" will be manifested in this circumstances as well. Although the creation of an innovation in the short term is not clearly understood in relation to the taxation process, it is obvious that there will be a change in the long term and new understandings will emerge. The importance of digital currencies as well as the transformation in the taxation process will be fully understood only after cash coins have been completely lost, as in the case of other exchange tools disappearing with the emergence of cash. Yet this does not mean that creative destruction has not been experienced even now. On the contrary, it provides an opportunity for both lawmakers and practitioners in the process of change to raise awareness of the process. Therefore, determining the effect of creative destruction in the taxation of crypto money, which is the aim of this study, also contributes to raising this awareness.

The practical determination of this destructive creation process, which is theoretically strongly predicted to be experienced in the long term due to the radical change and hypothetical conditions that cryptocurrencies will create in taxation, can be done by using qualitative and quantitative research methods such as questionnaires or interviews to be used for individuals who are not yet taxpayers but will become taxpayers in the future and individuals who are still taxpayers in the system in terms of the use of crypto money and its effects on the tax compliance process. Because, as an accepted reality in both domestic and foreign economic literature, the main factors shaping future behaviors are the personality traits, the environment they live in, their future prospects and the developments in the current period as well as the events that occur in the future.

Notes

1. As aforementioned, the general assumptions of the Schumpeter's theory have been listed by the help of not just one but several of his works since the development of the theory has spread more than one. The papers and associated pages which the assumptions are generated from are: 1927:25–26,82–84,287–294; 1934:6,35,56–67,1325–1334;1939:70–182; 1950:61–86, 215–218;1954:961–1008,1959;2005:106–112.

Bibliography

Andersen, Esben Sloth, (2004), "From Schumpeter's failed Econometrics to Modern Evometric Analysis: Creative Destruction as a Tale of Two Effects", *Conference of the International Schumpeter Society*, Bocconi University, Milan, 9–12 June.

Banco Central do Brasil, (2014), "Policy Statement No. 25–306 Supra note 1", Retrieved from http://www.bcb.gov.br/pom/spb/ing/IComunicado25306.pdf (07.03.2020).

Blackwell, Calvin, (2007), A Meta-analysis of Tax Compliance Experiments, *Georgia State Andrew Young School of Policy Studies, International Center for Public Policy Working Paper* (pp. 07–24)

Boucher, Philip, (2017), "How Blockchain Technology Could Change Our Lives: In-Dept Analysis", European Parliamentary Research Service, Retrieved from https://www.europarl.europa.eu/RegData/etudes/IDAN/2017/581948/EPRS_IDA(2017)581948_EN.pdf (25.02.2020).

Canada Revenue Agency, (2020), "Guide for Cryptocurrency Users and Tax Professionals". Retrieved from https://www.canada.ca/en/revenue-agency/programs/about-canada-revenue-agency-cra/compliance/digital-currency/cryptocurrency-guide.html (19.03.2020).

CH Alliance, (2019), "Investing n Cryptocurrency the Labyrinth of Tax Reporting", Annual Report of 2019, Retrieved from https://www.chappuishalder.com/wp-content/uploads/2019/02/3.-ARTICLE-02-Investing-1.pdf (15.02.2020).

Coricelli, Fabrizio and Dosi, Giovanni, (1988), "Coordination and Order in Economic Change and the Interpretive Power of Economic Theory", (G. Dosi, C. Freeman, R. Nelson, G. Silverberg and L. Soete Eds.), *Technical Change and Economic Theory*, New York: Printer, pp. 124–147.

Dogan, Zeki, Buyrukoglu, Selcuk and Kutbay, Huseyin, (2018), "Turkiye'de Bitcoin Islemlerinin Vergilendirilmesi ve Muhasebelestirilmesine Iliskin Oneriler", *Vergi Sorunlari Dergisi*, Vol. 41, N. 361, pp. 23–33.

Dolgun, Ugur, (2006), "Schumpeter'a Gore Ekonomik Gelismede Girisimcinin Dinamik Rolu", *Girisimcilik ve Kalkinma Dergisi*, Vol. 1, No. 1, pp. 173–176.

Farell, M. Ryan, (2015), "An Analysis of the Cryptocurrency Industry", UPENN Wharton School Scholarly Commons. 5–2015, Retrieved from https://repository.upenn.edu/cgi/viewcontent.cgi?article=1133&context=wharton_research_scholars (25.02.2020).

Fischer, M. Carol, Wartick, Martha and Mark M. Melvin, (1992), "Detection probability and taxpayer compliance: A Review of the Literature", *Journal of Accounting Literature*, Vol. 11, N. 1, pp. 1–46.

Ipek, Selcuk and Kaynar Ilknur, (2009), "Demografik Faktorlerin Vergi Uyumuna Etkisi: Bir Anket Calismasi SonuClari", *Maliye Dergisi*, Vol. 157, N. 1, pp. 116–130.

Inland Revenue Authority of Singapore – IRAS, (2019), "IRAS e-Tax Guide", https://www.iras.gov.sg/irashome/uploadedFiles/IRASHome/GST/Draft%20 e-Tax%20Guide%20_GST_Digital%20Payment%20Tokens.pdf (07.03.2020).

Jenner, Simon, (2017), "Blockchain: The Digital Tax Function's Leading-Edge Technology?", *Tax Notes International*, Vol. 88, N. 11, pp. 1087–1089.

Kirchler, Erich, (2007), *The Economic Psychology of Tax Behavior*, Cambridge: Cambridge University Press.

Krause, G. Eric, Velamur, K. Vivek, Burghardt, Tobias, Nack, Denny, Schmidt, Moritz and Treder, Tobias-Micha, (2016), *Blockchain Technology and the Financial Services Market State-of-the-Art Analysis,* HHL Leopzig Graduate School of Management Working Paper, Retrieved from https://www.infosysconsultinginsights.com/insights/blockchain-technology-and-the-financial-services-market/ (21.02.2020).

Kukrer, Ceyda, (2016), "Internet Ortaminda Bazi Faaliyetlerin Kavramsal Tanimlari ve Turk Vergi Sistemi Karsisindaki Durumlarinin Değerlendirilmesi", *Yonetim Bilimleri Dergisi*, Vol. 14, N. 27, pp. 583–604.

Laudonia, P. Mara, (2018), "The Legal Vacuum of the Bitcoin, Is It or Is It Not Money?", *TELAM*, http://www.telam.com.ar/notas/201702/180185-el-vacio-legal-del-bitcoin-eso-no-es-dinero.html, archived at https://perma.cc/P2WE-L8F9 (11.03.2020).

Legifrance, (2020), "Ordonnance n° 2017–1674 du 8 décembre 2017 relative à l'utilisation d'un dispositif d'enregistrement électronique partagé pour la représentation et la transmission de titres financiers", Retrieved from https://www.legifrance.gouv.fr/affichTexte.do?cidTexte=JORFTEXT000036171908&categorie (07.03.2020).

Nakamoto, Satoshi, (2008), "Bitcoin: A Peer-to-Peer Electronic Cash System", https://bitcoin.org/bitcoin.pdf (5.01.2020).

Ozturk, Nurettin and Koç, Asuman, (2006), “Elektronik Para, Diger Para Turleriyle Karsilastirilmasi ve Olasi Etkileri”, *Sosyal Ekonomik Arastirmalar Dergisi*, Vol. 6, N. 11, pp. 207–243.

Payment Services Act of Japan – PSA, (2009), “Act no. 59 Article 2 General Provisions Para 5”, http://www.japaneselawtranslation.go.jp/law/detail_download/?ff=09&id=2319 (07.03.2020).

Puviani, Amilcare, (1903), *Teoria della İllusione Finanziaria,* Milan: Independently Published.

Raiborn, Cecily, and Sivitanides, Marcos, (2015), “Accounting Issues Related to Bitcoins”, *The Journal of Corporate Accounting Finance,* Vol. 26, No. 2, pp. 25–34.

Schmolders, Gunter, (1959), “Fiscal Psychology: A New Branch of Public Finance”, *National Tax Association Journal,* Vol. 12, N. 4, pp. 340–345.

Schmolders, Gunter, (1976), *The Psychology of Money and Public Finance,* 2006 Edition, New York: Palgrave Macmillan.

Schumpeter, A. Joseph, (1927), “The Explanation of the Business Cycles”, *Economica.* Vol. 21, N. 4, pp. 286–311.

Schumpeter, A. Joseph, (1934), *The Theory of Economic Development,* Cambridge: Harvard University Press.

Schumpeter, A. Joseph, (1939), *Business Cycles: A Theoretical, Historical and Statistical Analysis of the Capitalist Process,* II Volume, New York: McGraw-Hill.

Schumpeter, A. Joseph, (1950), *Capitalism, Socialism and Democracy,* New York: HarperBrothers Publishers.

Schumpeter, A. Joseph, (1954), *History of Economic Analysis,* London: George Allen & Unwin.

Schumpeter, A. Joseph, (1956), *Ten Great Economists: From Marx to Keynes,* London: George Allen & Unwin.

Schumpeter, A. Joseph, (2005), “Development”, *Journal of Economic Literature,* Vol. 43, N. 1, pp. 108–120.

Schumpeter, A. Joseph, (2010), *The Nature and Essence of Economic Theory,* New Brunswick, New Jersey: Transaction Publishers.

Schumpeter, A. Joseph, (2012), *The Theory of Economic Development with a New Introduction by John E. Elliott.* Cambridge: Harvard University Press.

Senate Economic References Committee of Australia, (2015), “Digital Currency-Game Change or Bit Player”, Retrieved from ttps://www.aph.gov.au/Parliamentary_Business/Committees/Senate/Economics/Digital_currency/~/media/Committ ees/economics_ctte/Digital_currency/report.pdf (07.03.2020).

Solodan, Kateryna, (2019), "Legal Regulation of Cryptocurrency Taxation in European Countries", *European Journal of Law and Public Administration*, Vol. 6, N. 1, pp. 64–74.

Usta, Ahmet and Dogantekin, Serkan, (2017), *Blockchain 101*, Istanbul: Kapital Medya.

Unsal, Gulen, (2006), "08.03.2006 Tarihli ve 5472 Sayili Kanunla Yapilan Değisikliklere Gore Guncellenen Madde Gerekceleriyle Mulga 4389 Sayili Kanunun Ilgili Hukumleriyle Birlikte 5411 Sayili Bankacilik Kanunu", *BDDK Yayin No: YK/5*, Retrieved from https://tarama.uludag.edu.tr/cgi-bin/koha/opac-ISBDdetail.pl?biblionumber=75680 (17.02.2020).

Wingham, W. Dianne, (2004), "Enterpreneurship Through the Ages", (H.P. Welsch, Ed.), *Enterpreneurship: The Way Ahead*, New York: Routledge, pp. 27–42.

Zhang, Laney, (2018), "China – *Regulation of Crytocurrency in Selected Jurisdictions*", Retrieved The Law Library of Congress, https://www.loc.gov/law/help/cryptocurrency/regulation-of-cryptocurrency.pdf (07.03.2020).

Emre Kadir ÖZEKENCİ

"The Overview of Industry 4.0 and Society 5.0 Paradigms"

1 Introduction

Whether we accept or not, Industry 4.0 has already started to impact and change the world significantly. In the present, there are great opportunities for companies and people who are ready for/against the potential challenges of tomorrow. In other words, companies and people that adapt to the current changes and development won't face much difficulties when forming and implementing new business plans. As stated in The World Economic Forum (WEF) (2016), the world has become more and more competitive than ever before and people need to have critical skills for future work. One of the key critical skills is readiness for Industry 4.0. Without a doubt, technology and its components impact significantly each part of our daily lives. Nowadays, we are witnessing a huge change in almost every part of our life. In other words, we are living in a new world in which innovation is driven by fantastic technologies, and such technologies are bringing important changes to the economy and society. Thus, innovation is one of the key components of digitalization of industry and social infrastructure. Especially, after the beginning of the twenty-first century, globalization and rapid development in digital technology such as robotics, artificial intelligence, Internet of Things and big data cause significant changes not only in the industrial environment but also in social life (Savaneviciene et al., 2019:211–212). Japan is one of the pioneer countries in terms of innovation and sustainable development. Particularly, the Japanese government can be progressive in the field of education, agriculture, population and climate change with the help of latest innovations (Moss, 2019). Accordingly, the Japanese government was presented with "Society 5.0" as a core concept in the fifth Science and Technology Basic Plan in 2016 (Fukuyama, 2018:47). The main goal of the new economy and society increase the power of individuals. Without any discrimination, all individuals, including women and older people, can live securely and wealthy. Additionally, the new economy and society action plan has realized some issues such as environmental problems, terrorism, population and industrial competitiveness by creation of future (Keidanren, 2016:10–12). In this study, the paradigms of Industry 4.0 and Society 5.0 is investigated. The main objective of this study is to enhance the awareness of Industry 4.0 and Society 5.0 between individuals and businesses.

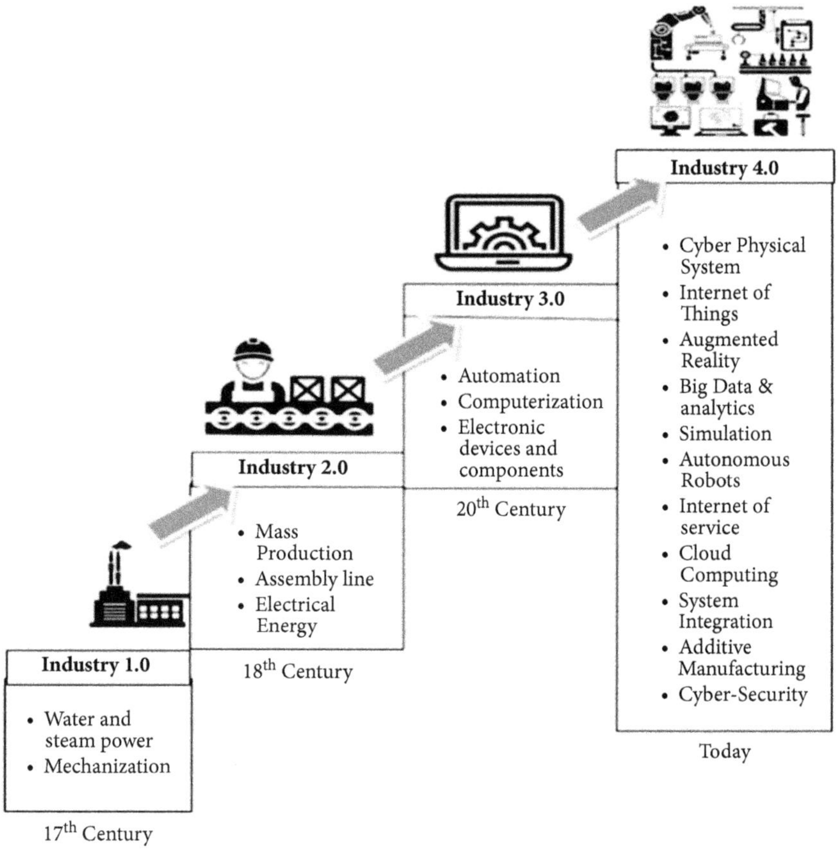

Fig. 1: The evolution of industry revolution. **Source:** Tay, et al., 2018

2 Industry 4.0 Paradigms

Until now, the world has witnessed three industrial revolutions. The diagram of evolution for industrial revolution is shown in Fig.1. The first industrial revolution started in the late eighteenth and in the beginning of nineteenth century (1760–1840). In the first industrial revolution, there was a transition from human and animal workforce to machinery,such as establishment of new chemical and iron manufacturing factories, enhanced steam power and improved water power efficiency (Mohajan, 2019:377). Also, in the first industrial revolution, textiles and iron industries played a crucial role (Ashton, 1997). The second industrial revolution covers the period from 1870 to 1914. In this period, the nature of production changed significantly. The world witnessed exponential growth in some industries. Industry 2.0 mainly used electric power for mass production. Chemicals, steel, transportation and agriculture industries played an important role in the second industrial revolution (Mokyr, 1999). The third industrial revolution started in the middle of the twentieth century (1960–2000). Great development in information technology led to new production processes. The invention of the Internet and smartphones and other major developments in information technology played a key role in the progress of the third industrial revolution (Akbaba, 2018:8). Also, in this period, electronics and information technology was used significantly to automate production (Schwab, 2017).

Today, we find ourselves in a new world called the "digital industrial revolution. "The fourth industrial revolution, also referred to as "Industry 4.0," can be described as "the modern and more sophisticated machines and tools with advanced software and networked sensors can be used to plan, predict, adjust and control the societal outcome and business models" (Mrugalska and Magdalena, 2016). Industry 4.0 was introduced as a new era of industrial revolution for the first time by Germany at the Hannover fair event in 2011. When the idea was shared with European manufacturing researchers and companies, the level of interest was quite high due to possibilities for increased productivity level and decreased cost-effectiveness (Tay et al., 2018:1380). Industry 4.0 consists of some disruptive technologies and paradigms, such as 3D printing, Smart Factory, big data, deep learning, Internet of Things, augmented reality, machine learning, cloud computing etc. (Catal and Tekinerdoğan, 2019: 100). With the help of the abovementioned new technologies, industrial systems become more and more secure, smart, effective, integral and reliable for the companies (Ahuett-Garza and Kurfess, 2018; Muhuri et al., 2019). Additionally, other major benefits and features of Industry 4.0 are as follows: enhanced operational efficiency and productivity, high level of automation, automated data exchanged, optimization, automation adaptation, value-added services, customization of production and human–machine interaction (Posada et al., 2015; Roblek et al., 2016; Thames and Schaefer, 2016; Lu, 2017).

In spite of numerous benefits of Industry 4.0, it also includes some negative consequences for companies and people. The biggest concerns are potential for job losses and "dehumanization" of our lives due to advanced technologies (Buytaert and Raj, 2016; Tanrıogen, 2018:164). However, there is no consensus about identifying "Industry 4.0 key technologies." In other words, organizations and academic researchers may express discrepancies on the key technologies of Industry 4.0 (Hermann et al., 2015; Chen et al., 2017). For instance, Boston Consulting Group (*BCG*) stated nine key disruptive technologies of Industry 4.0 (Rüßmann et al., 2015). On the other hand, Cheryl and Helena pointed out five key disruptive technologies (Cheryl and Helena, 2017). Another research conducted by McKinsey Global Institute explained twelve key disruptive technologies, whereas the WEF reported nine key disruptive technologies of Industry 4.0 (Manyika et al., 2013; World Economic Forum, 2016). The fourth industrial revolution is evolving at an exponential rate compared to previous industrial revolutions. Therefore, as stakeholders

Tab. 1: Demographics of Turkey with respect to the world (2020)

	Turkey	The world
Total population	83.88 million	7.75 billion
Urbanization	75 %	55 %
Internet users	62.07 million	4.54 billion
Internet users penetration	74 %	59 %
Active social media users	54.00 million	3.80 billion
Active social media users penetration	64 %	49 %
Mobile phone users	77.39 million	5.19 billion
Mobile phone users penetration	92 %	67 %

Source: Kemp, 2020

involved public and private sector, the academic world and the civil society, we must develop a comprehensive plan to be able to respond to it effectively (Krinis, 2016:12). Due to the several benefits of Industry 4.0, not only the industrial companies are interested in learning and implementing such new technologies, but also the academic world is interested in gaining insight about new paradigms in the educational curriculum (Catal and Tekinerdoğan, 2019:100).

2.1 The General Overview of Digital Transformation: Turkey and the Rest of the World

The number of Internet users in the world is approximately 4.6 billion and the same ratio is about 62 million in Turkey. The world average of Internet user penetration is 59 %, and the same ratio is 74 % for Turkey. The average daily time that Internet users aged 16 to 64 spend on different kinds of media and devices is about 7 hours 30 minutes. The active social media users are 54 million and 3.80 billion in Turkey and in the world, respectively. Of the total population, 64 % of people use social media platforms. The most-used social media platforms in Turkey are YouTube (90 %), Instagram (83 %), WhatsApp (81 %), Facebook (76 %) and Twitter (61 %). The number of mobile connections is about 5.20 billion in the world and the same ratio is about 78 million for Turkey. Of the total population, 92 % of people have mobile connections (Kemp, 2020).

Many countries around the world have started to establish or reorganize their manufacturing operations in the context of digitalization. In other words, the digital transformation process is adopted significantly by countries. For instance, the European Union announced a strategy called "Digitalization of European Industry" in2016 (European Commission, 2018). In addition, China declared its new strategy called "Made in China 2025" (McBride and Chatzky, 2019). A common point of both

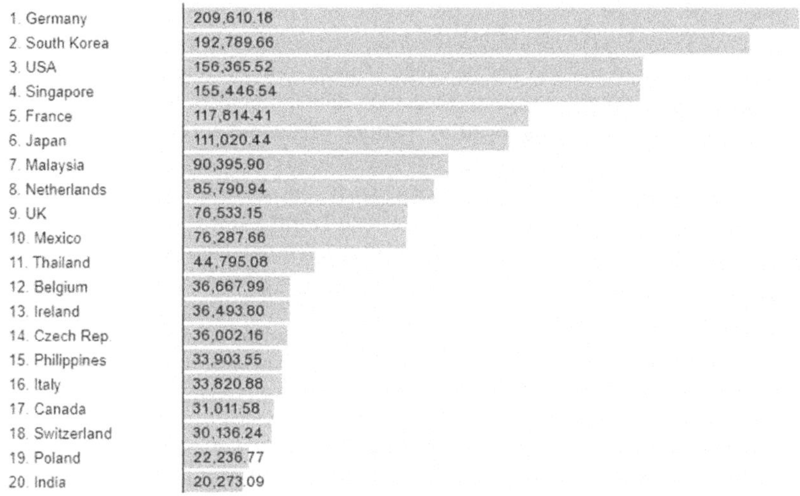

Fig. 2: Country rankings of high technology export in the period of 2018 (Million US dollars). **Source:** theglobaleconomy.com, 2018

strategies is that they transform their countries' low added value manufacturing into high added value manufacturing (Özlü, 2017:33–34). High-technology export can be described as "products with high R&D intensity, such as aerospace, computers, pharmaceuticals, scientific instruments, and electrical machinery" (theglobaleconomy.com, 2018). High-technology products' production and export is one of the key factors in terms of development and growth of countries (Yıldız, 2017:27). According to the technology and innovation report in 2018, science, technology and innovation are going to be play important roles in the achievement of sustainable development in the following years. Additionally, as indicated in 2030, the agenda for sustainable development, the role of technology and innovation are crucial to transform economies, increase productivity and living standards and decrease production cost and price. Especially, some emerging technologies such as big data, Internet of Things, artificial intelligence, 3D printing and drones are the greatest potential to enable the achievement of sustainable development goals (Unctad, 2018). Accordingly, at present, many developing countries allocate their resources to education and research activities to rise to the level of developed countries in terms of technological development (Göçer, 2013:216). The Fig.2 illustrates country-wise rankings of high technology export in 2018. Germany, South Korea and the United States are getting huge shares of high-technology export. Also, some developing countries such as India, Mexico and Malaysia try to get more shares in high-technology export by creating new strategy. Although, Turkey is the nineteenth largest economy in the world, the rank of high technology export of Turkey is 35 with 3116.83 million US dollars (theglobaleconomy, 2018).

After the decision of 24th January1980, Turkey opened its economy to rest of the world. In this regard, the new economic development policy called "export-oriented economy" was announced by Turkish policymakers. The main objective of the new strategy was to enhance the economic growth rate. However, the new economic development policy hasn't reached to the desired level, because many industrial productions have made mid-low technology instead of high-technology (Konak, 2018:71–72; Sahin, 2019:166). In other words, Turkey's export mainly consists of normal goods instead of products with high added value. In the past decade, the percentage of high-technology export hasn't changed significantly in Turkey; it increased from 3.4 % to 3.5 % in the period of between 2010 and 2018 (Ministry of Trade, 2019). Therefore, a new technology-oriented action plan has been prepared by the industry and technology ministry. According to the ministry, the new action program would be a critical milestone for the Turkish economy and it reduces the amount of current deficit in Turkey by using advanced technology (Dailysabah, 2019). At this point, the university–industry collaboration in the scope of Industry 4.0 is one of the key trigger factors for sustainable development. Research based on the WEF (2017) pointed out that information and communication technology (ICT) clearly has an influential role in innovation. It can be seen from Fig.3, there is a positive relationship between innovation and ICT use. For instance, Germany, Japan and the United States are relatively equal in terms of innovation level and ICT use. By contrast, some countries such as Brazil, Russia, Italy and Turkey don't support this statement. More clearly, although the number of Internet users, active social media users and mobile broadband subscriptions are quite high in the mentioned countries, the innovation rate is low compared to ICT use.

Not only in Turkey but also around the world, education of students (especially in higher education) has begun to change substantially due to digital transformation. Most universities are transforming their education plan to respond to the needs of the twenty-first century workforce (Rayner, 2015). At present, the nature of work is changing dramatically. Thus, one of the main missions of the academic world is to prepare students in accordance with the changing working environment. As stated by Tinmaz and Lee in 2018, by analyzing students' readiness for Industry 4.0, it will be possible to develop effective education or training courses for them. Accordingly, in recent years, there has been an increasing amount of literature on Industry 4.0. In the following section, the concept of Society 5.0 is discussed.

3 Society 5.0

Before discussing the Society 5.0 paradigm, it is better to understand previous society paradigms briefly. Fig.4 presents each stage of the society paradigms. Society 1.0 was based on agricultural activities. This term started with the

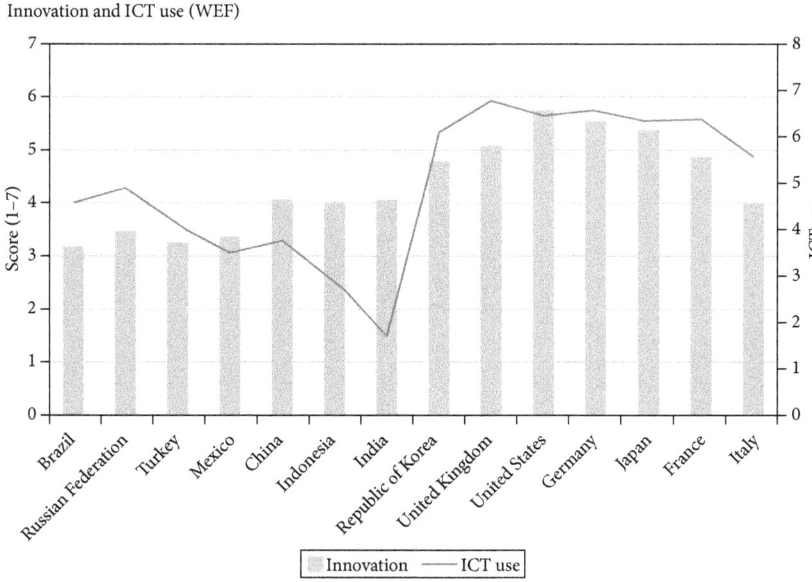

Fig. 3: Technology readiness and innovation analysis compared with the percentage of Internet users. **Source:** WEF, 2017

emergence of human beings. Society 1.0 can be also called as "hunting society" (Okamoto, 2019:4). In this manner, "Society 1.0" can be described as groups of people who have a strong ability of hunting and gathering crop, meat and other intermediates from nature. After the founding of Society 2.0, there were important developments in terms of organization and coordination. Society 2.0 or the "Agrarian Society" was more coordinated and evolved than the hunting society. For instance, irrigation techniques developed and farms were establishment during this period. Society 3.0, also referred to as the industrial society, continued until the end of the eighteenth century. With the help of industrial revolution, society produced huge amounts of products. At the same time, society had the chance to mass produce through this term. Then, society 4.0, or *information society*, came. The period of information society covers the twentieth century and the beginning of the twenty-first century. Society 4.0is best described as information, innovation, production and service ecosystem. Combining smart technologies with smart citizens—based on "man–machine symbiosis"—will lead to smarter societies (Helbing, 2016:9). Currently, we discuss Japanese revolution so-called Society 5.0.

The Japanese government announced a new development plan called Society 5.0 in 2016. According to the government of Japan, Society 5.0 is transformative and strategically quite important for the development of a society that will include

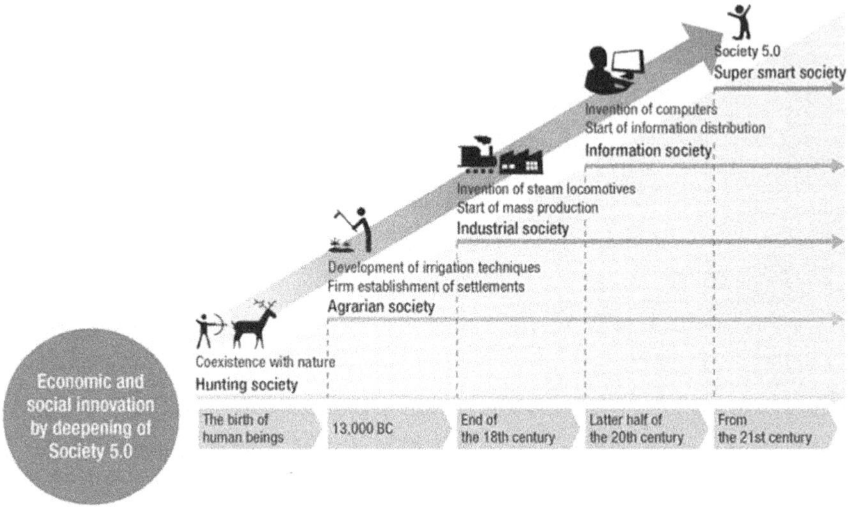

Fig. 4: The stages from Society 1.0 to Society 5.0

real world (physical space) and cyberspace by using ICT (Government of Japan, 2016:13). During the development and implementation of Society 5.0 paradigm, educational institutions, agencies and businesses are central to making effective collaboration (Harayama, 2017; Gladden, 2019:2). It is the goal of Society 5.0 to create a human-centric society in which both economic development and the resolution of societal challenges are achieved, and people can enjoy a high quality of life that is fully active and comfortable (Fukuyama, 2018:48). The additional purpose of Society 5.0 is to increased production and reduced loss of foodstuffs, promotion of sustainable industrialization and redistribution of wealth, and correction of regional inequality (Okamoto, 2019:6). With the help of this strategy, the Japanese government is going to be solve various societal challenges by leveraging the innovation of Industry 4.0, such as big data, artificial intelligence and robots and thus realize mid- to long-term growth. Therefore, the adoption of Society 5.0 becomes extremely important in a common strategy of implementation for the overall benefit for all level of parties that involve governments, citizens, businesses and universities (Manrique, 2019).

The Japanese government really attaches importance to this initiative because it believes that if the strategy is implemented effectively, Japan will be the first super-smart country in the world. Also, the government of Japan plans to share their experience with other nations who are interested in this development plan. Thus, nations can collaborate toward building a better and safer world (Cabinet Office, 2019). Parallel to this, Dr. Harayama, an Executive Member of the Council for

Tab. 2: The focus areas of Society 5.0 (economic and social perspective)

Sector	Policies
Cities and regions	Improved data sharing on energy, transportation, water, human traffic etc.
Energy	Development of affordable sustainable energy
Disaster prevention	Information sharing across organizations during natural disasters
Healthcare	Focus on prevention and individualized healthcare services
Agriculture and food	Utilization of technology for crop growth and optimization of the food value chain
Logistics	Data sharing across the whole supply chain
Manufacturing and services	Support for small businesses to produce high-quality goods
Finance	Diversification of financial services with the help of digital technologies, improved access to financial services.
Public service	Improved services by public administration based on digitalization and improved data sharing

Source: Adapted from Keidanren, 2018; Mavrodieva and Shaw, 2020

Science, Technology and Innovation, Cabinet Office of Japan, explained the Society 5.0 vision as follows:

> *a society where the various needs of society are finely differentiated and met by providing the necessary products and services in the required amounts to the people who need them when they need them, and in which all people can receive high-quality services and live a comfortable, vigorous life that makes allowances for their various differences such as age, sex, region, or language. It is called this to indicate a new society created by transformations led by scientific and technological innovation* (Harayama, 2017:10).

3.1 The Focus Areas of Society 5.0

In today's world, the effect of Society 5.0 can be felt in almost all areas of daily life. Nevertheless, Society 5.0 has mainly focused on nine economic and social sectors (Mavrodieva and Shaw, 2020:3). As can be seen from Tab.2, Society 5.0 will impact significantly various economic and social areas in the following years. The focus areas of Society 5.0 are discussed as follows (Keidanren, 2018:17–25):

- **Cities and Regions:** Society 5.0 will enable a high degree of living standard for both urban (high-density) and provincial (low-density) regions with the help of autonomous systems. For instance, transportation, energy and logistics flows will become more facilitating due to data sharing. Hereby, many financial and environment problems will be solved with high digitalization in the suburbs and rural areas. Such improvements will enhance alternative for places to work and live.

- **Energy:** Whether it is a supercity or rural area, effective energy networks are crucial for sustainable development. Society 5.0 will guarantee reliable energy with fair price to anyone. Thus, sustainable and diverse lifestyle will be available anywhere.
- **Disaster Prevention:** The number of natural disasters in the world has grown dramatically day by day, and the importance of efficient responses and resilience is increasing. With the help of some technological tools such as Internet of Things and social media, data will be collected during disasters and shared with the public and private sectors. Additionally, the latest technologies will be utilized for disaster mitigation and maintenance of infrastructures.
- **Healthcare:** Health and medicine sector will be transformed from traditional operational activities to more digitalized activities. Thus, anyone will be reach healthcare at the appropriate time. Society 5.0 will be offer high-quality healthcare services across regions and boundaries by leveraging innovative high-speed communication networks. Furthermore, individuals can manage their own lifecycle data with the help of artificial intelligence–based medical services.
- **Agriculture and Food:** Agricultural activities will be transformed into more attractive. For instance, agricultural robots, artificial intelligence and drones will control all processes of agricultural activities. Parallel to this, working hours will decrease significantly and productivity will increase exponentially. At the same time, in the food value chain, data and the latest technologies will be utilized to optimize all aspects of the production process, such as procurement, logistics and sales. According to data regarding consumer needs, production and sales information will be organized effectively. Hereby, there will be interactive communication between consumers and producers.
- **Logistics:** Logistics sector is one of the key indicators in economic growth and improved social infrastructure by facilitating flow of goods. With the effect of globalization and rapid growth in electronic commerce, logistics operations will require more sophistication and diversity in Society 5.0. For instance, the Internet of Things and Radio-frequency identification technology will be used to connect to networks during the transportation of cargos. Moreover, businesses will coordinate supply chain mechanism in real time by sharing data on all levels of business activities, such as procurement, production, logistics, sales etc. With the help of autonomous technologies such as drones, robots and artificial intelligence, logistics operation will be more automated and it will also increases efficiency.
- **Manufacturing and Services:** Advanced technologies such as artificial intelligence and 3D printing will provide effective instruments for manufacturing and services. Digital transformation in manufacturing and service industry will cause speedy creation of higher-quality goods. For instance, products regarding food, clothing and housing will be more customized and personalized. In other

words, consumers will be more effective during creation and other process by utilizing autonomous technologies such as 3D printers and artificial intelligence.

- **Finance:** Another important industry that is influenced by digital transformation significantly is the finance sector. At present, individuals and businesses are actively using digitalized financial services like Fintech due to security, convenience and low-cost. Such developments in the financial sector will allow individuals to live anywhere without cash. Along with this, innovative initiatives in the financial sector such as cryptocurrency and block-chain technology will create new forms of value exchange and lifestyle. In order to create greater income equality and living standards in emerging countries, improved financial services with advanced technology will be crucial.
- **Public Services:** After disruptive changes in public and private sector is more suitable, central and local governments will reorganize and rebuild their infrastructures based on digitalization. By establishing effective data management systems, they will provide higher value-added public services to their citizens.

In view of all that has been mentioned so far, we can state that Society 5.0 will be one of the crucial components behind the development of nations in the future. While Society 5.0 is developed by the Japanese Government, it's not limited to only Japan. The major objective of Society 5.0 is equivalent to the United Nations Development Program, entitled with "Sustainable Development Goals." According to the United Nations, current problems such as poverty, inequality and climate change will be solved by 2030 and innovation will play a central role in addressing challenges (United Nations, 2015; Harayama, 2017). Thus, Industry 4.0 and Society 5.0 are interconnected intensively. In the following section, Society 5.0's applications of the Industry 4.0 paradigm are discussed.

3.2 Industry 4.0 and Society 5.0

While the fourth industrial revolution is mainly focused on the development of industry with the digitalization of every step of the production process, Society 5.0 is a new concept which focuses on not only industry but also society. In other words, Industry 4.0 is the digital transformation of manufacturing; Society 5.0 is digitalization across all levels of the society. According to some researches, Society 5.0 is based on Industry 4.0 (Gorecky et al., 2014; Lasi et al., 2014). More clearly, it can be said that there is a strong and significant relationship between Industry 4.0 and Society 5.0 due to the use of common technology and similar mission. At present, we are living in the information society, and the importance of information is increasing day by day. However, creation and distribution of information are still human oriented in today's world. Society 5.0 will enable to create and distribute information by using latest technologies such as robots and artificial intelligence (Granrath, 2017). Consequently, both phenomena are taking the emerging

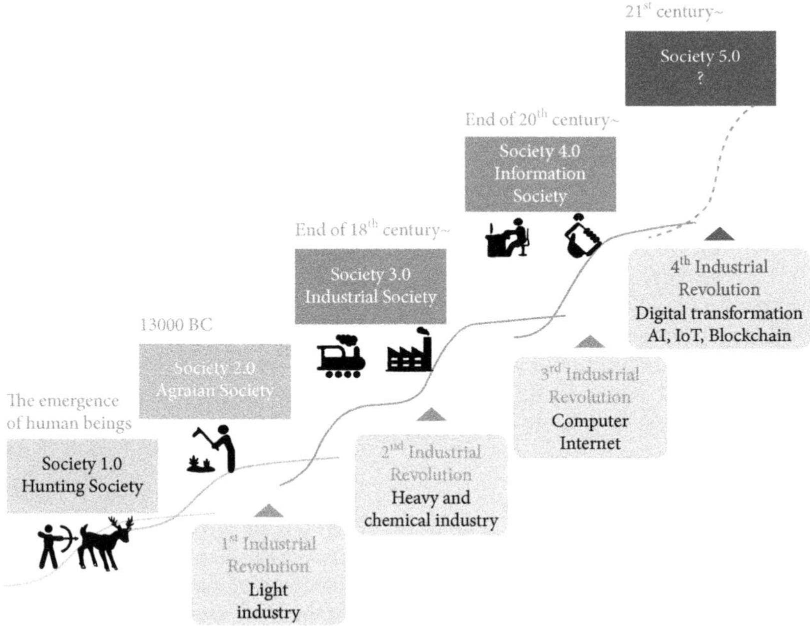

Fig. 5: Development of human society. **Source:** Keidanren Society 5.0 Co-Creation the Future, 2018

technologies as a driver and catalyst to enhanced general welfare of nations (Manrique, 2019). Also, the development of human society is illustrated in Fig.5.

The history of human society has been characterized by liberation from restrictions and acquisition of freedom through enhanced capabilities stemming from new tools and techniques. Moving from the hunting society to the agrarian society, humans gained and enhanced the ability to produce food, which freed them from hunger. In the industrial society, they increased production capabilities and mobility by harnessing motive power. In the information society digitalization enhanced the capacity of telecommunications and information processing, which drastically increased freedom to access information and communication of all kind and freely explore various opportunities for Internet-based services. Currently, we are witnessing Society 5.0, which is a human-focused society. In the future, humans will require imagination to change the world and creativity to materialize their ideas. Society 5.0 will be an imagination society (Keidanren, 2018:6). In the following section, literature review regarding Industry 4.0 and Society 5.0 is presented.

4 Previous Studies

4.1 Industry 4.0 Paradigm

One study by Lobo (2015) reviewed the Industry 4.0 revolution and the future of manufacturing execution systems. The findings stated that the manufacturing systems of the future, including manufacturing execution systems (MES) will have to be built to support this paradigm shift. Additionally, Qin et al. (2016) identified the improvement of Industry 4.0 in a production system and introduced the commonalities between of Industry 4.0 and manufacturing. The results emphasized that in the entire industry, there is a huge gap between recent industry and the achievement of Industry 4.0. In addition, researchers stated that the future of current manufacturing is developing in the direction of Industry 4.0. Another research made by Wyrwicka and Mrugalska (2017) showed the trends that are emerging in relation to the "4th generation industry" in the context of the consideration of its implementation to industrial enterprises in Poland. The findings indicated that many polish companies aren't prepared for Industry 4.0. In other words, many companies in Poland have low level of readiness about Industry 4.0. Also, Hofmann and Rüsch (2017) discussed the opportunities of Industry 4.0 in the context of logistics management. The findings demonstrated that potential Industry 4.0 implications in the context of Just-in-Time and cross-company Kanban systems in a precise manner positively. However, the concept of Industry 4.0 still lacks a clear understanding and is not fully established in practice yet. In addition, Xu et al. (2018) examined to introduce the communities within the industrial sectors to current developments and future opportunities. The results show that despite rapid development in Industry 4.0 both in academic and industry world, challenges still remain. Advanced ICT will contribute to the success of Industry 4.0. Furthermore, Branco et al. (2019) analyzed the Industry 4.0 readiness in manufacturing for the European Union. The analysis indicated that the implementation of the single digital market strategy with regard to the manufacturing sector demands the collection of more detailed information on Industry 4.0. A recent study by Tinmaz and Lee (2019) reviewed Korea University students' readiness level for Industry 4.0 Revolution. The finding showed that students are not well aware of Industry 4.0 fundamentals and stay skeptical and alert for its implementation, which is believed to bring more job opportunities while some current job titles disappear. Another recent research by Machado et al. (2019) investigated the digital readiness of a set of manufacturing industries. The results emphasized that Industry 4.0 processes can be hard for both small and larger companies due to high level of complexity. Also, lack of knowledge is still an obstacle for companies. Also, Rahmat et al. (2019) discussed how much Malaysian university students know about Industry 4.0 and how far they are ready to look for future careers. The results highlighted that some skill set were familiar to the respondents, but a number of the respondents indicated they were not familiar with many of those skill sets. Additionally, Ahmad et al. (2019) examined the factors that influence the readiness of university students on industrial revolution 4.0. The analysis explained that there is no relationship between the technical skills (ability to use technological tools) and nontechnical skills (problem-solving skills and leadership skills) of the student's readiness toward Industrial 4.0 revolution. Moreover, Aygün et al. (2019) determined the level of Industry 4.0's awareness of postgraduate students in Istanbul University. The findings presented that some concepts such as innovation, human-computer interaction and knowledge management are related to Industry 4.0. Moreover, most students believe that Industry 4.0 will cause unemployment and ethical problems. Most recent study by Chiarini et al. (2020) examined the Industry 4.0 technologies adopted in Italian manufacturing companies. The findings demonstrated that while these technologies do not contribute to any significant improvement in environmental sustainability, other strategies such as servitization, design-to-cost, supply chain integration and machinery—electronic equipment—and database integration, as well as lean strategy are enabled through the adoption of Industry 4.0 technologies.

4.2 Society 5.0 Paradigm

Prasetyo and Arman (2017) examined to get the right resource in implementing group management system in smart society platforms. The findings showed that conceptual design of group management systems have potential benefits to support Society 5.0 in the form of smart society platform. Additionally, Skobelev and Borovik (2017) analyzed modern technologies—from digital manufacturing to digital society—in developed organizations. The results highlighted that the growing popularity of digital economy has created a strong basis for the development of Industry 4.0 technologies already now and in the long term can serve as the launch pad for the creation of Society 5.0.One study by Ferreira and Serpa (2018) investigated Society 5.0 and Industry 4.0in the context of social development and sustainable innovation. Researchers indicated that this concept won't be restricted to only Japan; it will be extended to other nations. Another research made by Fukuyama (2018) discussed new human-centered societies, so-called Society 5.0. The researcher explained that Society 5.0 isn't just for the purpose of one country. It will contribute to resolving societal problems all over the world. Also, Takahashi (2018) examined potential risks and opportunities emerging from new technologies in the super-smart societies. The results emphasized that active and real-time data obtained in a continuous manner in a super smart society may help in the assessment of problematic and addictive behavior. However, ethical issues such as data and individual privacy should be resolved before wide acceptance of these innovations in the future societies. A recent publication by Onday (2019) reviewed Japan's *Society 5.0: Going Beyond Industry 4.0.* In the conclusion part, the researcher stated that Society 5.0 will reach different nations. However, there is still lack of knowledge about the concept of Society 5.0. Another recent research by Ellitan and Anatan (2020) discussed in detail the direction of change in corporate life from Industry 4.0 to Society 5.0. The findings demonstrated that there are two key different strategies—*Leading and Flexible*—which are very important to survive and achieve success in Industry 4.0 and Society 5.0.

In view of all that has been mentioned above, the importance of Industry 4.0 increases exponentially. Especially, in the manufacturing industry, there has been a growing interest in Industry 4.0 due to operational and financial gains. Many studies regarding Industry 4.0 investigated the manufacturing industry. There are a limited number of studies on university students' readiness level for Industry 4.0, as has been demonstrated above. Society 5.0 is a fairly new concept, as can be seen from the limited number of studies. Industry 4.0 and Society 5.0 are much interconnected paradigms. Researchers and decision makers have begun to adopt this new concept due to its various benefits to nations, since 2016. However, it still cannot be said to be at the desired level. Overall, previous studies highlight the need for more research on Industry 4.0 and Society 5.0 with different perspectives.

5 Conclusion

It can be seen from the information given above that the fourth industrial revolution began in the past decade. The impact of Industry 4.0 on every aspect of businesses is increasing dramatically each day. Toward the end of the nineteenth century, businesses started to reorganize their business function and operation by the effect of a new phenomenon called "globalization." Globalization allows businesses to expand their operation all over the world without any restriction. Although globalization seems to have affected businesses positively, it has also caused high competition in the global economy. Since the beginning of the twentieth century, in addition to globalization, disruptive changes in information technology have caused significant changes in the business environment. Parallel to this, businesses, especially industrial companies, are trying to adapt technological developments. Currently, many countries are investigating the concept of Industry 4.0 due to potential benefits to manufacturers. For instance, with the help of wireless connectivity and sensor data gathered from machines, manufacturers can monitor the current state of equipment and determine potential failures. Additionally, the number of industrial robots have increased exponentially in manufacturing sector. The price of industrial robots has decreased in recent years; therefore, the integration of automation systems is enhancing day by day. Especially Far East countries such as Japan, South Korea and China have highest adoption toward automation due to low cost of production. Moreover, big industrial companies such as General Electric (US), ABB Ltd (Switzerland), IBM (US), Mitsubishi (Japan), and Siemens (Germany) have changed their business models based on big data, Internet of Things and human–machine interface. As can be seen from the above, Industry 4.0offers great opportunities for industrial companies. However, there are still some challenges for small- and medium-sized enterprises (SMEs) in participating in the fourth industrial revolution. For instance, high infrastructure costs, perception and adaptation problems of new technologies and the inability to institutionalize are affecting negatively SMEs.

Nowadays, almost every day we hear "Industry 4.0 in manufacturing, marketing, management"; however, Japan is already moving onto its next step, which is called "Society 5.0." Society 5.0 is really a new and hot topic globally. However, we can positively state that the concept of Society 5.0 will be one of the crucial components in the sustainable development of future generations. Parallel to this, the major mission of Society 5.0 increases the general welfare of nations. Although, today, Japan is one of the most powerful nations on Earth, Japan has suffered many years in terms of social and economic issues due to atomic bombs. Therefore, Japan wants to create a "super-smart" society. According to Japanese authorities, Society 5.0 will enable possible to quickly solution in economy, education and society. In other words, the government of Japan plans to offer most suitable solutions to its citizens through the adoption of new technologies such as big data, artificial intelligence, drones etc. However, a super-smart society will not only relate to the Japanese nation, but also it will affect the entire world. At present, many countries are mainly focusing on the adoption of new technologies. Along with this, especially developed countries have created their long-term development plan customized through advanced technologies. It explains that why the rest of the world is paying close attention to this new concept. For instance, the Western countries announced a new strategy called "Digitalization of European Industry." Another strategy called "Made in China 2025"has been declared by the government of China. Additionally, many developed countries such as the United States, Germany and Japan have already started to allocate their resources to research and development activities on latest technologies. In the meantime, developing countries should be aware of the potential risk of new digital technologies. More clearly, new technologies will lead to job losses in developing countries due to integration of automation during production processes. Moreover, a new business model, which is characterized by advanced technologies, will change a competitive environment. It means that a single business may control or dominate the entire sector by using automated vehicles in one area. Thus, the impact of national governments will reduce in decisions and trends. Also, the algorithms of new technologies are mainly created by developed nations. If developing countries want to become more active in the global economy, they should be more concerned with development of algorithms. At present, some developing countries such as India, China and Russia are actively interested in the potential of new technologies. However, some developing nations such as Turkey, Brazil and Argentina are still far behind the potential of such technologies in sustainable development.

Overall, it can be said that business and social environments have drastically changed over the last decade. In recent years, not only private sectors but also public institutions have transformed their operation toward technology and innovation-based. The high level of potential benefits in the connection between people and things in the real and cyber worlds will provide effective solution in economy, healthcare, mobility, environmental issue, logistics, infrastructure and all aspects of social life. Whereas industrial companies create better productivity and resource efficiency, governments plan to decrease social inequality, poverty and environmental pollution by leveraging such technology. On the other hand, nations are facing several challenges such as data security, privacy, intellectual property and the supply of labor and skills during the adoption of these paradigms. Thus, it is very difficult to make definite judgments regarding these new phenomena. There are numerous arguments about the future of advanced technology. However, we are positive about two things. First, technological developments will continue exponentially. Second, nations that have adapted technological changes will be among the most powerful countries in the following years.

Last but not least, this chapter presents an overview of Industry 4.0 and Society 5.0. While the fourth industrial revolution is mainly focused on the development of industry, with digitalization in every step of the production process, Society 5.0 is a new concept which focuses on not only industry but also society. Recently, businesses and governments have started to adapt and implement Industry 4.0 and Society 5.0 into their long-term development plans. Especially, after disruptive changes in information technology, governments recognized that traditional strategies wouldn't be effective to solve the economic, social and environmental issues. Therefore, the importance and popularity of Industry 4.0 and Society 5.0 is increasing dramatically day by day. Both two new phenomena are taking emerging technologies as a driver and catalyst to enhanced general welfare of nations. Therefore, Industry 4.0 and Society 5.0 will be one of the crucial components behind the development of nations in the near future.

Bibliography

Ahmad, Rahman, Sagaran, Pannirchelvi, Kim-Soon, Ng, Sapry, Hairul R., and Omar, Siti S. (2019), "Factors Influence the Students' Readiness on Industrial Revolution 4.0", *International Journal of Recent Technology and Engineering*, Vol. 8, N. 2, pp. 461–468.

Ahuett-Garza, Horacio and Kurfess, Thomas, (2018), "A Brief Discussion on the Trends of Habilitating Technologies for Industry 4.0 and Smart Manufacturing", *Manufacturing Letters*, Vol. 15, pp. 60–63.

Akbaba, Ahmetİlker, (2018), *Dördüncü Endüstri Devrimine Geçiş Sürecinde Üç Boyutlu Yazici Kullaniminin Teknoloji Kabul Modeliyle Ölçümlenmesi: Otomotiv Endüstrisinde Bir Araştirma* (Doctoral Dissertation), Atatürk Üniversitesi, Sosyal Bilimler Enstitüsü,*Erzurum.*

Almada-Lobo, Francisco, (2015), "The Industry 4.0 Revolution and the Future of Manufacturing Execution Systems (MES)", *Journal of Innovation Management*, Vol. 3, N. 4, pp. 16–21.

Ashton, Thomas Southcliffe, (1997), The Industrial Revolution 1760–1830, *OUP Catalogue.*

Aygün, Doğuş, Yılmaz, İlker G., and Sati, Zümrüt, (2019), "Industry 4.0 Awareness: A Research on Postgraduate Students", *Conference: 6th*

International Management Information Systems Conference, Kadir Has University, İstanbul-Turkey, pp. 239–247.

Cabinet Office, (2019), "Society 5.0", https://www8.cao.go.jp/cstp/society5_0/index, (13.04.2020)

Castelo-Branco, Isabel, Cruz-Jesus, Frederico, and Oliveira, Tiago, (2019), "Assessing Industry 4.0 Readiness in Manufacturing: Evidence for the European Union", *Computers in Industry*, Vol. 107, pp. 22–32.

Catal, Cagatay and Tekinerdogan, Bedir, (2019), "Aligning Education for the Life Sciences Domain to Support Digitalization and Industry 4.0", *Procedia Computer Science*, pp. 99–106.

Chen, Baotong, Wan, Jiafu, Shu, Lei, Li, Peng, Mukherjee, Mithun and Yin, Boxing, (2017), "Smart Factory of Industry 4.0: Key Technologies, Application Case, and Challenges", *IEEE Access*, Vol. 6, pp. 6505–6519.

Chiarini, Andrea, Belvedere, Valeria, and Grando, Alberto, (2020), "Industry 4.0 Strategies and Technological Developments. An Exploratory Research from Italian Manufacturing Companies", *Production Planning & Control*, pp. 1–14.

Dailysabah, (2019), "Turkey's Latest Tech-Driven Industry Program to Prevent Imports Worth $30B", https://www.dailysabah.com/technology/2019/08/08/turkeys-latest-tech-driven-industry-program-to-prevent-imports-worth-30b, (04.06.2020).

Dries, Buytaert and DeveshRaj, (2016), "Could the Fourth Industrial Revolution Help Us Reach the Global Goals". *Global Agenda, Davos*, https://www.weforum.org/agenda/2016/02/davos-2016-and-the-fourth-industrial-revolution/, (20.04.2020).

Ellitan, Lena and Anatan, Lina, (2020), "Achieving Business Continuity in Industrial 4.0 and Society 5.0", *International Journal of Trend in Scientific Research and Development (IJTSRD)*, Vol. 4, N. 2, pp. 235–239.

European Commission, (2018), "Digitising European Industry", https://ec.europa.eu/digital-single-market/en/policies/digitising-european-industry, (28.05.2020).

Ferreira, Carlos M. and Serpa, Sandro, (2018), "Society 5.0 and Social Development", *Management and Organizational Studies*, pp. 26–31.

Fukuyama, Mayumi, (2018), "Society 5.0: Aiming for a New Human-Centered Society", *Japan Spotlight*, Vol. 1, pp. 47–50.

Göçer, İsmet, (2013), "Ar-Ge Harcamalarının Yüksek Teknolojili Ürün İhracatı, Dış Ticaret Dengesi ve Ekonomik Büyüme Üzerindeki Etkileri", *Maliye Dergisi*, Vol. 165, N. 2, pp. 215–240.

Gorecky, Dominic, Schmitt, Mathias, Loskyll, Matthias, and Zühlke, Detlef, (2014, July), Human-Machine-Interaction in the Industry 4.0 Era. In *201412th IEEE International Conference on Industrial Informatics (INDIN)*, pp. 289–294.

Government of Japan, (2016), "The 5th Science and Technology Basic Plan. Provisional Translation", https://www8.cao.go.jp/cstp/english/basic/5thbasicplan.pdf, (14.06.2020).

Granrath, Lorenz, (2017), "Japan's Society 5.0: Going beyond Industry 4.0", https://www.japanindustrynews.com/2017/08/japans-society-5-0-going-beyond-industry-4-0/, (07.06.2020).

Harayama, Yuko, (2017), "Society 5.0: Aiming for a New Human-Centered Society. Japan's Science and Technology Policies for Addressing Global Social Challenges. Interviewed by Mayumi Fukuyama", http://www.hitachi.com/rev/archive/2017/r2017_06/pdf/p08-13_TRENDS.pdf, (22.05.2020).

Helbing, Dirk, (2016), "Society 4.0: Upgrading Society, but How?", https://www.researchgate.net/profile/Dirk_Helbing/publication/304352735_Society_40_Upgrading_society_but_how/links/577e1a5a08aeaa6988af2cf8/Society-40-Upgrading-society-but-how.pdf, (28.04.2020).

Hermann, Mario, Pentek, Tobias, and Otto, Boris, (2015), *Design Principles for Industrie 4.0 Scenarios: A Literature Review*, Technische Universität Dortmund, Dortmund.

Hofmann, Erik Rüsch, Marco, (2017), "Industry 4.0 and the Current Status As Well As Future Prospects on Logistics", *Computers in Industry*, Vol. 89, pp. 23–34.

Keidanren, J.B.F. (2018), "Society 5.0 – Co-Creating the Future", http://www.keidanren.or.jp/en/policy/2018/095_proposal.pdf, (19.05.2020).

Keidanren, J.B.F. (2016), Toward Realization of the New Economy and Society – Reform of the Economy and Society by the Deepening of "Society 5.0", pp. 1–25.

Kemp, Simon, (2020), "Digital 2020: Turkey. *We Are Social, Hootsuite*",https://datareportal.com/reports/digital-2020-turkey, (02.06.2020).

Konak, Ali, (2018), "Yüksek Teknoloji İçeren Ürün İhracatının İhracat Hacmi ve Ekonomik Büyüme Üzerine Etkisi; Seçilmiş OECD Ülkeleri ve Türkiye Örneği", *JOMELIPS-Journal of Management Economics Literature Islamic and Political Sciences*, Vol. 3, N. 2, pp. 56–80.

Krinis, Dora, (2016), *VAT Challenges of the Digital Economy: An EU Perspective*(Doctoral dissertation), Universidade do Minho.

Lasi, Heiner, Fettke, Peter, Kemper, Hans, Georg, Feld, Thomas, and Hoffmann, Micheal, (2014), "Industry 4.0", *Business & Information Systems Engineering*, Vol. 6, N. 4, pp. 239–242.

Lu, Yang, (2017), "Industry 4.0: A Survey on Technologies, Applications and Open Research Issues", *Journal of Industrial Information Integration*, Vol. 6, pp. 1–10.

Machadoa, Carla G., Winrotha, Mats, Carlssonb, Dan, Almströma, Peter, Centerholtb, Victor, and Hallin, Malin, (2019), "Industry 4.0 Readiness in

Manufacturing Companies: Challenges and Enablers towards Increased Digitalization", *Procedia Cirp*, Vol. 1, N. 2, pp. 1113–1118.

Manrique, Christian, (2019), "Industry 4.0 and Society 5.0 by Christian Manrique", https://christianmanrique.com/2019/02/14/industry-4-0-and-society-5-0-by-christian-manrique/, (30.06.2020).

Manyika, James, Chui, Micheal, Bughin, Jacques, Dobbs, Richard, Bisson, Peter, and Marrs, Alex. (2013), *Disruptive Technologies: Advances That Will Transform Life, Business, and the Global Economy* (Vol. 180), San Francisco, CA: McKinsey Global Institute.

Martin, Cheryl and Leurent, Helena, (2017), "Technology and Innovation for the Future of Production: Accelerating Value Creation", World Economic Forum, Geneva, Switzerland, White Paper No. 00030896, 2017.

Mavrodieva, Aleksandrina, V. and Shaw, Rajib, (2020), "Disaster and Climate Change Issues in Japan's Society 5.0–A Discussion", *Sustainability*, Vol. 12, N. 5, pp. 1–17.

McBride, James and Chatzky, Andrew, (2018), Is 'Made in China 2025'a Threat to Global Trade, *Council for Foreign Relations.*

Ministry of Trade, (2019), "İhracat Ana Planı", https://ticaret.gov.tr/data/5d67a97a13b87799c4cc1fef/Ticaret_Sunum_29.08.19.pdf, (11.05.2020).

Mohajan, Haradhan, (2019), "The First Industrial Revolution: Creation of a New Global Human Era",*Journal of Social Sciences and Humanities*, Vol. 5, No. 4, pp. 377–387.

Mokyr, Joel, (1998), The Second Industrial Revolution, 1870–1914, *Storia dell'economia Mondiale, 21945.*

Moss, Jack, (2019), Society 5.0: How will Japan Approach "Data Capitalism", *Global Leadership Initiative.* G20 Summit – OSAKA.

Mrugalska, Beata and Wyrwicka, Magdalena, K. (2016), "Towards Lean Production in Industry 4.0",*Procedia Engineering*, pp. 466–473.

Muhuri, Pranab K., Shukla, Amit K., and Abraham, Ajith, (2019), "Industry 4.0: A Bibliometric Analysis and Detailed Overview", *Engineering Applications of Artificial Intelligence*, Vol. 78, pp. 218–235.

Okamoto, Masahide, (2019), "Standardization Activities on 'Society 5.0' in Japan. Society 5.0", *Standardization Promotion Committee, BSN Seminar*, Jakarta-Indonesia, pp. 1–19.

Onday, Ozgur, (2019), "Japan's Society 5.0: Going beyond Industry 4.0", *Journal of Business Economics*, Vol. 10, N. 389, pp. 1–7.

Özlü, Faruk, (2017), "The Advent of Turkey's Industry 4.0", *Turkish Policy Quarterly*, Vol. 16, N. 2, pp. 29–38.

Posada, Jorge, Toro, Carlos, Barandiaran, Inigo, Oyarzun, David, Stricker, Didier, de Amicis, Raffaele, ... and Vallarino, Ivan, (2015), "Visual Computing as a Key

Enabling Technology for Industrie 4.0 and Industrial Internet", *IEEE Computer Graphics and Applications*, Vol. 35, N. 2, pp. 26–40.

Prasetyo, YuliAdam, and Arman, Arry Akhmad, (2017), "Group Management System Design for Supporting Society 5.0 in Smart Society Platform", *International Conference on Information Technology Systems and Innovation (ICITSI)*, pp. 398–404.

Qin, Jian, Liu, Ying, and Grosvenor, Roger, (2016), "A Categorical Framework of Manufacturing for Industry 4.0 and Beyond", *Procedia Cirp*, Vol. 52, pp. 173–178.

Rahmat, Ainol, Mardhiyah, Adnan, Airil, Haimi, Mohd, and Mohtar, Nor Marini, (2019), "Industry 4.0 Skillsets and 'Career Readiness': Can Malaysian University Students Face the Future of Work", In *International Invention, Innovative & Creative Conference (InIIC Series 2/2019)*, Kuala Lumpur, Malaysia.

Rayner, Gerry and Papakonstantinou, Theo, (2015), "Student Perceptions of Their Workplace Preparedness: Making Work-Integrated Learning More Effective", *Asia-Pacific Journal of Cooperative Education*, Vol. 16, N. 1, pp. 13–24.

Roblek, Vasja, Meško, Maja, and Krapež, Alojz, (2016), "A Complex View of Industry 4.0", *Sage Open*, Vol. 6, N. 2, DOI: 2158244016653987.

Rüßmann, Micheal, Lorenz, Marcus, Gerbert, Philipp, Waldner, Manuela, Justus, Jan, Engel, Pascal, and Harnisch, Micheal, (2015), "Industry 4.0: The Future of Productivity and Growth in Manufacturing Industries", *Boston Consulting Group*, Vol. 9, N. 1, pp. 54–89.

Sahin, Begum E. (2019), "Impact of High Technology Export on Economic Growth: An Analysis on Turkey", *Journal of Business Economics and Finance*, Vol. 8, N. 3, pp. 165–172.

Savanevičienė, Asta, Statnickė, Gita, and Vaitkevičius, Sigitas, (2019), Individual Innovativeness of Different Generations in the Context of the Forthcoming Society 5.0 in Lithuania", *Inžinerinė Ekonomika*, pp. 211–222.

Schwab, Klaus, (2017), "The Fourth Industrial Revolution: What It Means, How to Respond". *World Economic Forum*, https://www.weforum.org/agenda/2016/01/the-fourth-industrial-revolution-what-it-means-and-how-to-respond, (01.04.2020).

Skobelev, Petr Olegovich and Borovik, S. Yu, (2017), "On the Way from Industry 4.0 to Industry 5.0: From Digital Manufacturing to Digital Society", *Industry 4.0*, Vol. 2, N. 6, pp. 307–311.

Takahashi, Taiki, (2018), "Behavioral Economics of Addiction in the Age of a Super Smart Society: Society 5.0", *Oukan (Journal of Transdisciplinary Federation of Science and Technology)*, Vol. 12, N. 2, pp. 119–122.

Tanrıogen, Zeynep M. (2018), "The Possible Effects of 4th Industrial Revolution on Turkish Educational System", *Eurasian Journal of Educational Research*, Vol. 18, N. 77, pp. 163–184.

Tay, ShuIng, Lee, TeChuan, Hamid, NorZ. A., and Ahmad, AhmadNurA. (2018), "An Overview of Industry 4.0: Definition, Components, and Government Initiatives", *Journal of Advanced Research in Dynamical and Control Systems*, Vol. 10, N. 14, pp. 1379–1387.

Thames, Lane, and Schaefer, Dirk, (2016), "Software-Defined Cloud Manufacturing for Industry 4.0", *Procedia Cirp*, Vol. 52, pp. 12–17.

The Global Economy, (2018), "High-Tech Exports-Country Rankings", https://www.theglobaleconomy.com/rankings/High_tech_exports/#Turkey, (03.04.2020).

Tinmaz, Hasan and Lee, Jin H. (2019), "A Preliminary Analysis on Korean University Students' Readiness Level for Industry 4.0 Revolution", *Participatory Educational Research*, Vol. 6, N. 1, pp. 70–83.

United Nation, (2015), "Transforming Our World: The 2030 Agenda for Sustainable Development", https://sustainabledevelopment.un.org/post2015/transformingourworld, (12.05.2020).

United Nations Conference on Trade and Development, (2018),"Technology and Innovation Report 2018: Harnessing Frontier Technologies for Sustainable Development", https://unctad.org/en/PublicationsLibrary/tir2018_en.pdf, (27.06.2020).

World Economic Forum, (2017), "The Global Competitiveness Report 2017–2018". *World Economic Forum*, 2017, http://reports.weforum.org/global-competitiveness-index-2017-2018/, (20.05.2020).

World Economic Forum, (2016), "The Future of Jobs: Employment, Skills and Workforce Strategy for the Fourth Industrial Revolution". In *Global Challenge Insight Report*,World Economic Forum, Geneva.

Wyrwicka, Magdalena K. and Mrugalska, Beata, (2017), "Industry 4.0" towards Opportunities and Challenges of Implementation, *DEStech Transactions on Engineering and Technology Research* (ICPR).

Xu, Li, Da, Xu, EricL., and Li, Ling, (2018), "Industry 4.0: State of the Art and Future Trends", *International Journal of Production Research*, Vol. 56, N. 8, pp. 2941–2962.

Yıldız, Ümit, (2017), *Brıcs Ülkeleri ve Türkiye'de Yüksek Teknoloji İhracatı ve Ekonomik Büyüme İlişkisinin Panel Veri Analizi.*

Mehmet ÖZYÜREK

The State, Tech Companies and Surveillance in the Psychopolitical Era

1 Introduction

The Internet is now the most significant platform for communication, shopping, education, health, dating, etc. While surfing the Internet or using it as a communication tool, individuals share myriad types of personal data and information with tech companies such as Internet providers, social media platforms, video sharing websites and browsers. Today, the number of individuals using smart home appliances is increasing noticeably. The Internet of Things is becoming ordinary with smartphones, smart watches, smart TVs and other smart appliances that are used in daily life. Once individuals have one of these devices, the devices start reporting any information that they see and hear since most of these devices are connected to a cloud system on the Internet (Balkin, 2017: 1169–1170). The information is shared with the provider tech companies. From that point, the companies know who the individuals are, where they are and what they are doing. There is no doubt this environment seems like the one demonstrated in the book *Nineteen Eighty-Four* (McMullan, 2015).

The analogy made between today's world and *Nineteen Eighty-Four*, however, is not enough to depict today's surveillance culture. In *Nineteen Eighty-Four*, it was only Big Brother, the leader of Oceania, watching people (Orwell, 2007). Today, tech companies, along with the state, conduct surveillance. In effect, some of the tech companies have more information about citizens than the national intelligence agencies (Han, 2017: 76). However, unlike the traditional surveillance techniques conducted by states, tech companies also trade the information about the users with third parties such as advertising companies and other states. In some circumstances, tech companies cooperate with states (Rozenshtein, 2018: 104). Companies not only monitor individuals but also manipulate their behaviour. They use psychologists and data scientists to do this (Cadwalladr, 2017). These companies focus on influencing and modifying the behaviour of individuals for more profit (Zuboff, 2016). In other words, tech companies use their *psychopolitical* steering power over the users (Han, 2017).

Furthermore, states themselves, undoubtedly, continue to be important actors in terms of surveillance. States, who are supposed to protect the privacy of individuals from the *psychopolitical* steering of the tech companies, use the platforms of these companies to conduct surveillance on individuals. In this respect, most states justify mass Internet surveillance on the grounds of national security (Michaels, 2008).

In response to the pressure by national governments, some companies call on governments around the world to be more transparent and adhere to human rights principles when conducting surveillance (Reform Government Surveillance, 2020). For this purpose, some tech companies formed a coalition called Reform Government Surveillance (RGS) in 2017. The coalition focuses on urging governments to reform their surveillance laws and practices (RGS, 2020). In addition to this coalition, some individual companies claim to be responsible in respecting the privacy of their users (Pollach, 2011: 92).

Yet, can tech companies protect the privacy of personal data and information? To what extent can they reduce their *psychopolitical* steering? Should tech companies regulate themselves on their own in the present *psychopolitical* era? What role should governments play in this regard? The following chapter answers these questions. In order to answer the above questions, the chapter uses a qualitative interdisciplinary approach, which draws on a number of different disciplines and sub-disciplines such as philosophy, different fields of law, political economy and corporate responsibility. This normative analysis has various objectives: the first is to show the change in the nature of the surveillance society. Even though there have been numerous studies to investigate surveillance societies (such as Bauman, 2000; Clarke, 1988; Lyon, 2001; Marx, 2015; Norris and Armstrong, 1999; Torpey, 2007; Wright et al., 2008; Zuboff, 2016), no study, to our knowledge, has analysed the *psychopolitical* steering of the tech companies in surveillance from the perspective of Byung-Chul Han's theories (2017). The second aim of the chapter is to show the competition between the tech companies and states in the *psychopolitical* era. In this regard, the chapter not only gives a glimpse into the relationship between the states and companies in surveillance but also shows how they race each other in the manipulation of individuals through big data.

The rest of the chapter is divided into five parts: after this introduction, in order to describe how the surveillance mechanisms alter in the *psychopolitical* era, section 2, references Byung-Chul Han's (2017) *Psychopolitics.* In order to contemplate the background of Han's *Psychopolitics*, section 2 briefly introduces Michel Foucault's thoughts on *biopolitics as well.* This philosophical summary and Han's theoretical perspective constitute the philosophical background for the following analysis of surveillance society. In section 3, the spotlight is upon the key elements of the surveillance power of the tech companies. In this respect, section 3 gives some of the well-known examples depicting how the tech companies use personal data about individuals. Then, in section 4, the focus is on the governments' role in the surveillance of individuals. The section demonstrates that even though states are fundamentally responsible for protecting the privacy of their citizens, they seem reluctant to perform this task. In section 5, self-regulation and the assertion by the tech companies of being responsible in protecting users' data are explored.

2 From Bentham's Panopticon to Psychopolitics

French Philosopher Michel Foucault borrows the concept of *panopticon* from Jeremy Bentham in his book called *Discipline and Punish* (Foucault, 2012: 196–231). *Panopticon* is an architectural design used for buildings such as prisons, factories or schools (Bentham, 1995). It is a type of complex building with a central watchtower surrounded by cells. The main aim of this design is that even though the people in the cells are thoroughly visible to the watchman in the central tower, the watchman cannot be seen from the cells. Even if there is no watchman in the tower, people who are separated from each other in their cells cannot see inside of the tower. Therefore, people in the cells always feel they are being watched.

Foucault uses *panopticon* to depict how individuals are controlled in disciplinary societies. In order to understand the role of *panopticon* in Foucault's philosophy, it may be helpful to mention the general features of Foucault's depiction of *biopolitics*. According to Foucault, the sovereign power, which comes from the threat of the sword, altered in the seventeenth and eighteenth centuries (Foucault, 1981). As Foucault says, the power over society became "more [sophisticated] than the threat of death" in this process (Foucault, 1981: 143). The sovereign power of the state was enhanced by the combination of direct and indirect coercion on society (Mythen & Walklate, 2006: 385). The state began to pay more attention to correct and train the body rather than torture it (Muller, 2004: 52). According to Foucault, this constituted a radical change in the nature of power since "diverse techniques for achieving the subjugation of bodies and the control of populations" became possible (Foucault, 1981: 140). Foucault describes this alteration by using the term *biopower*. *Biopower* has become more common in clinics, prisons, schools, factories and hospitals (Nadesan, 2008: 26). In this development, social welfare and the administration of life have become more important for the authorities, such as states. In short, biological existence has become political (Agamben, 1998).

With *biopolitics*, states become more active monitoring the mortality, life expectancy, and fertility of their populations (Nadesan, 2008: 24). Power is dispersed throughout society (Gutting, 2005: 87). Concurrently, *biopolitical* authorities such as psychiatrists, psychologists, social workers, etc. emerge (Nadesan, 2008: 26). After a while, individuals internalize the surveillance of these *biopolitical* authorities. This internalization constitutes one of Foucault's explanations about *biopolitics*, which refers to a type of social control that is based upon individual self-governance (Lemke, 2011: 34). In this type of society, the surveillance capability of *panopticon* can have a broader horizon as individuals monitor their own behaviour (Gutting, 2005: 96).

In today's digital world, however, *biopolitics* seems obsolete (Jeffries, 2017). In this way, Han (2017: 34–39) claims that Foucault's account needs updating. In order to update Foucault's philosophy, Han starts with the concept of *panopticon*. Unlike the traditional illustration of *panopticon*, used by Bentham and Foucault, in which people are to be separated from each other in their cells, Han depicts today's digital *panopticon* as a system that exploits freedom (Han, 2017: 12). He portrays this new environment as *Psychopolitics*, in which people can be seen as free to take their own decisions without any intervention. Free people are described as "auto-exploiting labourers in his or her own enterprise"(Han, 2017: 72) In this society, rather than external pressure to make people visible to *panopticon*, people themselves race each other to be visible. Individuals would like to expose anything they possess to others (Han, 2015: 9). In the *psychopolitical* era, social media and big data alter the disciplinary *biopower* into *psychopower*. *Psychopower* manipulates people "... by pleasing and fulfilling. Instead of making people compliant, it seeks to make them dependent" (Han, 2017: 25). Social media and personalized search engines destroy the public sphere. In this new environment, in which "[o]ne encounters only oneself and one's own life," individuals lose their ability for critical thinking (Han, 2015: 35).

However, digital *panopticon* may appear more efficient than the traditional *panopticon* since the former does not forget anything (Han, 2017: 62). Furthermore, digital *panopticon* makes predictions about the future behaviour of the individuals as well (Han, 2017: 73). It uses algorithms to predict what individuals will do in the future. More importantly, it uses micro-targeting to manipulate individuals (Han, 2017: 74).

The truth of the matter is that in today's societies surveillance is usually grounded in manipulation rather than coercion (Marx, 2002: 15). Tech companies do not force individuals to share their personal data and information with them, but individuals are usually eager to do that by their own free will. Individuals agree to be more transparent by their own choices. In this era, individuals are their own broadcasters who are expected "to express themselves as much as possible" since this is the way in which the tech companies obtain more data about individuals (Balkin, 2017: 1162). Once individuals start sharing, they let tech companies control and exert surveillance on themselves. Unlike any other surveillance systems, individuals themselves sustain the digital system of surveillance with their contribution. Facebook, Twitter, Instagram or YouTube may be good examples in this regard. However, even though individuals give consent to terms of service agreement with tech companies when they create an account on these platforms, they usually do not know how the companies will use the data they share (Vincent, 2017). Once an individual become a member, "[i]ntelligence [of the individual] has no access to outside, because it makes a choice between options in a system. Therefore, intelligence does not really exercise free choice: it can only select among the offerings the system affords" (Han, 2017: 95). Then the individual becomes a self-surveilling subject (Han, 2017: 72).

3 Beyond Surveillance: The Companies with Psychopower

As mentioned above, in today's *psychopolitical* era, tech companies not merely monitor individuals but also predict what individuals may do in the future. Companies categorise and judge individuals according to data and information (Balkin, 2017: 1164). They sell this data on behaviour and emotion to business customers

(Zuboff, 2016). In this process, tech companies influence and manipulate the behaviour of individuals (Han, 2017: 56). Furthermore, tech companies and their business customers who buy data from these companies make decisions about people by using algorithms. These decisions affect the profiles and the identities of individuals in the digital world (Balkin, 2017). Some of the individuals are being classified as inferior to others. It creates "a new digital class society" (Han, 2017: 77).

"Big Data is a highly efficient *psychopolitical* instrument" (Han, 2017: 21e). Google is "the largest and most successful 'big data' company" (Zuboff, 2015: 79). Therefore, Google may be the most significant actor in *psychopolitics*. There may be several examples to show how Google plays its role as today's digital *panopticon*. Google's Google Map, for example, takes people's private pictures on the street without their consent (Knight, 2007). The company that owns YouTube and the operating system Android are accused of violating users' privacy (Miller, 2013). Google knows everything what individuals search for on the web and which apps they use (Curran, 2018). Google is also known as one of the cookie makers to track people on the Internet (Fowler, 2019). Google's browser Chrome appears to allow more cookies to carry out surveillance of people on the web than any other browser (Fowler, 2019). These cookies help Google track the things that the users search for and the websites they visit, then the company shows advertisements according to this surveillance. On Android devices, Chrome also sends individuals' locations to Google when they search for anything on their phones (Fowler, 2019). In the end, all the data that individuals share enables Google to provide a more personalized digital environment.

The above examples may be enough to call Google today's *panopticon*. Yet the surveillance that Google conducts is beyond that. The company, for example, is accused of accessing sensitive medical records of patients (Vincent, 2019). Google accessed the records of the University of Chicago Medicine between 2009 and 2016 (Vincent, 2019). More recently, Google acted similarly in another health project called "Project Nightingale" (Copeland, 2019). In accordance with this project, even though private data on patients, such as names, addresses, dates of birth and medical conditions of patients, are transferred to Google, none of the patients or doctors is informed about this transfer (Copeland, 2019). Likewise, in the United Kingdom, the company was slammed for accessing information about British patients without their consent (Boiten, 2020). Alphabet, Google's parent company, bought DeepMind, which had information about 1.6 million British patients in 2014 (Boiten, 2020). Nevertheless, at the time, DeepMind promised to share the data only to develop an application called Stream to alert doctors at risk (Vincent, 2017).

Lastly, but most importantly, individuals now give their free consent to share their health data with Google. Individuals create live data for Google as the company bought Fitbit in 2019 (Phelan, 2019). With Fitbit, which produces wearable products for health and fitness, Google can monitor individuals' daily health and fitness data, such as temperature, heart rate, calories burned, sleep, etc. In brief, Google gradually makes more individuals monitor themselves on different subjects.

Google, however, is not the only tech company that can be investigated from the perspective of *psychopolitics*. Facebook may be another *psychopolitical* actor. Individuals become their own broadcasters on Facebook for free. During this broadcasting they do not only share personal information such as their names and relationship status but also share their reactions to social and political issues. Then, Facebook shares some of this data with other third parties such as advertisers and political data analytics firms (Rosenberg et al., 2018). Facebook also uses this data in experiments in order to manipulate the users' emotional and psychological responses (Booth, 2014). For example, the company gathers psychological insights on young people and shares these insights with advertisers (Levin, 2017).

The scandal of the Cambridge Analytica (CA), a political data analytics firm, proved how Facebook becomes a determining factor in manipulating people's political choices. In this scandal, the information on the Internet was used to manipulate swing voters in the United States and the voters who were inclined to vote for leave in the process of Brexit in the United Kingdom (Rosenberg et al., 2018). The scandal showed that Facebook "likes" of many people might play a role in election campaigns. Psychology professor Dr. Aleksandr Kogan developed an application called "thisisyourdigitallife" on Facebook (Rosenberg et al., 2018). This app helped him gather data of Facebook users and their friends on Facebook. Facebook users agreed to share their data for a small payment. However, they were not informed that the data that they agreed to share would be used for political manipulation in the 2016 US Presidential Election (Rosenberg et al., 2018). The data made it possible to reach potential Trump voters. Similarly, the company influenced the result of the EU referendum by helping the Leave campaign in the United Kingdom (Cadwalladr, 2017). The information about the voters was in the hands of the CA's scientists. As the CA scandal shows, while the users enjoy Facebook's free platform, they create free data for the company. This helps Facebook, and the other companies that Facebook provides with data, in manipulation of individuals.

In addition to Facebook and Google, companies like Apple, Amazon, Microsoft, etc. would also be added into the list of companies that have data-driven business models (Amnesty International, 2019). Most of these companies are well-known global companies. There are, however, some other tech companies such as "Rubicon Project, AdSonar, Quantcast, Undertone, Traffic Marketplace" that most people may not have heard of (Schneier, 2015). Companies such as Dataminr, Geofeedia, PATHAR, and TransVoyant help the US authorities to carry out surveillance (Fang, 2016). As the next section shows, Geofeedia, for example, helps the US law enforcement agencies to track protesters on Twitter and Instagram (Bromwich et al., 2016). Another example may be the big data company called Acxiom, which collects information about consumers in the United States (Turow & Draper, 2014: 137). It is claimed that this company has more information than the Federal Bureau of Investigation (FBI) (Han, 2017: 107). Companies doing business in automotive, telecommunication and banking sectors are amongst the customers of

Acxiom. Acxiom evaluates individuals for those companies as to whether they are worth targeting (Turow & Draper, 2014: 137). Acxiom helps its business customers have "360-degree customer view" (Han, 2017: 67). Another example may be the data mining company Palantir Technologies. This company has contracts worth millions with the US Immigration, Customs and Enforcement Agency (O'Carroll and Franco, 2017). Los Angeles Police Department and the Central Intelligence Agency (CIA) also use the data mining tools of the company (Peretti, 2017).

Doubtless there are more examples that could be mentioned. However, listing all of these tech companies and investigating how they carry out *psychopolitical* surveillance on individuals are undeniably beyond the scope of this chapter. The purpose of this study is not a comprehensive critique of these companies. The above examples show that individuals lose "the ability to form reasonable expectations and assessment who has access to [their] information" (Lanzing, 2016: 14). In this new digital version of *panopticon*, the watchtower becomes invisible. Unlike the traditional version of *panopticon*, where inmates were aware of the threat of being watched, the surveillance by the digital *panopticon* is no longer visible (McMullan, 2015). Tech companies such as "Google and social networks, which present themselves as spaces of freedom, are assuming panoptic forms" (Han, 2015: 49). As highlighted in the example of Acxiom, tech companies now have a 360-degree view. This makes today's digital *panopticon* a more efficient surveillance mechanism than Bentham's *panopticon* since, as underlined above, Bentham's *panopticon* had some blind spots to inspect secret wishes and thoughts of the inmates (Han, 2017: 73).

4 Race for Psychopower: States vs Tech Companies

One can argue that human rights are violated in the *psychopolitical* era (Sieber, 2019). Surveillance-based business model of the tech companies may constitute a threat to human rights (Amnesty International, 2019). Even though *Psychopolitics* is a broader concept than human rights, the autonomy of the individuals seems to be in danger since the surveillance of these companies is based upon "active steering" rather than "passive surveillance" (Han, 2017: 21–22). "[I]ndividuals face threats of control and surveillance by Big Data"(Balkin, 2017: 1153). Social rights, cultural rights and political rights may be seen as some of the rights violated by the *psychopolitical* steering of the tech companies (Sieber, 2019). Yet, who should be responsible for protecting individuals in the *psychopolitical* era? Governments? Or should tech companies regulate themselves?

4.1 State Reluctance for Regulation

With respect to international human rights law, states have obligations to protect human rights in accordance with the Universal Declaration of Human Rights (UDHR), the International Covenant on Civil and Political Rights (ICCPR) and the International Covenant on Economic, Social and Cultural Rights (ICESCR). In this respect, states are also expected to have regulatory arrangements to protect human rights against the impact of tech companies. Some of these arrangements are also listed in the UN Guiding Principles on Business and Human Rights (UNGP) and the OECD Guidelines for Multinational Enterprises.

Tech companies constitute a threat to the freedom of individuals in the *psychopolitical* era. *Psychopolitical* steering of tech companies breaches social rights, cultural rights and political rights of individuals (Sieber, 2019). Presumably, data privacy may be the main element in the protection of these rights. With respect to data privacy, one can assert that "government regulation and enforced laws are the only effective methods" to make tech companies respect data privacy (Vila et al., 2003). In some countries, such as in Australia, Brazil and Germany, authorities started to address the negative impact of tech companies over individuals (Amnesty International, 2019: 44). Some national governments already have institutions and regulations to protect privacy of individuals (Regan, 2012: 399–400). Even in the United States, where the government prefers "hands-off approach to protecting online privacy," the state authorities have begun investigating tech companies (Pisa & Polcari, 2019: 14; Amnesty International, 2019). Individual states seem to pay more attention to data privacy now, such as California passed the California Consumer Privacy Act of 2018.

However, there are some obstacles in the regulation of tech companies by national regulatory mechanisms. First, even if a government passes a privacy law, it may not be able to enforce it due to lack of adequate resources. Small developing countries, for example, with limited resources may not be able to regulate giant tech companies such as Google, Facebook, etc. (Pisa, 2019). The legal and technical capacity of these countries may not be adequate to enforce possible stricter regulations.

Second, even if a state has adequate resources, it may not be eager to pass strict laws for data privacy. It is argued that regulating tech companies may reduce the competitiveness of the tech companies in a certain country (Tiku, 2019). For example, possible future regulations on the US tech companies are seen as a factor to weaken the competitiveness of American companies against Chinese companies (Tiku, 2019). American companies claimed that putting extra burden on them might help Chinese companies. Therefore, a state which considers economic outcomes of a possible regulation on its national tech companies may not be eager to pass strict laws to regulate tech companies in terms of data privacy.

The third obstacle may be the lobbying activities of tech companies. Tech companies lobby governments, such as the United Kingdom, the United States, Canada, India, Vietnam, Argentina, Brazil, Malaysia and all 28 states of the EU, to weaken regulation (Cadwalladr & Campbell, 2019). In this respect, lobbying activities of Google and Facebook in the EU may be the best example (Kayali, 2019). Google, for instance, spent more than €8 million to lobby the EU in 2018 (Amnesty International, 2019: 45). Both Google and Facebook lobbied against the European General Data Protection Regulation (GDPR). These and other tech companies not only lobby governments but also fund non-governmental organizations to act in favour of their arguments (Fang, 2019).

The fourth obstacle to regulating tech companies may be related to the national governments' interests in surveillance. Internet platforms created and controlled by the tech companies constitute an avenue for governments to conduct surveillance on individuals, which cannot be otherwise possible (Balkin, 2017: 1179). In addition to the online advertisers mentioned above, the intelligence agencies of the states are also interested in the data that the individuals share on the Internet (Shahbaz & Funk, 2019). National security threats, such as terrorism, may be one reason in the collaboration of states and tech companies (Michaels, 2008). Furthermore, as the Covid-19 pandemic showed, another reason for a collaboration between the states and tech companies may be for the surveillance of pandemic patients (Foer, 2020). For this reason, in many countries, governments are collaborating with tech companies to track individuals (Amnesty International, 2020).

In the *psychopolitical* era "[r]epressive forces don't stop people from expressing themselves but rather force them to express themselves" (Han, 2017: 94). China may be a major example in this regard. The Chinese government and the tech companies already started using *psychopolitical* steering to alter the behaviour of citizens (Foer, 2020). Nascent social-credit system in the country uses big data to design good citizens through penalties and rewards. Besides making a great deal of investment in digital surveillance, the country has a close relationship with the tech companies to monitor individuals (Freedom House, 2019). The surveillance activities that the Chinese government conducts on WeChat, the Chinese version of WhatsApp, may be a good example in this regard. According to an allegation, the Chinese authorities detained a person from the Uighur Muslim minority because a friend of his on his social media checked in from Mecca (Shahbaz & Funk, 2019).

Another example can be the US government. When Edward Snowden leaked the documents related to surveillance activities of the American intelligence, one of the revelations was about the National Security Agency's (NSA) PRISM programme (Bernal, 2018: 179). In accordance with this programme, the NSA accessed data on the users of tech companies such as Facebook, Google, Microsoft, YouTube, Skype and Yahoo. More recently, according to the report by the American Civil Liberties Union (ACLU), the government organs in the United States used the social media monitoring firm Geofeedia's services for surveillance (Leetaru, 2016). The police used the data provided by Geofeedia to arrest the Baltimore protestors, after the death of Freddie Gray in 2016 (Bromwich et al., 2016). Furthermore, in the United States, some agencies in the Department of Homeland Security (DHS) purchase tools to bypass surveillance of individuals' private information shared on the Internet (O'Carroll & Franco, 2017). As underlined above, the data mining company Palantir Technologies, which has contracts worth millions with the US Immigration, Customs and Enforcement Agency, may be the most important tech company as far as this is concerned (O'Carroll & Franco, 2017).

Fifth, for any nation, it may be difficult to regulate the activities of transnational tech companies since most tech companies which use online cloud platforms cross the national borders (Manyika et al., 2016). The geographical scope of the Internet is beyond the scope of any national regulation. Therefore, an international set of rules may be more effective to regulate tech companies (Smith, 2019). Nevertheless, it is difficult to say that there is an international coordination in the world today. Many countries do not even have any regulation for data privacy (Eggers et al., 2018). When it comes to countries which have privacy regulation, most of them use regulatory techniques that differ from each other.

Nevertheless, the European Union passed a unique comprehensive regulation, the General Data Protection Regulation (GDPR), in 2016. The next section briefly examines the role of GDPR in the protection of individuals against the *psychopolitical* steering of tech companies.

4.2 General Data Protection Regulation (GDPR)

The GDPR, which came into force on May 25, 2018, is a significant international step to regulate tech companies. GDPR encapsulates rules, such as how to collect and handle data, for tech companies doing business in the EU. All tech companies that market goods or services to the EU residents are subject to the GDPR (2018: article 3). GDPR has some important articles to protect the privacy of individuals. For example, in accordance with Article 3, tech companies have to explain how they process users' data in "a concise, transparent, intelligible and easily accessible form, [by] using clear and plain language." This article focuses on reducing the opacity of how the tech companies process data. According to Article 6 of the GDPR, on the other hand, tech companies are required to request the consent of the users to process their personal data. The next important article of the GDPR may be Article 15, which requires tech companies to give users an access to their personal data. If the user requests, the tech company is required to disclose the reason why it is processing the user's data and for how long the data will be stored. Lastly, Article 83 of the GDPR states that if a company breaches the requirements, it may be fined up to $20 million or 4 % of annual turnover. In 2019, for example, the supervisory authority in France fined Google €50 million for breaching the GDPR (Rosemain, 2019).

The GDPR may be one of the unique and comprehensive regulations for protection of data privacy. However, it may not be a panacea for protection of individuals against *psychopolitical* steering. First of all, even though the GDPR is an international regulation, national governments have a key role in the enforcement. In accordance with Article 52(4) of the GDPR, member states are responsible for providing "the human, technical and financial resources" of the supervisory boards called data protection authorities (DPAs). Nonetheless, according to the report published by Brave, governments seem unsuccessful in supporting DPAs (Brave, 2020). According to the report, DPAs do not have financial budget and personnel and resources to investigate tech companies. DPAs seem weak to regulate giant tech companies. Secondly, even though the GDPR seems to require tech companies to ask the consent of people for data processing, it may also be seen as a factor forcing people to give consent to the tech companies to violate their privacy (Hare, 2019). Even if people consent to tech companies, this may not mean that most people take rational decisions. People may not even understand the privacy policies of tech companies, which are long and complex (van Ooijen & Vrabec, 2018: 95–96). Thirdly, state regulations mostly focus on the tech companies. The GDPR, for example, focuses on the tech companies rather than the state authorities. As underlined above, states may process personal data by claiming national security concerns (Human Rights Watch, 2018). Thus, the GDPR may not be strong enough to fight against the surveillance power of the tech companies and governments (Swant, 2019).

In short, international laws also seem insufficient to protect individuals against *psychopower*. As the GDPR case showed, even though states pass strict international regulations, the enforcement of these regulations seem extremely difficult. More importantly, as Snowden stated, "regulating protection of data presumes that the collection of data in the first place was proper, that it was appropriate and that it doesn't represent a threat or danger" (Swant, 2019). Therefore, from this perspective, the GDPR, or other regulations, may merely be interpreted as a competition for control of surveillance. In Snowden's words, "[t]hat it's okay to spy on your customers or your citizens so long as it never leaks, so long as only you are in control of what it is" (Swant, 2019).

4.3 Reform Government Surveillance Coalition, Self-Imposed Privacy Policies and Other Privacy Tools by Tech Companies

Even though states are reluctant to regulate tech companies – and constrain themselves – in surveillance of individuals, some tech companies call on governments around the world to reform their current surveillance laws and practices (Reform Government Surveillance Coalition, 2020). They ask governments to adhere to the principles of protecting human rights when conducting surveillance. In addition to putting pressure on governments they even claim to regulate themselves beyond law to protect privacy. Yet, can tech companies improve their behaviour to respect the privacy of personal data? Is it possible for them give up their *psychopolitical* steering?

Some tech companies announce that they will pay more attention to the protection of data privacy (Wong, 2019). Even the companies criticised above started promising to respect privacy. For example, immediately after the CA scandal, Facebook declared it would investigate the misuse of the data of its users (Rosenberg et al., 2018). It changed its policy with respect to apps. The company also claimed to improve transparency in its advertisement policy (Shukla, 2019). Its founder Mark Zuckerberg underlined the importance of the protection of the privacy of its users (Bell, 2019).

In fact, in addition to privacy policies, some companies focus on designing products and services for greater privacy. In this respect, Apple, advertising its products by highlighting their high privacy features, may be one example. In 2019, one of Apple's slogans was "What happens on your iPhone stays on your iPhone" (Shaban, 2019). Tim Cook, the CEO of Apple, highlights the importance of privacy and the necessity of comprehensive privacy laws (Cook, 2019).

Furthermore, Google introduced End-to-End, an extension to its Chrome, for better users' privacy as well (Somogyi, 2014). The company claims that they compete with other companies in protection of the users' privacy (Wong, 2019). Google also introduces new privacy features. Google's incognito mode for using Google Maps, for example, shows that the tech company pays more attention to privacy now (Wong, 2019).

As Pollach (2011) demonstrates, according to some tech companies, privacy is seen as a part of their social responsibility. For this purpose, some tech companies even become members of certification networks to show that they attach importance to users' privacy (Mendel et al., 2018: 93). The Direct Marketing Association (DMA) and TRUSTarc may be good examples in this regard. These networks provide companies with private certification (Tang et al., 2008). In short, tech companies, either individually or within an association, now claim that they will improve the privacy of the users without a stricter government regulation.

Furthermore, when governments request data from some companies, companies such as Yahoo, Google and Microsoft, such companies share the information about this request with public (Schneier, 2014). For example, Facebook received nearly 140,000 requests from government authorities for user data in the second half of 2019 (Facebook, 2020). Tech companies complain about the requests by governments on their websites (See Apple Inc., n.d.). In some circumstances, tech companies may also refuse to help governments; Apple, for example, refuses to unlock the iPhones of criminals (Rozenshtein, 2018). The company discloses the information about the requests of governments regarding its customers' information and data in its website (Apple Inc., n.d.).

Tech companies also seem to focus on constraining government surveillance (Rozenshtein, 2018: 105). In 2018, tech companies such as Apple, Dropbox, Facebook, Google, LinkedIn, Microsoft, Snap Inc., Twitter and Verizon Media created a platform called the Reform Government Surveillance Coalition. These companies call on governments around the world to adhere to the principles of protecting human rights when conducting surveillance (RGS, 2020). In addition to putting pressure on governments to reform their current surveillance laws and practices, the RGS wants governments to ensure transparency and privacy. The RGS criticises the governments' efforts for weakening the encryption systems that tech companies provide to their users (Whittaker, 2018). Moreover, the RGS Coalition also expects governments to solve possible conflicts among jurisdictions with mutual legal assistance treaty processes (RGS, 2020).

All these developments show that even though tech companies conduct surveillance on individuals, they also claim to improve policies and services to protect the privacy of individuals. They resemble national governments that both carry out surveillance and claim to protect the privacy of their people. Similar to governments, tech companies, which have millions of users in different parts of the world, possess their own governance structures. Mark Zuckerberg defines Facebook as follows: "In a lot of ways Facebook is more like a government than a traditional company. We have this large community of people, and more than other tech companies we're really setting policies" (Klonick, 2017). In fact, one of the interpretations that may be extracted from Zuckerberg's words is that Facebook is not only a tech company but also a governance structure. It has its own governance mechanisms to control its own society. Likewise, other tech companies such as YouTube and Twitter also have their own governance mechanisms to control the content of the posts that their users share on their platforms (Balkin, 2017: 1183).

But why would tech companies pay attention to the privacy of their users? There is no doubt the first reason may be the current regulations and possible future regulations. As discussed in the previous section, for example, the GDPR may be an important factor in this regard. On the other hand, some governments expect the tech companies to regulate themselves (Balkin, 2017: 1183). According to this view, the relationship between the tech companies and the individuals may be built upon a fiduciary relationship (Balkin, 2017). In other words, tech companies may act as doctors, who maintain the privacy of the information about their patients. It is asserted that tech companies can be self-regulated in order to be socially responsible as well (Klonick, 2017: 1615). Furthermore, international efforts to curb the negative impact of companies on human rights may be another reason for the changing attitude of tech companies towards privacy. The UN Guiding Principles on Business and Human Rights and the OECD Guidelines for Multinational Enterprises may be examples putting pressure on companies with respect to privacy. Lastly, but most importantly, the business interests of tech companies in promoting privacy may be another reason. Privacy is seen as one of the subjects in corporate strategies of tech companies (Pollach, 2011: 91). Privacy plays a role in their competitiveness, as companies are ranked by private initiatives according to their privacy commitments and policies (Ranking Digital Rights, n.d.). Tech companies that pay attention to reputation may improve their privacy policies as well (Pollach, 2011: 91). For example, after the Snowden revelations, tech companies focused on reforming their privacy policies to fix their damaged reputations. (Timm, 2014). Some of these companies see the revelations as an opportunity to differentiate their products (Rozenshtein, 2018: 116).

Unlike governments, tech companies can cross national borders. For instance, Facebook has more users outside the United States than in the United States (Rozenshtein, 2018: 117). The governance structures of the tech companies described above may regulate all the activities of the company in different parts of the world (Balkin, 2017: 1182). Even if governments are reluctant to regulate tech companies, tech companies may choose to pay more attention to privacy for competitiveness. Thus, for example, even in countries where the privacy laws are weak, tech companies may respect privacy. Moreover, as Rozenshtein (2018: 105) highlighted, they may help to constrain governments' surveillance.

However, self-imposed privacy policies of tech companies may have some disadvantages. First, these policies are shaped by the market. As Hirsch (2010: 455) highlighted, this market "is characterized by highly imperfect and asymmetrical information, firms can collect and use far more personal data than they could in a hypothetical perfect market." Thus, the privacy of individuals is mostly left to the mercy of the tech companies. Second, customer preferences shape market mechanisms (Hirsch, 2018: 455). But, customers may not pay attention to long privacy policies of tech companies (van Ooijen & Vrabec, 2018: 95–96). Even if they read these policies, they may have difficulties in understanding them because of their complexity. Therefore, if individuals are not able to understand privacy policies, it may be optimistic to expect them to punish or reward a tech company according to its privacy policy. Similarly, individuals may not be able to punish or reward a tech company according to the company's privacy policy if the company has monopoly. Thus, the tech company may avoid implementing its privacy policy, which may increase the company's liability (Mendel et al., 2012: 94). Third, even though some tech companies such as Apple, Google and Facebook may pay attention to reputation, some tech companies which mostly have business relationships with state agencies may not have strict privacy policies in force to protect individuals. Acxiom or Palantir, mentioned above, may be two examples. In fact, individuals may not even have heard of the names of some of these tech companies, such as "Rubicon Project, AdSonar, Quantcast, Undertone, Traffic Marketplace" (Schneier, 2015). Privacy protection may even be against the nature of these companies. Fourth, although government regulations are shaped by democratically approved laws in democracies, self-imposed privacy policies of tech companies are governed by a private elite bureaucracy, which is likely to pay attention to the interests of the company, rather than those of the citizens. Fifth, even if tech companies now design products and services for greater privacy, they seem not to change their surveillance-based business models (Amnesty International, 2019). One reason for sustaining the surveillance-based business models of tech companies may be that people like free and convenient products and services (Schneier, 2015). These features may also be seen as the main factor in the sustainability of the *psychopolitical* power of most tech companies.

Lastly, when it comes to constraining the surveillance of national governments, companies may not resist the pressure of national governments due to their financial interests. A company may merely focus on a stable environment for its business in that specific country (Balkin, 2017: 1179). Therefore, it may be difficult to resist the requests of governments. Moreover, governments may use indirect techniques to push tech companies to share information with the state authorities. For example, when WikiLeaks revealed diplomatic and military documents to public in 2010, the US authorities put an indirect pressure on the website. The government merely used patriotic values to make the public put pressure on tech companies (Balkin, 2017: 1180). Immediately after the government's reaction, companies such as Amazon, hosting WikiLeaks on its cloud service, and companies such as PayPal, MasterCard and Visa, letting people make donations to WikiLeaks, ceased their business relationship with WikiLeaks (Bambauer, 2012: 892).

As portrayed above, the services of some of the tech companies, such as search engines and social media companies, claim that they provide their platforms for free. They expect the users share as much information as they can. Then, they sell the users' data (Zuboff, 2016). Therefore, as the Amnesty International's report highlighted, even though tech companies have privacy policies, they continue their surveillance-based business model (Amnesty International, 2019). This surveillance-based model constitutes the main element of the *psychopower* on people. No matter what privacy policy tech companies have, they seem never stop collecting users' data.

Conclusion

This chapter has shown that surveillance in the *psychopolitical* era is no longer mere information collection by nation states. Individuals are controlled by the manipulation of big data rather than coercion as Foucault theorised decades ago. Surveillance in this era is based upon the exploitation of freedom.

As Krasmann (2017: 15) clearly highlights, today, it is difficult to realize "whether it is state agencies or commercial companies that are invested in the processing of data; and we do not know whether and when they actually watch, and what they might see." Companies, nation states and most importantly individuals themselves play a role in today's digital *panopticon*, which exploits the psychic realm of individuals.

In the *psychopolitical* era, neither states nor tech companies seem to surrender their surveillance activities. Rather than stopping data collection, these actors try to dominate each other in the control of surveillance. In this competition, on the one hand, some tech companies collaborate with national governments for their business interests. National governments, on the other hand, appear to take advantage of the facilities of the products and services provided by the tech companies.

Moreover, the chapter has also shown that states and tech companies compete with each other in order to control surveillance. Some tech companies, for example, reject the requests made by governments for the data of their users. Some tech companies emphasise their responsibility to protect their users. The tech companies which pay attention to their reputation claim not to share personal data and information with governments. New policies, services and products of tech companies for more privacy and their campaigns for more transparency show how some states fall behind the tech companies in control of *psychopolitical* steering.

However, the chapter also highlighted that the protection of individuals against *psychopolitical* steering necessitates more than self-imposed privacy policies and corporate responsibility because the business models of tech companies, based upon data processing and targeted advertising, constitute the main element in *psychopolitical* steering. Furthermore, as it is also highlighted, tech companies, which do not pay attention to corporate reputation, may not even hesitate to share information with governments. Therefore, self-imposed privacy policies and corporate responsibility are unlikely to be a solution in protecting individuals in the *psychopolitical* era.

The analysis leads to the following conclusions. The first conclusion may be related to the role of states and national regulations. It is noticeable that strong data protection by government regulation may help individuals to protect themselves from *psychopolitical* steering. As underlined in terms of the GDPR, regulations can help individuals to ask tech companies what kind of personal data they collect about users and how they share this information. Yet, as Snowden clarified faultlessly, regulations for data protection do not stop data collection (Swant, 2019). The second conclusion may be on the voluntary role of tech companies. Even if some tech companies have better privacy policies, products and services, they are unlikely to stop surveillance. In other words, as Schneier (2015) rightly states, "choosing among providers is not a choice between surveillance or no surveillance."

Moreover, in the *psychopolitical* era, as stated above, "[i]ntelligence does not really exercise free choice: it can only select among the offerings the system affords." Therefore, in today's intelligence, "[t]he law is not the only thing that can protect you," as Snowden said. "Technology [created by tech companies] is not the only thing that can protect you. We are the only thing that can protect us" (Swant, 2019). Therefore, one option against *psychopolitical* steering may be "the right to say nothing, because only then is there a chance of framing the rare, and ever rarer, thing that might be worth saying" (Han, 2017: 94).

Whether tech companies will stop *psychopolitical* steering or be stopped by nation states is an open question. Even though this chapter shows the changing nature of surveillance in today's world and the competition for the control of surveillance in this new era, it has some limitations. Most of the arguments regarding the *psychopolitical* steering in this chapter rely upon normative justifications. Yet, these assumptions may be supported by future empirical research. Future research could also examine the role of tech companies, or international laws, in redesigning the Internet to end *psychopolitical* steering.

Bibliography

Agamben, Giorgio, (1998), *Homo Sacer: Sovereign Power and Bare Life*, Stanford: Stanford University Press.

Amnesty International, (2019), "Surveillance Giants: How the Business Model of Google and Facebook Threatens Human Rights", https://www.amnesty.org/en/documents/pol30/1404/2019/en/, (21.11.2019).

Amnesty International, (2020), "COVID-19: Surveillance and the Threat to Your Rights", https://www.amnesty.org/en/latest/news/2020/04/covid-19-surveillance-threat-to-your-rights/, (12.06.2020).

Apple, (n.d.), "Apple Privacy", https://www.apple.com/privacy/government-information-requests/, (09.04.2020).

Balkin, Jack M., (2017), "Free Speech in the Algorithmic Society: Big Data, Private Governance, and New School Speech Regulation", *UC Davis Law Review*, Vol. 51, pp. 1149–1210.

Bambauer, Derek E., (2012), "Orwell's Armchair", *The University of Chicago Law Review*, Vol. 79, pp. 863–944.

Bauman, Zygmunt, (2000), *Liquid Modernity*, Cambridge: Polity.

Bell, Emily, (2019), "Mark Zuckerberg's Facebook Mission Statements Hide His Real Aim", *The Guardian*, https://www.theguardian.com/media/commentisfree/2019/mar/10/mark-zuckerberg-facebook-mission-statements-hides-his-real-aim, (26.03.2020).

Bentham, Jeremy, (1995), "Introduction", (Bozovic, Miran Ed.), *The Panopticon Writings*, London: Verso.

Bernal, Paul, (2018), *The Internet, Warts and All: Free Speech, Privacy and Truth*, New York: Cambridge University Press.

Boiten, Eerke, (2020), "Our Personal Health History is too Valuable to be Harvested by the Tech Giants more Essential than ever", *The Guardian*, https://www.theguardian.com/commentisfree/2020/feb/16/our-personal-health-history-is-too-valuable-to-be-harvested-by-tech-giants, (16.02.2020).

Booth, Robert, (2014), "Facebook Reveals News Feed Experiment to Control Emotions", *the Guardian*, https://www.theguardian.com/technology/2014/jun/29/facebook-users-emotions-news-feeds, (30.06.2020).

Brave, (2020), "Europe's Governments are Failing the GDPR: Brave's 2020 Report on the Enforcement Capacity of Data Protection Authorities", https://brave.com/wp-content/uploads/2020/04/Brave-2020-DPA-Report.pdf, (02.04.2020).

Bromwich, Jonah E., Victor, Daniel and Isaac, Mike, (2016), "Police Use Surveillance Tool to Scan Social Media, A.C.L.U. Says", https://www.nytimes.com/2016/10/12/technology/aclu-facebook-twitter-instagram-geofeedia.html, (04.05.2020)

Cadwalladr, Carole, (2017), "The Great British Brexit Robbery: How Our Democracy was Hijacked", *The Guardian*, https://www.theguardian.com/technology/2017/may/07/the-great-british-brexit-robbery-hijacked-democracy, (07.05.2020).

Cadwalladr, Carole and Campbell, Duncan, (2019), "Revealed: Facebook's Global Lobbying against Data Privacy Laws", *The Guardian* https://www.theguardian.com/technology/2019/mar/02/facebook-global-lobbying-campaign-against-data-privacy-laws-investment, (18.04.2020).

Clarke, Roger, (1988), "Information Technology and Dataveillance", *Communications of the ACM*, Vol. 31, N. 5, pp. 498–512.

Cook, Tim, (2019), "You Deserve Privacy Online. Here's How You Could Actually Get It", *Time*, https://time.com/collection/davos-2019/5502591/tim-cook-data-privacy/, (13.06.2020).

Copeland, Rob, (2019), "Google's 'Project Nightingale' Gathers Personal Health Data on Millions of Americans", *the Wall Street Journal*, https://www.wsj.com/articles/google-s-secret-project-nightingale-gathers-personal-health-data-on-millions-of-americans-11573496790, (11.03. 2020).

Curran, Dylan, (2018), "Are You Ready? Here is all the Data Facebook and Google have On You", *The Guardian*, https://www.theguardian.com/commentisfree/2018/mar/28/all-the-data-facebook-google-has-on-you-privacy, (30.03. 2020).

Eggers, William D., Turley, Mike and Kishnani, Pankaj, (2018), "The Future of Regulation Principles for Regulating Emerging Technologies", *Deloitte Insights*, https://www2.deloitte.com/us/en/insights/industry/public-sector/future-of-regulation/regulating-emerging-technology.html, (04.06.2020).

Facebook, (2020), "Facebook Transparency: Government Requests for User Data", (July–December 2019), https://transparency.facebook.com/government-data-requests/jul-dec-2019, (10.06.2020).

Fang, Lee, (2016), "The CIA is Investing in Firms that Mine your Tweets and Instagram Photos" *The Intercept*, https://theintercept.com/2016/04/14/in-undisclosed-cia-investments-social-media-mining-looms-large/, (14.05.2020).

Fang, Lee, (2019), "Silicon Valley-Funded Privacy Think Tanks Fight in D.C. to Unravel State-Level Consumer Privacy Protections", *The Intercept*, https://theintercept.com/2019/04/16/consumer-privacy-laws-california/, (13.04.2020).

Foer, Franklin, (2020), "What Big Tech Wants Out of the Pandemic: The Firms are all too Eager to Help the Government Manage the Coronavirus Crisis", *The Atlantic*, https://www.theatlantic.com/magazine/archive/2020/07/big-tech-pandemic-power-grab/612238/, (14.06.2020).

Foucault, Michel, (1981), *The History of Sexuality, Vol. 1: An Introduction*, (Robert Hurley, Trans.), Harmondsworth: Penguin.

Foucault, Michel, (2012), *Discipline and Punish: The Birth of the Prison*, (Alan Sheridan, Trans.), New York: Vintage.

Fowler, Geoffrey A., (2019), "Goodbye, Chrome: Google's Web Browser has Become Spy Software", *Washington Post*, https://www.washingtonpost.com/technology/2019/06/21/google-chrome-has-become-surveillance-software-its-time-switch/, (21.06.2019).

Freedom House, (2019), "Freedom on the Net 2019: China", https://freedomhouse.org/country/china/freedom-net/2019, (20.03.2020).

Gutting, Gary, (2005), *Foucault: A Very Short Introduction*, Vol. 122, Hampshire: Oxford University Press.

Han, Byung-Chul, (2017), *Psychopolitics: Neoliberalism and New Technologies of Power*, London: Verso Books (eBook Version), https://search.ebscohost.com/login.aspx?direct=true&scope=site&db=nlebk&db=nlabk&AN=1468203.

Han, Byung-Chul, (2015), *The Transparency Society*, Stanford: Stanford University Press.

Hare, Stephanie, (2019), "These New Rules were Meant to Protect our Privacy. They don't Work", *The Guardian*, https://www.theguardian.com/commentisfree/2019/nov/10/these-new-rules-were-meant-to-protect-our-privacy-they-dont-work, (18.05.2020).

Hirsch, Dennis D., (2010), "The Law and Policy of Online Privacy: Regulation, Self-regulation, or Co-Regulation", *Seattle University Law Review*, Vol. 34, pp. 439–480.

Human Rights Watch, (2018), "The EU General Data Protection Regulation: Questions and Answers", https://www.hrw.org/news/2018/06/06/eu-general-data-protection-regulation, (10.05.2020).

Jeffries, Stuart, (2017), "Psychopolitics: Neoliberalism and New Technologies of Power by Byung-Chul Han", *The Guardian*, https://www.theguardian.com/books/2017/dec/30/psychopolitics-neolberalism-new-technologies-byung-chul-han-review, review, (30.04.2020).

Kayali, Laura, (2019), "Inside Facebook's Fight against European Regulation" *Politico*, https://www.politico.eu/article/inside-story-facebook-fight-against-european-regulation/, (26.05.2020).

Klonick, Kate, (2018), "The New Governors: The People, Rules and Processes Governing Online Speech", *Harvard Law Review*, Vol. 131, pp. 1598–1670.

Knight, Sam, (2007), "The Times all Seeing Google Street View Prompts Privacy Fears", *The Times*, https://www.thetimes.co.uk/article/all-seeing-google-street-view-prompts-privacy-fears-t6m0th6sqv3, (01.06. 2007).

Lanzing, Marjolein, (2016), "The Transparent Self", *Ethics and Information Technology*, Vol. 18, N. 1, pp. 9–16.

Leetaru, Kalev, (2016), "Geofeedia is Just the Tip of the Iceberg: the Era of Social Surveillence", *Forbes*, https://www.forbes.com/sites/kalevleetaru/2016/10/12/geofeedia-is-just-the-tip-of-the-iceberg-the-era-of-social-surveillence/#50050d0f5b90, (02.04.2020).

Lemke, Thomas, (2011), *Biopolitics: An Advanced Introduction*, New York: New York University Press.

Levin, Sam, (2017), "Facebook told Advertisers It Can Identify Teens Feeling 'Insecure' and 'Worthless", *The Guardian*, https://www.theguardian.com/technology/2017/may/01/facebook-advertising-data-insecure-teens, (15.05.2020).

Lyon, David, (2001), *Surveillance and Society: Monitoring Everyday Life*, Buckingham and Philadelphia: Open University Press.

McMullan, Thomas, (2015), "What Does the Panopticon Mean in the Age of Digital Surveillance?", *The Guardian*, https://www.theguardian.com/technology/2015/jul/23/panopticon-digital-surveillance-jeremy-bentham, (13.04.2020).

Manyika, James, Lund, Susan, Bughin, Jacques, Woetzel, Jonahan R., Stamenov, Kalin, and Dhingra, Dhruv, (2016), *Digital Globalization: The New Era of Global Flows*, Vol. 4, San Francisco: McKinsey Global Institute.

Marx, Gary T., (2015), *Windows into the Soul: Surveillance and Society in an Age of High Technology*, Chicago: University of Chicago Press.

Marx, Gary T., (2002) "What's new about the "New Surveillance"? Classifying for Change and Continuity", *Surveillance and Society*, Vol. 1, N. 1, pp. 9–29.

Mendel, Toby, Puddephatt, Andrew, Wagner, Ben, Hawtin, Dixie, and Torres, Natalia, (2012), *Internet Privacy and Freedom of Expression*, UNESCO Series on Internet Freedom, Paris: UNESCO.

Michaels, Jon D., (2008), "All the President's Spies: Private-Public Intelligence Partnerships in the War on Terror", *California Law Review*, Vol. 96, pp. 901–966.

Miller, Claire Cain, (2013), "Google to Pay $17 Million to Settle Privacy Case", *New York Times*, https://www.nytimes.com/2013/11/19/technology/google-to-pay-17-million-to-settle-privacy-case.html, (18.03.2020).

Muller, Benjamin, (2004), "Globalization, Security, Paradox: Towards a Refugee Biopolitics", *Refuge: Canada's Journal on Refugees*, Vol. 22, N. 1, pp. 49–57.

Mythen, Gabe and Walklate, Sandra, (2006), "Criminology and Terrorism: Which Thesis? Risk Society or Governmentality?", *British Journal of Criminology*, Vol. 46, N. 3, pp. 379–398.

Nadesan, Majia Holmer, (2010), *Governmentality, Biopower, and Everyday Life*, New York: Routledge.

Norris, Clive and Armstrong, Gary, (1999), *The Maximum Surveillance Society: The Rise of CCTV*, New York: Oxford University Press.

O'Carroll, Tanya and Franco, Joshua, (2019), " 'Muslim registries', Big Data and Human Rights" , *Amnesty International*, https://www.amnesty.org/en/latest/research/2017/02/muslim-registries-big-data-and-human-rights/, (27.03.2020).

Orwell, George, (2019), *Nineteen Eighty-Four*, Singapore: Penguin Random House.

Peretti, Jacques, (2017), "Palantir: the 'Special Ops' Tech Giant that Wields as Much Real-world Power as Google", *The Guardian*, https://www.theguardian.com/world/2017/jul/30/palantir-peter-thiel-cia-data-crime-police, (30.05.2020).

Phelan, David, (2019), "Google Buys Fitbit for $2.1 Billion: Here's What It Means", *Forbes*, https://www.forbes.com/sites/davidphelan/2019/11/01/

google-buys-fitbit-for-21-billion-heres-what-it-means/#5d527432732f, (12.03.2020).

Pisa, Michael, (2019), "Developing Countries Seek Greater Control as Tech Giants Woo the "Next Billion Users", *Center for Global Development,* https://www.cgdev.org/blog/developing-countries-seek-greater-control-tech-giants-woo-next-billion-users, (22.04.2020).

Pisa, Michael and Polcari, John (2019), "Governing Big Tech's Pursuit of the "Next Billion Users", *Center for Global Development,* https://www.cgdev.org/publication/governing-big-techs-pursuit-next-billion-users, (20.03/2020).

Pollach, Irene, (2011), "Online Privacy as a Corporate Social Responsibility: An Empirical Study", *Business Ethics: A European Review,* Vol. 20, N. 1, pp. 88–102.

Ranking Digital Rights, (n.d.), About Ranking Digital Rights, https://rankingdigitalrights.org/about/, (09.03.2020).

Reform Government Surveillance Coalition $(RGS)_{a}$, (n.d.), "Putting Principles into Action", https://www.reformgovernmentsurveillance.com/principles/, (20.03.2020).

Reform Government Surveillance Coalition $(RGS)_{b,}$, (n.d.), "Purpose and Members", https://www.reformgovernmentsurveillance.com/about/, (27.03.2020).

Regan, Priscilla M., (2012), "Regulating Surveillance Technologies: Institutional Arrangements", (Kirstie Ball, Kevin D. Haggerty and David Lyon, Eds.), *Routledge Handbook of Surveillance Studies,* New York: Routledge.

Rosenberg, Matthew, Confessore, Nicholas and Cadwalladr, Carole, (2018), "How Trump Consultants Exploited the Facebook Data of Millions", *New York Times,* https://www.nytimes.com/2018/03/17/us/politics/cambridge-analytica-trump-campaign.html, (17.05. 2020).

Rozenshtein, Alan Z., (2018), "Surveillance Intermediaries", *Stanford Law Review,* Vol. 70, pp. 99–190.

Schneier, Bruce, (2014), "Don't Listen to Google and Facebook: The Public-Private Surveillance Partnership is Still Going Strong", *The Atlantic,* https://www.theatlantic.com/technology/archive/2014/03/don-t-listen-to-google-and-facebook-the-public-private-surveillance-partnership-is-still-going-strong/284612/, (17.05.2020).

Schneier, Bruce, (2015), "How we Sold our Souls – and More – to the Internet Giants", *The Guardian,* https://www.theguardian.com/technology/2015/may/17/sold-our-souls-and-more-to-internet-giants-privacy-surveillance-bruce-schneier, (01.05.2020).

Shaban, Hamza, (2019), "Apple Stars at Giant Tech Confab CES – without Actually Being There", *Washington Post,* https://www.washingtonpost.com/technology/2019/01/07/apple-burns-google-giant-billboard-touting-privacy-ces/, (11.06.2020).

Shahbaz, Adrian and Funk, Allie, (2019), "Freedom on the Net 2019 Key Finding: Governments Harness Big Data for Social Media Surveillance", *Freedom House,* https://freedomhouse.org/report/freedom-on-the-net/2019/the-crisis-of-social-media/social-media-surveillance, (13.04.2020).

Shukla, Satwik, (2019), "A Better Way to Learn About Ads on Facebook", https://newsroom.fb.com/news/2019/03/a-better-way-to-learn-about-ads/, (28.03.2019).

Sieber, Alexander, (2019), "Does Facebook Violate its Users' Basic Human Rights?", *NanoEthics,* Vol. 13, N. 2, pp. 139–145.

Smith, Brad, (2019), "Tech Firms Need More Regulation, The İndustry Must Cooperate to Solve Problems – but Government Must Take a More Active Role as Well", *The Atlantic,* https://www.theatlantic.com/ideas/archive/2019/09/please-regulate-us/597613/, (06.04.2020).

Somogyi, Stephan, (2014), "Making End-to-end Encryption Easier to Use", *Google Security and Privacy,* https://security.googleblog.com/2014/06/making-end-to-end-encryption-easier-to.html, (07.05.2020).

Swant, Marty, (2019), "Edward Snowden Thinks Even the EU's Sweeping Privacy Law Is too Weak", *Forbes,* https://www.forbes.com/sites/martyswant/2019/11/05/edward-snowden-thinks-even-the-eus-sweeping-privacy-law-is-too-weak/#47fcd80b67ff, (07.05.2020).

Tiku, Nitasha, (2019), "Big Tech: Breaking Us Up Will Only Help China", *Wired,* https://www.wired.com/story/big-tech-breaking-will-only-help-china/, (12.05.2020).

Timm, Trevor, (2014), "Four Ways Edward Snowden Changed the World – and Why the Fight's not Over", *The Guardian,* https://www.theguardian.com/commentisfree/2014/jun/05/what-snowden-revealed-changed-nsa-reform, (15.04. 2020).

Torpey, John, (2007), "Product Review: Through Thick and Thin: Surveillance after 9/11", *Contemporary Sociology,* Vol. 36, N. 2, pp. 116–119.

Turow, Joseph and Draper, Nora, (2012), "Advertising's New Surveillance Ecosystem", (Kirstie Ball, Kevin D. Haggerty and David Lyon, Eds.), *Routledge Handbook of Surveillance Studies,* New York: Routledge.

van Ooijen, Iris and Vrabec, Helena U., (2019), "Does the GDPR Enhance Consumers' Control over Personal Data? An Analysis from a Behavioural Perspective", *Journal of Consumer Policy,* Vol. 42, N. 1, pp. 91–107.

Vila, Tony, Greenstadt, Rachel and Molnar, David, (2003), "Why We Cannot Be Bothered to Read Privacy Policies: Models of Privacy Economics as Lemons Market", (Norman Sadeh, Ed.), *Proceedings of the Fifth International Conference on Electronic Commerce,* New York: ACM Press.

Vincent, James, (2015), "Google's DeepMind and UK Hospitals Made Illegal Deal for Health Data, says Watchdog", *The Verge,* https://www.theverge.

com/2017/7/3/15900670/google-deepmind-royal-free-2015-data-deal-ico-ruling-illegal, (03.03.2020).

Vincent, James, (2019), "Google Accused of Inappropriate Access to Medical Data in Potential Class-action Lawsuit", *The Verge*, https://www.theverge.com/2019/6/27/18760935/google-medical-data-lawsuit-university-of-chicago-2017-inappropriate-access, (17.04.2020).

Whittaker, Zack, (2018), "Tech Giants Hit by NSA Spying Slam Encryption Backdoors", *ZDNet*, https://www.zdnet.com/article/coalition-of-tech-giants-hit-by-nsa-spying-slams-encryption-backdoors/, (12.06.2020).

Wong, Julia C., (2019), "My Data Security is Better than yours: Tech CEOs Throw Shade in Privacy Wars", *The Guardian*, https://www.theguardian.com/technology/2019/may/09/google-sundar-pichai-privacy-apple-facebook-data, (10.03.2020).

Wright, David, Gutwirth, Serge, Friedewald, Michael, Vildjiounaite, Elena and Punie, Yves (Eds.), (2008), *Safeguard in a World of Ambient Intelligence*, Dordrecht: Springer.

YouTube, (n.d.), "Copyright Management Tools: Content Verification Program", https://support.google.com/youtube/answer/6005923?hl=en, (15.04.2020).

Zuboff, Shoshana, (2015), "Big Other: Surveillance Capitalism and the Prospects of an Information Civilization", *Journal of Information Technology*, Vol. 30, pp. 75–89.

Zuboff, Shoshana, (2016) "Google as a Fortune Teller: The Secrets of Surveillance Capitalism", *Frankfurter Allgemeine*, https://www.faz.net/aktuell/feuilleton/debatten/the-digital-debate/shoshana-zuboff-secrets-of-surveillance-capitalism-14103616.html?printPagedArticle=true, (14.03.2020).

Seda Gökçe TURAN

Evaluating Perception of Youths Toward Digital Identity

1 Introduction

It is noted by researchers that the twenty-first century is the age of technology and has witnessed many transformations and changes not only in the social but also on an individual aspect. Undoubtedly, this change and transformation is felt in many different fields, from art to culture and education, in societies. Researchers believe that the main reason we experience these effects of the technological era so much is the widespread use of information technology tools in daily life (Eklund and Roman, 2017; Hodkinson, 2017; Loh and Lim, 2019). According to researchers, this innovation and usage of technology shapes how adult and youth get in touch with other people (Flanagin and Metzger, 2008; Goldman et al., 2008). Especially for youths, the usage of digital platforms as a second place of life has had important consequences not only for social transformation but also for the digitalization of their identity. According to Subrahmanyam and Greenfield (2008:120), social media and digital media are highly effective on youths. Moreover, they stated that physical world and virtual world are psychologically linked to each other. On the other hand, Coleman (2011, cited in Sarıyar, 2019:93) argued that digital environments were not a second "life place," but people could easily change their mind and experience different identities in the virtual world, which means that those places are kind of role-playing areas.

"Identity," as a concept, is a trait that differentiates individuals, while at the same time it defines concepts such as national identity, gender identity, cultural identity and individuals' common characteristics (Buckingham,2008: 1). Youths' identity development process is generally characterized by identity depression, conflicted behaviors, changes in mood and absence of fear to risky behaviors or even enthusiasm about it (Buckingham, 2008: 2). During the process of identity construction in the adolescence period, youths try to take new roles and responsibilities and distinguish the consequences of their actions and behaviors (Loh and Lim, 2019:2). Moreover, during the adolescence period, views of others and an attraction for other people are very important for youths. In terms of digital environments, it seems that youths have a chance to be in an environment with audiences where they can present their identities. Unlike mass media, digital media allow users interact and be an active user instead of a passive audience. In addition, users can present their digital identities to other people as they can produce their own content (Özdemir, 2015:113).

Within the frame of literature, the present study aims to encapsulate youths' digital identities in the context of Goffman's Dramatugy Theory. This study focuses on how the digitalized world affects youths' perceptions toward identity, risks to negative perceptions of digital identity, consequences it can cause and suggestions for the construction of a healthy identity in the digital society for youths.

2 Digital Media and Digital Identity

Especially for understanding and analyzing youths' engagement in the digital world, the term "identity" has become important more than ever. Because the term "identity" involves physical, cognitive and social dimensions, in literature, finding a common definition for it is not easy. According to Erikson, the term identity depends on an emotional or psychological structure, but Marcia noted that identity is composed of one's drives, beliefs and experiences (cited in Atak, 2011: 164). Chen (2016: 521) noted that individuals' identity is composed of two parts, frontstage and backstage. On the frontstage, individuals' performances, which include their emotions, appearance and style, could develop their "idealized" identity. The vital component of being frontstage audiences who support or watch their acting. Individuals' identities also have a backstage part where not only they prepare themselves for acting but also escape from responsibilities or roles related with the frontstage. On the other hand, Cho and Jimerson (2017: 886) stated that identity is characterized by how individuals internalize social structures. In terms of this structure, identity has two dimensions, which are personal sense of self and negotiating roles and social structures in different situations. For a healthy and stable identity, feedbacks from other people, especially peers, seem vital for youths, which is also one of the characteristics of digital environments (Middaugh, 2019:2). As most of the researchers stated, construction of an identity is not a fixed possession or label, but it is a process mostly shaped by social interaction (Weber and Mitchell, 2008:43). Loh and Lim (2019: 4) noted that peer relations provide the lens for self-consciousness, which is a great source for identity development. As underlined several times for identity development, youths need to involve and engage in the social context, which is important for gaining new perspectives, opportunities for new negotiations and experience their digital identities (Loh and Lim, 2019: 4).

Digital identity could not be evaluated separate from real life identity or social identity. But some researchers (Işıklı and Küçükvardar, 2016: 103) argue that digital identity is a kind of extended real identity. In addition, Baudrillard (2015:75) stated that when users join the digital world, the difference between the human and the machine disappears. This disappearance specifies the user's digital identity and how they present themselves. Because it is much fuller of freedoms than in the real world, people prefer digital environments more and experience their identities there (İlhan, 2013:236).But other users see these identities with doubt. Especially in terms of cyber-relationships, users only deal with the other users' "digital identity" and they perceive them only this identity they see. Moreover, they are not sure whether the other user, with whom they communicate, is a troll or not. From a digital identity perspective, just like how people in the real world try to get to know each other through verbal or body language, in digital environments, users try to know each other through social media profiles, content sharing or status. Srivastava and Roychoudhury (2020:1) stated that digital identity includes a variety of features that identify the user within the given online social network, which is distinct from the other users. These digital identities allow the users to interact with other users while being present themselves.

Of course, digital identity and real identity could not separate from each other. But it is a common fact that online environments differ from being face-to-face (Cho and Jimerson,2017:886); thus, the interaction between digital identity and real identity would be different. In digital environments, users experience new identities which are linked to their real-life identity. But before discussing digital media and digital identity, we should talk about mass media effects on youths. Mass media has long been an important source of youths' identity development in terms of TV characters and their roles in contents such as soap operas, movies or news (Middaugh, 2019: 1). Mass media has a different effect on youths, but mostly it affects individual perception toward justice and expectation about the real world. In mass media, often a "fair" world where the evil is punished and the good people always win is presented and idealized (Mora, 2008: 11). This presented world of course is not real and affects youths' perception toward the real world. After rise of the Internet and digital environments, researchers stated that (Middaugh, 2019:1) there are new dimensions to the relationship between media and youths' identity development and structure. In terms of digital environments and digital identity, it is the same with mass media effect. In a digital environment, youths are mostly exposed to different disclosure or hate speeches which are related to a fair world, justice or discrimination. Consequently, being exposed to this type of disclosure affects the construction of their digital identity.

In recent years, studies about youths' self-presentation on social media networking sites have increased significantly. Most researchers discuss self-presentation with online sharing activities (Gregson and Rose, 2000).Other researchers underline that people present their identity through images, status, social media profile and friend lists to other users (Hogan, 2010: 377). Expressing or presenting themselves with a different identity is not always a "bad" thing. As part of their developmental stage, sometimes adolescents have difficulty in talking about themselves or establishing healthy communication. They could be a bit shy (Middaugh, 2019:2). But on digital platforms (social media, online games etc.), youths could pretend like an idealized person, become role model for themselves. They behave as if they were at a comfort zone. In addition to this, since for some youths face-to-face communication, especially in a romantic relationships, could be very difficult, they feel more comfortable with digital communication. But the key point is how digital environments affect their digital identities. Because if they get used to digital communication and pretend as if their digital identity is their only identity, daily life communication and real relationships would be more difficult for them.

Today, researchers also focus on different self-presentations on different digital platforms. As variety of digital platforms researchers (Schwartz and Halegoua, 2014) define new identity forms, such as "Spatial Self." According to Schwatz and Halegoua (2014:2), people tend to display their physical activities on social media; they record, publish or archive their daily experiences or mobility to present their identity to audiences; this is called their "spatial self." Researchers claim that this spatial presentation is counted as same as the status messages, photos or videos that are shared via social media. Users who share their location on social media network sites not only share their locations but also create a life map with their cultural perspectives (Schwatz and Halegoua, 2014:4).

Furthermore, there is another dimension to digital identity: social identity. According to Erikson's theory, social identity constitutes interaction with other people in the social field and observing and testing an individual's identity through other peoples' views (Atak, 2011:168). Tajfel and Turner (1979) noted that the social identity theory underlines both intergroup relations and intragroup conflicts and corporations (cited in Hogg, 2016: 3). In terms of these group identities, "Social media identity bubbles" is term used by Keipi et al. (2017) to refer to people using social media to interact with other users and validate their identity (Cited in Oksanen et al.,2020: 2). Oksanen et al. (2020:2) classified social media identity bubbles into three factors, which are closeness to digital social networks (social identification), tendency to interact with similar-mind users (homophily) and confidence toward information from the like-minded. These tendencies shape both digital identities and digital experiences of youths. Davis (2013: 2283) argued that Internet usage indirectly affect youths' sense of identity by reducing the quality of offline friendships. Because it's believed that the time spent engaging in online activities reduces the quality of their face-to-face relationships. But for online gamers, this argument could be different. According to Eklund and Roman (2017: 284), online gamer pupils mostly prefer other online gamers for friendship, and mostly they prefer friendships at school with other online gamers. So, we can conclude like there is a specific digital culture in cyberspace, and youths mostly create and develop their own identity, both digital and real world. In order to explain how youths present their digital identities, Eric Goffman's Dramaturgy Theory would be appropriate.

3 Erving Goffman: Dramaturgy Theory and Digital Identity

The construction of identity in youths takes place not only on the family and community basis but also on the axis of media and the digital world. Mora (2008:1) noted that mass media is accepted as a determinant for societal values since most of the individuals are affected by it. Regarding online behaviors, Zeng et al.

(2020: underline "prosocial" behaviors and note that online prosocial behaviors could allow individuals to act as if they are not shy or believe in a just world at individual and contextual factors. Moreover, moral perfectionism, which is one of characteristics of identity, could motivate social media users to online prosocial behavior. As most of the researchers noted that in terms of constructing digital identity, bystanders and other users' existence and their applause to moral and prosocial behaviors is called the "audience effect." The term "audience," which basically affects the construction of youths' identity and is also a basic characteristic of their developmental stage, stands out in digital environments. This audience effect influences youths' determination of their social roles, which, as Goffman stated (1995; cited in Boz, 2015), shapes the social interactions of individuals. Goffman (1959:22; cited in Hogan 2010: 378) defined these interactions and performances as" activity of an individual which occurs during a period marked by his continuous presence before a particular set of observers and which has some influence on the observers." According to Goffman, individuals' social roles are shaped by and affected other people's behavior and the feedback they receive from them. This shaping process occurs through social interaction among the people, which depends on the flow and circulation of information according to the social understanding of these people. Goffman describes the "audience" effect in his Dramaturgy Theory, which explains social interaction with theatrical scene analogy. Just as actors and actresses want to make a positive impression on audiences to get applause, individuals want to make such good impression on other people too. Goffman also remarks that people express their feelings, thoughts, desires and beliefs using gestures, words or nonverbal communication pathways just like actors and actresses do in their role-plays. Within the frame of Goffman's Dramaturgy Theory, digital environments are considered as environments where youths present themselves both verbally and nonverbally, experience the audience impact and configure their identity (Bullingham and Vasconcelos, 2013: 101). Although the concept of identity is primarily considered on an individual basis, that is physical appearance, character structure values, judgments and attitudes, in digital environments, social identities become more visible. It is one of the early stages of building identity for youths to regulate their behavior according to group norms and group behaviors. Rather than a different structure, digital environments are a different "culture." Moreover, in the twenty-first century, it is seen an indispensable part of everyday life. In this context, digital environments are very important not only for digital identities but also for self-development, because these environments affect the social identity of youths. Youths' self-esteem and self-confidence would increase when accepted or approved in digital environments (i.e., social media, blogs, vlogs, online gaming, etc.), just like the person who is affiliated with a group feels like they belong there. Unfortunately, this does not happen always in a positive way. Especially risky and dangerous cyber behaviors like perception of body image, sharing every activity continuously, comparing oneself with everybody all the time, cyberbullying to others while expressing their ideas and suffering from these behaviors may lead to youths developing a negative perception toward their identity. In order to analyze youths' digital identity, it should be emphasized that individuals in a digital environment could be "anonymous." As Middaugh (2019:2) noted, anonymity has two sides, it is both risky and beneficial. For identity construction and development of youths who have limited opportunities for exploration in their communities, trying to explore being anonymous could be possible. This view about the benefits of being anonymous is based on early researches and observations that state that online chatrooms allow users to experience different personas with little risks. But also, this anonymity allows youths to behave in risky ways such cyberbullying or indulge in hate speech because they assume that social rules cease to function in the digital world and hence they would not be punished for their inappropriate behaviors. Although digital identity and real identity are not interchangeable concepts, their interaction depends on how often youths use digital environments and how the line between real life and the digital world is defined by them. For example, using the digital world as a "bolt hole" by youths who cannot be part of peer groups in daily life or feel unendorsed will make them give more importance to their digital identity. Digital environments are mostly seen as a safe place for youths in order to experience different roles, but Livingstone (2003) stated that identity development is shaped by social interaction, and because of this, motivation behind using digital platforms should be analyzed. For this reason, social interaction should be emphasized to understand youths' digital identity and social media usage habits.

Youths who socially interact tend to present themselves in accordance with social norms and acceptable labels, such as conscious, ethical, moral, and responsible. This causes them to build an identity in a digital environment and experience their "idealized" identity (Yüzüncüyıl, Buluş and İşman, 2016: 613). As Goffman stated in his dramaturgical approach, people present an idealized rather than an authentic identity of themselves (Hogan, 2010: 378). According to Manago (2015), this idealization is not always in a positive way. Research findings noted that problematic Internet usage, online gaming and Internet addiction are very high among youths who use the Internet for identity exploration; such individuals have difficulty in communicating, are shy and have large differences between their ideal and actual selves (cited in Middaugh, 2019:2).

One of the most important dimensions of self-presentation is "facemask," which is defined as a persona in identity literature. This face-mask acting has opportunities and risks on online world where users could self-share and self-expose. Due to human beings always have risk while presenting themselves in different ways and as different personas, especially in digital world, which is free from reliable or complete boundaries (Couldry and Hepp, 2017: 146). For a stable and healthy identity construction, youths mostly need stable and healthy identity role models. But on online platforms, especially social media, youths mostly face with different personas who could be even troll or wrong role model (racist, discriminatory or extremist users). Of course, how to present themselves and presentation of other users affects youths' construction of digital identity. As DeFleur (2010: 135) noted that with audience effect a user's perception and online behaviors. In other words if users feel they were watching by others their online behaviors would be different. By taking into consideration the construction of digital environments and audience effects, it could be said that digital environments have huge effect on youths' identity development and present this identity as "digital identity." In the next section, the digital environment's effect on youths' identity perception is discussed.

4 How Digital Environments Affect Youths' Perception of Identity?

The adolescence period is often characterized by big changes in human life, which is vital for identity development, which is mostly shaped by social relationships (Middaugh, 2019:1; Atak,2011: 163). Besides this, youths are using digital environments in order to facilitate social interaction, which seems very important in constructing an identity (Goldman et.al., 2008: 185). The most obvious characteristic of a digital environment is the existence of unlimited, wide variety of opportunities for learning, social interaction and individual entertainment (Flanagin and Metzger, 2008: 5). For understanding the effect of digital environments on youths, we must consider physical, social, and cognitive factors about identity, which are important for youths' digital environment experiences for risky and beneficial sides. Oksanen et al. (2020: 3) noted that identity dynamics have a vital role for peoples' way to handle negative or bullying experiences ondigital platforms. Moreover, if people feel threatened about their identity on online platforms, they react more strongly. So, digital identity could affect online behaviors also.

According to Işıklı and Küçükvardar (2016: 91), noted that at digital environments people and things had identity. As normal, after gaining this identity users and things have a different relationship and interaction. In terms of social network sites, Stuztman (2006:10) noted that popular social networks communities simplify sharing of identity information, and these identity information are more subjective and holistic. Also, from the point of users, especially digital media platforms have different norms and rules. For example, for Tweeter, the number of followers is considered as an indicator of status; moreover, the number of retweets and likes are also very important (Sarıyar, 2019: 64).

While discussing about the relationship between digital identity and digital environments, we should also specify cyber behaviors such as cyber harassment and its effects on digital identity construction of youths (Loh and Lim, 2019: 4).

Cyberbullying or online harassment, which are so common among youths, is a kind of harassment which involves intentional behaviors and acts to harass someone or groups by using any technological devices (Lee and Shin, 2017:352). Youths whose identity is generally based on social media or digital environment profiles would be much more cyberbullied than other people (Oksanen et al., 2020: 1–3). Digital environments allow users to be anonymous. This characteristic would be both beneficial and harmful for the users. Being anonymous gives freedom to the users to hide their real identity and present themselves as different persons. In addition, digital platforms allow users to explore each other both digital and human beings. Normally, one's identity is associated with their past, their experiences, relationships, ethnicity etc., but for a digital identity, users do not need any of them (Sarıyar, 2019: 84).

In social media, there are some groups that discriminate against people or share racist disclosures which affect people's beliefs about themselves or others. Youths who are exposed to too much hate speeches about "others" perceive others as they are presented on social media or digital environments. Both victims and perpetrators of hate speeches construct their digital identity on social media and present themselves as racist or nonracist. Moreover, by taking into consideration race or gender issues, scholars claim that youths defined digital environments as not only how race and gender issues are visible but also how it could be presented (Bullingham and Vasconcelos, 2013: 103). This presentation could be evaluated by social identity theory as group members in other words social media users determine the other users' identity, people tend to identify themselves and group norms also online (Oksasen et al, 2020:3). On social media, it is not so difficult to find users who disclose their beliefs, ideas or opinions even in a friendly or unfriendly way (hate speech, hate crime, rude online behaviors etc.) and who influence youths' perception toward the real or virtual world and indirectly their digital identities. Digital identities are characterized by different dimensions such as searching for controlling to uncertainty, self-esteem, social belonging and experience for different identities (Oksasen, 2020: 3). Evaluating the relationship between self-concept and digital environment usage, Davis (2013) stated that youths who use digital platforms for identity exploration tend to have less apparent self-concepts. Middaugh (2019) pointed out social media's effects on youths' identity in terms of marginalization, adaptation of negative identity and risk for fragmentation. As Hodkinson (2017:273) noted, all those identity discussions focused on the youths' privacy perspective while using digital environments, social media networks and sharing or posting any documents related with their private life.

Construction of healthy digital identity in terms of these dimensions is discussed in the concluding part.

5 Conclusion

For the digital age, there is a strong relationship between digital platforms and users' digital identities. Within the frame of characteristics of digital environments and youths' developmental stages, it is obvious that interactive digital platforms had a strong influence on constructing and experiencing identity. Youths see digital platforms as if a place to express themselves both verbally and nonverbally, and they often need to present themselves appropriate to social expectations and cultural values of society (Zhao et al., 2013:2), which means that they experience idealized digital identity. Moreover, they have a chance to recognize other users in terms of their identities. But, if youths see digital platforms and online interactions as a determinate for their identity, the dark side of the digital environments, such as cyber harassment, hate speech etc., would be more destructive. To encourage youths to construct a healthy digital identity as researchers, teachers, parents should help youths to be aware of cyber-harassment, digital violence, fake news and media manipulation. As researchers stated, digital citizenship is described as the ability to identify media and mass media messages, create media messages in a responsible way using responsible and conscious technology and adapt to ethical rules of digital world (Xu et al., 2019) is considered as a most effective tool for protect youths from dark side of digital environment. Moreover, digital media literacy skills would also make youths stronger to construct a digital identity.

Bibliography

Atak, Hasan, (2011), "Identity Development and Identity Formation: A Theoretical Assessment", *Current Approaches in Psychiatry*, Vol. 3, No. 1, pp. 163–213.

Baudrillard, Jean, (2015), Le Pacte De Iucidite Ou I'intelligence Du Mal, (Oğuz Adanır, Trans.), Ankara: Doğu-Batı Publication.

Boz, Nevfel, (2012), Dijital Identity and Self-Presentation in the New Media, (Unpublished Doctoral Dissertation), University of Marmara, İstanbul.

Buckingham, David, (2008), "Introducing Identity", (David Buckingham, Ed.), Y*outh, Identity, and Digital Media,* The John D.and Catherine T., MacArthur Foundation Series on Digital Media and Learning, Cambridge, MA: The MIT Press, pp. 1–24. doi:10.1162/dmal.9780262524834.001.

Bullingham, Liam and Vasconcelos, Ana C. (2013), "The Presentation of Self in the Online World: Goffman and the Study of Online Indentities", *Journal of Information Sciences*, Vol. 39, No. 1, pp. 101–112.

Chen, Chih-Ping, (2016), "Playing With Digital Gender Identity and Cultural Value", *Gender, Place and Culture*, Vol. 23,No. 4, pp. 521–536.

Cho, Vincent and Jimerson, Jo Beth, (2017), "Educational Management", *Administration and Leadership*, Vol. 45, No. 5, pp. 884–900.

Couldry, Nick and Hepp, Andreas, (2017), *The Mediated Construction of Reality*, Polity Press, USA.

Davis, Katie, (2013), "Young People's Digital Lives: The Impact of Interpersonal Relationships and Digital Media Use on Adolescents' Sense of Identity", *Computers in Human Behavior*, Vol. 29, pp. 2281–2293.

Eklund, Lina and Roman, Sara, (2017), "Do Adolescent Gamers Make Friends Offline? Identity and Friendship Formation in School", *Computers in Human Behavior*, Vol. 73, pp 284–289.

Flanagin, Andrew J. and Metzger, Miriam J. (2008), "Digital Media and Youth: Unparalleled Opportunity and Unprecedented Responsibility", (Miriam J. Metzger and Andrew J. Flanagin Eds.), *Digital Media, Youth and Credibility*,

The John D. and Catherine T. MacArthur Foundation Series on Digital Media and Learning, Cambridge, MA: The MIT Press, pp. 5–28.

Goldman, Shelly, Booker, Angela and McDermott, Meghan, (2008), "Mixing the Digital, Social, and Cultural: Learning, Identity, and Agency in Youth Participation", (David Buckingham, Ed.), *Youth, Identity, and Digital Media*, The John D. and Catherine T. MacArthur Foundation Series on Digital Media and Learning. Cambridge MA: The MIT Press, pp. 185–206.

Gregson, Nicky and Rose, Gillian, (2000), "Taking Butler Elsewhere: Performativities, Spatialities and Subjectivities", *Environment and Planning D: Society and Space*, Vol. 18, pp 433–452.

Hodkinson, Paul, (2017), "Bedrooms and Beyond: Youth, Identity and Privacy on Social Network Sites", *New Media and Society*, Vol. 19, No. 2, pp. 272–288.

Hogan, Bernie, (2010), "The Presentation of Self in the Age of Social Media: Distinguishing Performances and Exhibitions Online", *Bulletin of Science, Technology and Society*, Vol. 30, N. 6, pp. 377–386.

Hogg, Michael A. (2016), "Social Identity Theory", (S. McKeown, Ed.), *Understanding Peace and Conflict Through Social Identity Theory*, Peace Psychology Book Series, Switzerland: Springer International Publishing.

Işıklı, Şevki and Küçükvardar, Mert, (2016), *Information Revolution*, Ankara: Birleşim Publication.

İlhan, Süleyman, (2013), "Identity Construction in Liquid Society: Transitive, Eclectic, Self-Centered Identities", *Fırat University Journal of Social Sciences*, Vol. 23, No. 2, pp. 233–246.

Lee, Changho and Shin, Namin, (2017), "Prevalence of Cyberbullying and Predicors of Cyberbullying Perpetration Among Korean Adolescence", *Computers in Human Behavior*, Vol. 68, pp. 352–358.

Livingstone, Sonia, (2003), "Children's Use of the Internet: Reflections on the Emerging Research Agenda", *New Media and Society*, Vol. 5, No. 2, pp. 147–166.

Loh, Renea S.M. and Lim, S. Sun, (2019), "Youth Digital Culture", (Renee Hobbs and Paul Mihailidis, Eds. In Chief), (Gianna Cappello, Maria Ranieri, and Benjamin Thevenin, Assoc. Ed.), *The International Encyclopedia of Media Literacy*, Wiley.

Middaugh, Ellen, (2019), "Media and Adolescent Identity Development", (Renee Hobbs and Paul Mihailidis, Eds., in-Chief), (Gianna Cappello, Maria Ranieri, and Benjamin Thevenin, Assoc. Eds.), *The International Encyclopedia of Media Literacy*, Wiley.

Mora, Nazan, (2008), "Media and Cultural Identity", *International Journal of Humanities*, Vol. 5, No. 1, pp. 1–14.

Oksanen, Atte, Oksa, Reetta, Savela, Nina, Kaakinen, Markus and Ellonen, Noora, (2020), "Cyberbullying Victimization at Work: Social Media Identity", *Computers in Human Behavior*, Vol. 109, pp. 1–11.

Özdemir, Zafer, (2015), "New Trend In Identity Achievement In Social Media: Using Selfies", *Maltepe University Journal of Communication Faculty*, Vol. 2, No. 1, pp. 112–131.

Sarıyar, Hatice, (2019), *The Conceptualization of İdentity in The Digital Era and The Reality: Twitter Parody Accounts*, (Unpublished Master's Dissertation), Marmara University, İstanbul.

Schwartz, Raz and Halegoua, Germanie R. (2014), "The Spatian Self: Location-Based Identity Performance on Social Media", *New Media and Society*, pp. 1–18.

Srivastava, Deepesh K. and Roychoudhury, Basav, (2020), "Words Are Important: A Textual Content Based Identity Resolution Scheme Across Multiple Online Social Networks", *Knowledge-Based Systems*, Vol. 195, pp. 1–17.

Stutzman, Frederic, (2006), "An Evaluation of Identity– Sharing Behavior in Social Network Communities", *Journal of the International Digital Media and Arts Association*, Vol. 3, No. 1, pp. 10–18.

SubrahmanyamKaveri and Greenfield, Patricia, (2008), "Online Communication and Adolescent Relationships", *The Future of Children*, Vol. 18, No. 1, Children and Electronic Media, pp. 119–146.

Weber, Sandra R. and Mitchell, Claudia, (2008), "Imagining, Keyboarding, and Posting Identities: Young People and New Media Technologies", (David Buckingham, Ed.), The John D. and Catherine T. MacArthur Foundation Series on Digital Media and Learning, Cambridge MA: The MIT Press, pp. 25–48.

Xu, Shun, Yang, Harrison H., Macleod, Jason, and Zhu, Sha, (2019), "Social Media Competence and Digital Citizenship among College Students", *Convergence: The International Journal of Research into New Media Technologies* 2019, Vol. 25, No. 4, pp. 735–752.

Yüzüncüyıl, Kübra S., İşman, Aytekin and Buluş, Berkay, (2016), "Transformation of Socializing into Digitalization and Representation of Digital Identity", *TRTAkademi*, Vol. 1, No. 2, pp. 608–619.

Zeng, Pan, Zhao, Xue, Xie, Xin, Long, Jie, Jiang, Quianyun, Wang, Yuhui, Qi, Lin, Lei, Li and Wang, Pengcheng, (2020), "Moral Perfectionism and Online Prosocial Behavior: The Mediating Role of Moral Identity and the Moderating Role of Online Interpersonal Trust", *Personality and Individual Differences*, Vol. 162, pp. 1–6.

Zhao, Xuan., Salehi, Niloufar, Naranjit, Sasha, Alwaalan, Sara, Voida, Stephen and Cosley, Dan, (2013), "The Many Faces of Facebook: Experiencing Social Media as Performance, Exhibition, and Personal Archive", CHI'13, April 27–May2, Paris France.

Eren KESER and Fatih USLU

Digitalization and Augmented Reality Applications: Whether Augmented Human or Lost Humanity?

1 Introduction

Even though communication has started with the existence of humanity, the mass communication researches, which consisted of the impact of social change and transformation after the industrial revolution, emerged as a result of efforts to understand social processes about society and people. Since the seventeenth century, an industrial revolution has taken place over an average of 100 years; however, with the introduction of the third industrial revolution as the Internet, everything has started to be unlike before and industrial revolutions have also accelerated. From now on, by completing the industrial revolutions, we have passed to social revolutions. In this context, Society 5.0 has the characteristics of a valuable social revolution. Revolutions led by digital technologies such as Industry 4.0 and Society 5.0 have contributed to the name of our age as digital age. When digital revolution was experienced in the twenty-first century, as a result of the penetration of the Internet in all areas of life, it is seen that mass communication turns toward an individual-oriented communication by changing/transforming. Although it is commonly thought that communication started with the interpersonal communication process in hunter and collector tribes, and then there was a tendency toward mass communication as a result of the formation of mass society with the transition to urban life and urbanization, it is actually cave paintings, which are located in the caves, which are the living areas of social groups in the prehistoric period, that are in some way the first examples of mass communication. However, the discussion of how the mass society can be controlled and directed has started with the development and spread of communication science studies in social scientific disciplines with the invention of communication technologies such as printing technologies, photography, telephone, cinema, radio, television, and the Internet; and with the effect of industrialization. As a result of this, it is seen that mass communication theories are formed by laying the foundation of thoughts, efforts, and communication studies aimed to understand mass communication. First of all, theoretical approaches on communication/media studies are based on two basic theoretical distinctions: (1) mainstream media studies, which laid its foundations based on the advocacy of "laissez-faire, "the liberal saying of Adam Smith, and free market economy, which legitimizes the existing social order and prioritizes the liberal economic order and individual entrepreneurship, and (2) critical media studies, which continue with the tradition of cultural studies in the following process and argue that mainstream media studies serve to current system, based on Marxist thought, examining communication and the system with a critical approach. In conclusion, with the fact that the competence of these theoretical approaches formed in the early twentieth century became questionable, the approach in which the social changes in the cultural and social field, which are called "technological determinism" in the literature with a reductionist perspective, are determined only by the effect of technology. Apart from this approach, with the transformation in political organizations, the economic acceleration that occurs because of the change in the production models and the technological developments that occur with the convergence of distances make social change/transformation and the traditional mass media tools are affected by this and turn into digital communication tools. In other words, new communication environments have been created to interact instantly with the content of the TV, which is regarded as one of the traditional communication tools by social media environments with its network infrastructure. These communication environments, which are defined in the literature as new media, new communication media, digital communication media etc., can be said to have created a new "digital culture" by playing an active role in changing the society. These new communication environments, which are effective in the formation of digital culture, manifest themselves by the development and spread of the Internet. Initially, the process had started with Web 1.0 only for the purpose of spreading one-way information and based on text; later, it continued with Web 2.0, where new communication environments such as social media, where interactions can take place, became more common; finally, in today's society, the transition to Web 3.0, where virtuality and mobility are at the forefront, has become a multi-multimedia form that includes elements such as text, visual, video, virtual spaces, and augmented reality (AR), which can also be called interactive media. In addition, in the coming years, it is imaginable that with the use of Web 4.0, screen-based communication tools will be threw background out of focus and all communication environments can turn into a single-tool-multimedia form that can be reached through AR as a single channel. M. Castells, who reduces the integration of communication tools realized in the form of multi-multimedia to the concept of mass self-communication by linking the formation of the digital community to the development process of the Web, expresses that:

> *Arisen with the development of Web 2.0 and Web 3.0, in other words, with the increased broadband capacity, innovative open source software and avatar interaction in three-dimensional virtual spaces, strengthening of computer graphics and interfaces, devices and applications that support the spread of social spaces on the internet.* (Castells, 2016: 102)

It is also seen that the computer was a closed book before the Internet and but later became interactive with the Internet and gained multimedia feature; the Internet became interactive with Web 2.0 as well. In relation to that, in his study, Ryan (2019: 157), who also touches on the limited aspects of Web 1.0, which is seen as revolution in terms of information technologies, emphasizes that Internet web browsers were a communication medium which did not allow the content producer and the target group to communicate with each other and focused only on dissemination of information, and the Internet has been prevented from being a written text when it gains bidirectional communication by means of Web 2.0. Thus, by means of Web 3.0 in today's digital culture age, it can be assumed that there is a digital community that surrounded by new media tools that inholds digital communication, integrated with traditional communication environments, virtual environment, interactive communication, interactive, and networked communication.

2 Communication and New Media Tools in Digital Culture

Communication has hundreds of definitions. It can be said that the simplest definition that can be made by taking into consideration all the theoretical researches, studies, suggested theories, and sections of life in order to understand the communication process is actually a persuasion. If we look at the definitions in the literature, according to the Turkish Language Association (TDK, 2020)[1] dictionary, communication can be defined as "Conveying emotions, thoughts, or information to others in every conceivable way, information, intercommunication and communication. Information exchange carried out by using tools such as Telephone, telegraph, television, radio etc., information, intercommunication, correspondence, communication." In other words, "Communication is a result of the process of human survival and a human phenomenon that undergoes changes according to the developments in the manner of human survival" (Oskay, 2017: 15). In addition, "Communication is face-to-face speaking, television, information dissemination, our hairstyle, literary criticism: an infinite number of additions can be made to the list" (Fiske, 2003:15). Thereby, in line with these definitions, it can be said that communication, which started with the existence of human and socialization process, is also a basic structure that forms cultures. In his study, M. Castells addresses that culture can only exist through communication and reveals that "Because culture is transmitted and initiated through communication, cultures themselves – our historically produced belief systems and traditions – have transformed fundamentally with the new technological system, and will evolve even more over time" (Castells, 2008: 441). As indicated in Tab. 1, in addition, with this transformation, the Digital Culture formed by new forms of social organization, Jan Van Dijk mentions about a social structure which has passed from the mass society to the Network Society as a result of the modern society form and organization and defines as the network society. According to Van Dijk, who calls digital media users' practices as cyberculture, while performing their daily activities with the proliferation of smart portable devices such as mobile phones and tablets in today's media environment,

1 Turkish Language Association, "Turkish Language Association Dictionary", https://sozluk.gov.tr/ (13.06.2020).

Tab. 1: Typology of mass society and network society

Characteristics	Mass society	Network society
Basic elements	Collectivities (Groups, organizations, communities)	Individuals (connected by networks)
Structure of elements	Homogeneous	Heterogeneous
Scale	Extended	Extended and Aged
Scope	Local	"Glocal" (global and local)
Connectivity and connectedness	High in elements	High between Elements
Intensity	High	Less
Centralization	High (few centers)	Low (multicenter)
Comprehensiveness	High	Low
Community type	Physical and indivisible	Virtual and disconnected
Organization type	Bureaucracy vertical integrated	Infocracy Horizontally differentiated
Household type	Large with extended family	Small with diverse family relationships
Basic communication type	Face to Face	Increasingly mediated
Media type	Broadcast mass media	Narrowcast interactive media
Number of media	Low	High

Source: Van Dijk, 2018: 69.

which have been developed with the computer and the Internet, individuals can now make artistic works as a feature of digital culture and information sources in different forms in the mass society can coexist under the name of multi-media as a result of this era. Thereby, screen-oriented communication tools can be seen everywhere in the network society. He mentions an era in which this situation has prevented paper and face-to-face communication, has a rapid flow of information with the effect of computer technologies, the consumer can also broadcast on his own as a manufacturer, produce artistic video content and is easier to specialize (Van Dijk, 2018: 293–298).

After the industrial revolution and the use of steam machines technology and then the space races that started in the middle of the twentieth century, developments in communication technologies (technological tools such as telegraph, camera, phone, radio, television, video camera, wireless, satellite technologies, GSM, computer with modem, three-dimensional (3D) printers, virtual reality (VR) and AR), which have progressed with great acceleration, are increasingly taking place in daily life. Accordingly, communication technologies in today's digital societies; In environments including public spaces, organizations and private living spaces, it also develops to include situations that facilitate the daily activities of the individual, shorten the time, and where information flow is at the highest level, and it is compulsory to meet the basic needs required to sustain life. In particular, the use of AR tools such as *Hololens,*[2] one of the wearable computer technologies used especially in the field of health, is becoming more and more widespread before and after the operation. Similarly, it offers a wide range of work within the engineering, academia, and digital marketing/advertising sectors, bringing the distances closer. Also, the very fast flow of information can be realized by the combination of the virtual and the real, and the interaction/meeting of different cultures creates a new digital culture/community, which can be provided in a much more interactive, faster, and more realistic way than other communication tools, and virtual is considered as real. It can be mentioned that this age of digital culture, which has started to form with new media tools, along with intense information on itself, creates an audience/target group, both active and passive, by accommodating the active audience of new communication tools by changing/transforming from the passive audience situation seen in traditional media tools (radio, television) to interactive communication and virtual interaction. While any content on TV, which is the traditional communication tool, is in the passive state in the news flow, it can also interfere with the program with the help of network, Internet data-based computers, and mobile devices and affect the content instantly when necessary (Gungor, 2018: 140). With all that, both active and passive audience/target group status also affect manipulation to a higher rate than ever before. Briefly, while individuals in the digital community are exposed to disinformation that they are not aware of as a passive audience as well as manipulation, it causes both to accelerate and become more widespread by reproducing and sharing this destruction process, which also came under the influence of an active audience. This process is realized through indistinguishable modified photos, digital video recordings created with an artificial intelligence (AI) system, distorted and reproduced news, and digital games that allow the historical process to be reconstructed with fantastic items and imaginary events, increasingly affecting this digital community day by day. Regarding the digitalization process, Ayhan (2019: 184) mentions that with Web 3.0, control has passed from human to AI, and this digitalization process

2 The Hololens device is an augmented reality application tool developed by Microsoft. It is a new communication tool with multi-media feature and provides digital image transfer in the physical space. It is used in entertainment, education, engineering, architecture, and health fields, and with the software developed for the device, it facilitates and accelerates the daily workflow in different business sectors. For further details, see: https://www.microsoft.com/en-us/hololens/hardware (10.06.2020).

has caused confusion. AR applications, suggestion systems in content monitoring, and software that contains certain algorithms can be shown as examples. In addition, with the algorithms included in various programs and with the help of AI, the new image obtained from two different photographs can be given a real photo impression and it cannot be understood whether there is a photomontage when examined. The originality of this photograph can be understood with the help of AI. rather than human expertise. During the ongoing process, by means of the visual archives in the Internet database, with the help of AI, works can be done as if they were artisanal by placing elements such as mountains, forests, streams, and houses on a blank digital canvas. Digital community members can increase the manipulation by adding their own speeches and mimics to the video recording of any opinion leader, artist, and scientist with the help of AI. Lastly, the history flow that exists in digital game stories can be presented as an alternative history. There are productions like digital games Mass Effect 2 and Witcher 3[3] which show pornographic violence as the natural flow of life, with more than 20 million sales figures to date. In the digital culture era, especially online digital games that active almost every hour of the day and player individuals who can be seen as their virtual community, become virtual individuals who live as if they were real individuals of a universe where unattainable and desired is achieved and pleasure is fulfilled and did not actually exist with game flows, scenarios and ways of playing. These new communication environments, which are formed as a result of digital change/transformation and whose usage is increasing day by day, are becoming indispensable to life. Regarding all these, the annual Global Digital (January 2020)[4] report, in collaboration with We Are Social and Hootsuite, reveals this increasing and expanding digital change/transformation process. In the report, as of January 2020, mobile users constitute 67 percent of the world population of 7.75 billion, which is 5.19 billion people. Of them, 4.54 billion are Internet users and this is almost two thirds of the world's population, that is 59 percent. As indicated in Fig. 1, it is also seen that half of the world's population uses social media.

It is observed in Fig. 2, the number of digital users in Turkey is located above the world average. According to the report, there are 77.39 million Internet users in Turkey, which is 92 percent of the population; accordingly, almost the entire population uses the Internet. In addition, active social media users constitute 64 percent of the population with 54 million people and are above the world average. These data are constantly increasing at an unstoppable speed, and it can be said that life without the Internet is almost impossible in the age of digital culture.

3 For a detailed review of the Witcher 3: Wild Hunt's explicit content, see: https://www.ign.com/wikis/the-witcher-3-wild-hunt/Sex_and_Romance (08.06.2020).

4 For Digital Around the World in Jun 2020 Turkey, see: https://datareportal.com/reports/digital-2020-turkey (12.06.2020).

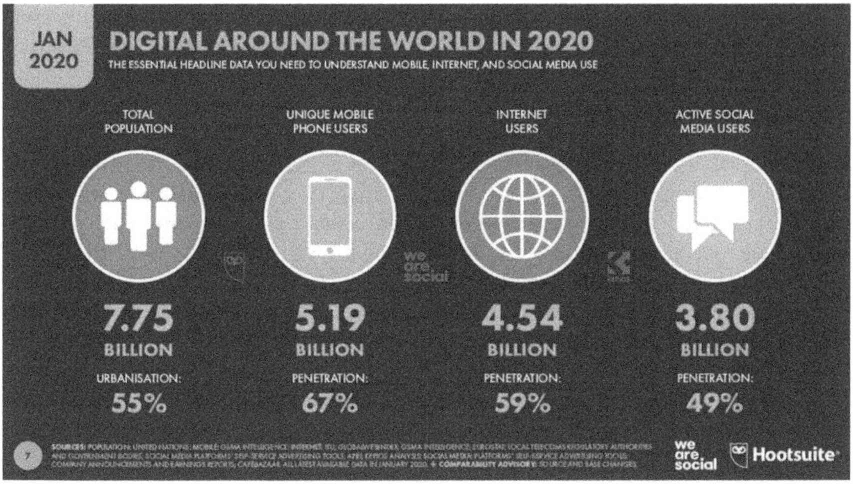

Fig. 1: Digital around the world in 2020. Source: https://datareportal.com/reports/digital-2020-turkey (12.06.2020)

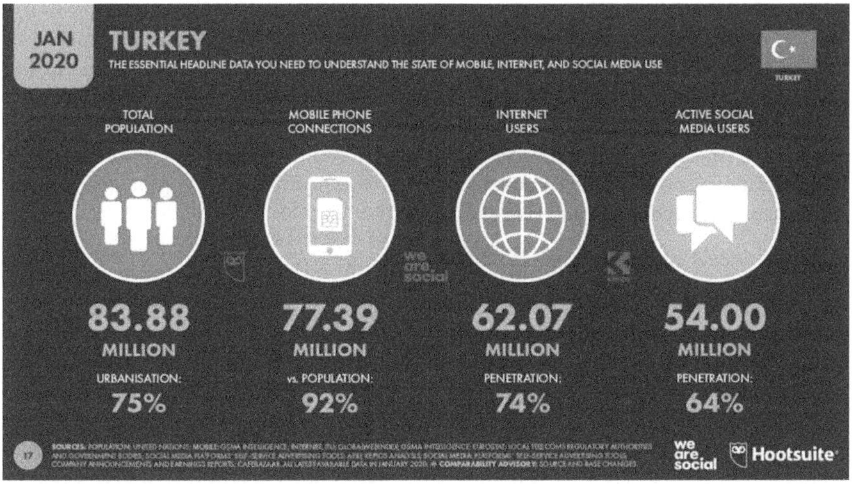

Fig. 2: Digital around the world in 2020 (Turkey). Source: https://datareportal.com/reports/digital-2020-turkey (12.06.2020)

To briefly look at the new communication environments that effective in accelerating this digitalization process, which can also be seen as updated variations of existing new media tools:

- **Virtual Reality Applications**

It is the creation of artificial/virtual spaces that provide a sense of reality to the brain, with the help of 3D glasses, devices such as various software and touch sensors, designed to be completely closed in order not to see the real space, to replace the physical field of view in the real life environment, with the aid of various software and touch sensors. In other words, the VR user experiences the physical reality in a different environment, which is created virtually by ignoring real-life space with the help of the device. Although VR software is widely used in the entertainment industry today, it is also used in the fields of education, military technology, engineering, and medicine[5]. With help of this technology, a person in any real-estate office can see almost the same physical condition of the house in a very remote part of the world. In addition, it can be used in the field of marketing communication for international meetings of a large company, through a virtual avatar, as if everyone is together, and the sense of reality is very high (Kelly, 2016: 240).

Although it has become widespread in recent years, this technology dating back to the 19th century, the use of it as a new communication medium is possible by means of the devices become accessible to all segments, with the reduction of the cost, as a result of the advancement of computer technology, the development of the Internet and the spread of free circulation. While reaching individuals in the field of entertainment through game consoles such as PlayStation, Xbox etc., and have gained functionality in other areas through such as HTC Vive and Oculus Rift devices over PC. Although the use of virtual reality technology is seen as new today, it also has a long development process and position that can be accepted in the historical process as can be seen in Tab. 2.

- **Augmented Reality Applications**

AR, which is developed on VR technology, is the transfer of digital objects and environments on physical environments in real-life spaces, and the transfer and feeling of reality is further strengthened. In other words, sound, images, graphics, and animations produced by computer software are transferred to the real-world environment instead of the screen, so that physical images consisting of digital codes are obtained. According to Milgram and Kishino, it is actually an artificial reality environment based on VR and using digital objects without physical reality

5 It is able to contribute to the process and minimize the risk with the help of 3D ultrasound, where the vessels or the brain can be seen, which facilitates the work of the surgeon in difficult operations during the operation and by replacing the mock-ups in surgical trainings. For further details, see Harvey (2020).

Tab. 2: Development process of virtual reality technology

Year	Technological development
1838	Wheatstone stereoscope
1849	Brewster stereoscope
1903	Parallax effect
1915	The first 3D movie
1929	Link flight simulator
1946	The first computer(ENIAC)
1948	Holography
1956	The Sensorama
1960	Heileg's HMD system
1965–68	Ian Sutherland – the ultimate display (FIPS)
1967	Traub – Varifocal mirror
1970'ler	Computer-based stereo images
1972	Digital pong game
1973	Evans & Sutherland Computer Co-operation
1976	Videoplace
1977	Apple, Commodore and Radio Shack computers
1979	The first Data Gloves
1979	LEEP optics
1981	SpaceGraph developed by Larry Sher at BB&N
1981	SGI was founded
1983	The first book on virtual reality: "Artificial Reality," written by Myron W. Krueger
1985	Commercial Lc container images
1985	NASA AMES
1985	The first virtual reality company VPL was established
1986	Chapel Hill's UNC System
1986–89	Super Cockpit program
1987	Virtual Reality published on Scientific American with Jim Foley's article "VR Comes to the Public's Attention"
1990	SIGGRAPH panel discussions
1990'lar	FakeSpace Boom Display System

(continued)

Tab. 2: Continued

Year	Technological development
1991	International Conference on Artificial Reality and Telexistence in Japan
1992	CAVE system
1995	Virtual Workbench
1995	IEEE Virtual Reality Annual International Symposium (VRAIS 95)
1995	Beginning of clinical virtual reality
1998	Disney Quest's opening
1999	IEE VR conference (Instead of VRAIS)
2001	First computer-based cube rooms SAS3/SAS Cube, designed by Z-A Production
2003	Los Angeles IEEE VR conference
2004	Chicago IEEE VR conference
2007	Announcement of Google Street View application
2010	The first Oculus Rift Prototype announced by Palmer Luckey
2013	Valve developed the platform where content is stored to virtual reality technology
2014	Valve announced SteamSight Prototype and 2016 as the release date
2014	Facebook bought Oculus VR worth 2 billion USD
2014	Sony announced the Morpheus project as PlayStation VR for the game consoles
2015	Kickstarter company exceeded 150,000 USD in its investment campaign for Gloveone and succeeded
2015	HTC and Valve companies announced HTC Vive and controls
2016	At least 230 companies have focused on developing VR technology
2017	Sony has patented wireless VR technology similar to its VR technology

Source: Yengin & Bayrak, 2017: 96–97.

(Milgram et al., 1994: 283). AR devices and applications, whose development and use are considered to be more widespread than VR, can find more usage areas as they take place within the physical world. Unlike Microsoft's Hololens device, devices like Nreal and Google Glass[6] seem more probable that they will become more functional both in the flow of daily life and in business life as if using normal

6 https://www.nreal.ai/ (10.06.2020); https://www.google.com/glass/start/ (10.06.2020).

Tab. 3: Development process of augmented reality applications

Year	Augmented reality application
1957	Cinematographer Morton Heilig created a simulator named "Sensorama," and a very perceptual experience that will activate the sense organs
1968	Ivan Sutherland from Harvard University designed the first-worn viewer named "Sword of Democles" and guided the development of AR
1990	Thomas Caudell and David Mizell provided technicians with a head-mounted digital viewer for Boeing in the 1990s to guide technicians to make cable connections correctly
1992	Starting to use the digital glasses that Professor Steve Mann has been working on since the 1980s under the name "Eye Tap"
2000	Accelerating the development of applications for mobile devices and starting to create new technologies since 2000s
2004	Launch of the first video-based augmented reality system for mobile phones by Mathias Möhring
2008	Using the Wikitude Program called "AR Travel Guide," which is a vision and location-based AR application developed by Mobilizy Company
2008	The introduction of bionic contact lenses in the terminator movie and the award-winning short film *Sight* by Babak Parviz and his team
2009	Putting into practice, "Sixth Sense" AG project being developed at Pranav Mistry MIT Media Laboratory
2012	Introducing the first examples of smart glasses named "Glass" that Google has been working on for many years
…	Many companies such as Vuzix, Recon, Oculus, Meta, and Nreal continue to develop their augmented and virtual reality glasses

Source: Ozgunes & Bozok, 2017: 146–160.

glasses, since they are lighter and more portable. Even though AR applications may seem new, they are actually thought to have started in 1962 by Morton Heilig with a device called Sensorama. Also, unlike virtual reality applications, the historical development process of augmented reality technology that has progressed until today is seen in Tab 3.

By help of the AR applications that are still being developed, it is possible to make life easier. It can be listed as: (1) smart roads to see directions on glasses instead of getting directions while walking in the city; (2) video-animated menus showing how the foods are while passing a road with cafes and restaurants when looking at the signs; (3) applications that reflect the conversation of the other party simultaneously when needed to speak in a foreign language; (4) applications where the engineer can see a problem in the construction installation interactively by looking at the problematic place; (5) applications that enables you to see the route and stop in an interactive way without having to write any information on public transportation vehicles (Franklin, 2019: 169).

- **3D Printers**

The process started in 1859, when François Willème performed the first "3D scanning" by photographing objects from different angles at the same time by using 24 cameras, has become finding areas of usage in engineering, medicine, industrial production, arts, small businesses, food industry and personal uses by using PLA (Polylactic Acid) based on corn starch, which has no harm to health today, and ABS (Acrylonitrile butadiene styrene), which is a very solid petroleum-based plastic etc. By using AR device, digital printing of digital objects and products that are interacted with the help of various software can be provided as desired. In particular, procedures such as prosthesis, implant, and artificial organs[7] production in the medical field are included in the literature. Gerrand (2014: 230–231) states that the process may progress when working with more patients by mentioning the benefits of the implant design they produce for a patient and by expressing his experience with the 3D printing model.

- **Interactive Video Content Environments**

In parallel with the development of digital media, TV and its programs, which are traditional media tools, are also changing/transforming. This transformation process has created digital platforms such as Netflix, Amazon TV, Disney Plus, and Apple TV , and their number is increasing day by day. These communication environments are accessible from any communication device with the screen and also become the center of life. As stated by Postman (2014: 101), the users place the products, the services that obtained, at the focal point of their life. In this way, viewers, who are consumers in the postmodern lifestyle, are determined by a digital lifestyle where their needs are created by triggering their desires through the media channel (Baudrillard, 2010: 60). Specially applications such as Netflix, which can be watched in more than 190 countries, provide unlimited content access, ad-free, binge-watching (watching in rapid succession) and suggestion systems[8] created based on the popular content, trigger this desire further and increase the addiction and direct it toward a self-centered and self-alienated society model. As stated by Castells, it is a society model that contains an abundance of information and data within the digital community/network society, where Internet, social

7 With the help of various programs, it is possible to produce prosthesis and surgical instruments by making 3D designs (Harvey March 2020).

8 https://help.netflix.com/tr/node/100639 (10.06.2020).

media, mobile access, excessive consumption, and communication technologies are changing.

- **Video Games and Consoles**

The process that started as the first home-type game console with the device released in 1972 under the name of Magnavox Odyssey by hosting the pong game and its derivatives, in the following years, continued with companies such as Atari, Sega, and Nintendo. One of the best-selling home video consoles in the 1990s, Nintendo has been the pioneer of the technologies we see in video game consoles right now. Although it has become widespread and developed, the beginning of being seen as a new communication medium has progressed in parallel with the development process of the Web; transformation in digital media has also had an impact on digital gaming and has changed/transformed from single player games into the type of online gaming that use broadband internet infrastructure. Online games such as Playerunknown's Battlegrounds (PUBG), World of Warcraft, Guild Wars 2, League of Legends, Dota 2, Conan Exiles, etc. have become new communication environments, where gender can be determined, appearance can be adjusted, clothing style can be determined, and virtual identities can be created by choosing the style of war, race and faction; at the same time, friendships created by connecting with users at the back of beyond, which are never seen in real life, can be struggled as if they were real by performing various tasks in the virtual world. In addition, by creating teams-clans, these friendships are maintained through communication technologies in the real-life environment. Virtual IDs can be created by using subscriptions on the PC or by downloading the game application; PlayStation Network and Xbox Live user accounts can be opened via game consoles such as PlayStation and Xbox, and these communication environments can be used like social media tools. Since these communication environments are difficult to monitor and capture, they can be used by various criminal organizations. Recently, the allegation that the ISIS terrorist organization carried out the Paris attacks planning by communicating via PlayStation 4.[9]

3 The Age of Digital Tribes in Network Society

In the literature, terms such as "web society," "network society," "digital community," and "digital human network" can be used for the people of the digital culture age. In this age, we need electronic communication networks based on radio signals, such as fiber cables, which can provide Internet data exchange and data exchange from satellite as well as electricity and water lines. For the new generation of young people, named as "digital natives" of today's society, a world without the Internet is seen as being deprived of everything and abstracting. As a result

9 https://www.birgun.net/haber/paris-katliaminin-playstation-ile-duzenlendigi-iddia-edildi-95228 (09.06.2020).

Tab. 4: McLuhan's Periodic Media Society Classification Table

Age	Medium Type	Medium	Era
Tribes	Verbal	Speaking	Before AD 1500
Detribalization	Mechanic	Press	1500–1900
Tribalization	Electrical	Television	After AD 1900s

Source: Laughey, 2007: 35.

of the developments in communication technologies, the information flow is very fast, and it can be provided with new multi-spatial environments created by the combination of the virtual and the real; the interaction of different cultures/data exchange takes place and creates a new digital culture/digital community. Citing that the digital culture started with the creation of the digital community by the Internet in the twenty-first century, Castells emphasizes that the organizational structure of the digital community is in the form of an Internet-based network society/digital community, referring to the fact that digital community individuals produce, spread, and store information through communication technologies (Castells, 2000: 110).

Based on McLuhan's periodic media society classification and today's society model, which Van Dijk called the "Age of Networks" for the post–twenty-first century, the era that communication technologies surrounds every moment of our lives can be named as "Age of Digital Tribes" society, which uses the Internet as a tool. As indicated in Tab. 4, in his periodical media society work, McLuhan considers the post-1900s AD as the electronic age and mentions that human life in a "global village" environment in the society model he describes as Tribalism, and adding that the visual sense is pushed into the background and the auditory world comes to the fore again.

Especially, it can be said that since the 2000s, as a result of the current developments in the flow of life and the digitalization of everything, digital organizations, in which the interaction has become increasingly widespread, have begun to form. Thereby, it can be added to McLuhan's periodic media society classification. It is possible to talk about an era that can be defined as the "Age of Digital Tribes," where audio, visuality, and touch are at the forefront and the Internet can be seen as a tool where wireless connection/radio signals are used as tools. In this society model where organizations, communication, and lifestyles are digitalized, virtuality is dominant in its nature. In this society model that is appealing to all senses, integration of traditional and new media communication mediums also in this period, internet pages and browsers started to lose their importance with the introduction of communication environments where virtual spaces become widespread by help of developments on VR devices, and with the AR technology where real environment conditions and digital can be used together. Inspired by

Tab. 5: The New Era to Be Added to the Periodic Media Community of McLuhan

Age	Medium Type	Medium	Era
Digital Tribes	Wireless Connection (Fiberoptic and Satellite Infrastructure)	Internet	After AD 2000s

the research of Marshall McLuhan, Kelly (2016: 75–100) divides the IT process into three periods and mentions that each new environment replaces the previous one. By pointing out the period when computers started to be used in places like offices, as the initial phase of the IT process; emphasizes a period when browsers dominated by the internet spreading and storing information, and lastly the third period in that we live in, where the importance of social media has increased and web pages have started to lose importance. In addition to all this, he states that with the influence of the printing press, there has been a transition from verbal culture to written culture, and today, a digital era dominated by billions of all kinds of screened mobile devices and video game consoles. Based on Mcluhan's periodic media society classification, the new society model that may be added as a result of the listed technological developments is seen in Tab. 5.

While evaluating the classification and theoretical philosophies made by philosophers, it is also necessary to take into consideration the period they lived in and the political and socio-cultural structure of the societies. Therefore, it can be considered to contribute to the literature with the analysis to be made by taking into consideration the social structure that transformed with the developing technology in the period we live. One of the points to be considered here is the renewal and new classification caused by continuous technology. When talking about the evaluations, the analysis, which made by considering the preferences, perceptions, opinions and attitudes of the generation, regarded as a new generation within the society, without reducing it to determinist approaches that make technology fetishism, will be more consistent in specification and identification of problems. In this way, it can be thought that errors can be minimized. On the background of technological determinist approaches, as Herbert Marcuse stated, for post-modern societies, technology also has the function of suppressing and directing the individual under the name of technological rationality and making it a one-dimensional thinking pattern. In conclusion, Williams (1985) points out, those who produce and spread technology are sovereign powers that are actually determinants of the political and economic order. Opposes the view that technology is the main factor in the change of societies and underlines that technology is identified by societies, not the societies by technology. Finally, the ones listed throughout the article are seen in Tab. 6, the periodic media society where Mcluhan's table has been updated and changed as a result of communication technologies.

Tab. 6: Periodic Media Society by the Age of Digital Culture

Age	**Medium Type**	**Medium**	**Era**
Tribes	Verbal	Speaking	Before AD 1500
Detribalization	Mechanic	Press	1500–1900
Tribalization	Electrical	Television	After AD 1900s
Digital Tribes	Wireless Connection (Fiberoptic and Satellite Infrastructure)	Internet	After AD 2000s

Conclusion

In today's world, with the transition from the "modern" process to the "post-modern" process in production, it can be stated that these new social life forms called digital community/network society are transformed by taking part in "post-modernism." As a result of this, in the postmodern period, although digital culture/digital community communication environments are considered as digital/virtual, it is seen that they have actually changed/transformed based on the means of traditional communication (TV, radio, newspaper etc.). This digitalization process of communication can be accepted as a communication age where the audiences can instantly interact with the content with the help of a mobile device while watching TV channels; where online player communities/groups formed by virtual structuring in digital video games; While Netflix, which is one of the video content watching environments such as Disney Plus, Amazon Prime, Puhu Tv, Blu Tv, had been operating as a video tape/CD/DVD rental company before the 2000s as a movie/series viewing platform, it has been transformed with the effect of digitalization and the content is presented under the name of suggestion system with the help of AI, and the continuity of consumption is ensured; where audiences can also create newspaper or TV news, such as news agencies in the environments that integrate the old and the new. In addition to these, in the age of digital culture, with information, virtual interaction is so much that manipulation is higher than ever before. With the Internet database, with the help of AI, photos can be made as if they were artisanal and indistinguishable from the truth, manipulation can be made with the help of AI by adding others' speeches and mimics to the video recording of any opinion leader, and digital videos can be seen from bulletins with virtual announcers. Digital games that show pornographic violence as the natural flow of life by presenting the flow of the current history are presented as if there was an alternative time. All these put more influence on the digital community. Therefore, it increases manipulation even more as an effect of the digitization process. Since accessing the right information becomes increasingly difficult in the digital culture era, filtering systems, which can be seen as one of the applications that can be made, are not only limited to the Internet. On the social media, digital games, mobile applications, interactive applications, AR applications, and protective and preventive software systems become operable to parents, or those whom is responsible, in the form of "Do you want to download this application, do you approve?" or "Do you want to login to this application?" etc. For example, YouTube, the video sharing platform, can be seen as a beginning to the application model, where the video category can be selected as education, comedy or child, with the category option that the content owner being asked for information about who will be his target audience. In particular, Web 3.0 and the subsequent Web 4.0 are used by X and Y generations born before 1980 and 1999 to get more information and to meet their information needs at certain times of the day; however, it can be stated that Z generations, seen as digital natives born after 1999, are more affected by misinformation, considering that they use the Internet as a means of entertainment and spending with its visual and multi-media applications as a means of entertainment and time. In conclusion, technology-based entertainment and communication tools such as all kinds of social media, digital game environments, and AR in postmodern life offer a virtual, incomplete, and manipulated lifestyle that actually removes its users from real life and reproduces the existing sovereignty structure (Oskay, 1998: 169). The most important issue to be considered in the process of building artificial human modeling is to ensure that the human will not be defeated by technology, and the human being, the most important element of humanity, should be placed at the center of these digital transformation and developments. Consequently, the use of technological data should be permitted, without harming humanitarian and moral values, and by keeping human beings in the center. Finally, with the changing social structure, acceptance, and opinions, the individuals of the digital culture age society feel ready for their orientation toward the artificial human.

Bibliography

Ayhan, Ahmet, (Ed.), (2019), *Dijitalleşen İletişim* [*Digitalizing Communication*], Konya: Literaturk Academia.

Baudrillard, Jean, (2010), *Tüketim Toplumu: Söylenceleri/Yapıları* [The Consumer Society], (H. Deliceçaylı and F. Keskin, Trans.), İstanbul: Ayrinti Press.

Birgün, (2015), "Paris katliamının PlayStation ile düzenlendiği iddia edildi", https://www.birgun.net/haber/paris-katliaminin-playstation-ile-duzenlendigi-iddia-edildi-95228, (9.6.2020).

Castells, Manuel, (2000), *Globalization and Identity in the Network Society*, Prometheus.

Castells, Manuel, (2008), *Eformasyon Çağı: Ekonomi, Toplum ve Kültür / Ağ Toplumunun Yükselişi* [*Information Age: Economy, Society and Culture / Rise of Network Society*] (Edition 2 b., Vol. 1), (E. Kılıç, Trans.), İstanbul: İstanbul Bilgi University Press.

Castells, Manuel, (2016), *İletişim Gücü* [*Communication Power*] (E. Kılıç, Trans.), Istanbul: İstanbul Bilgi University Press.

Datareportal, (2020), "Digital 2020: Turkey", https://datareportal.com/reports/digital-2020-turkey, (12.06.2020).

Fiske, John, (2003), *İletişim Çalışmalarına Giriş* [*Introduction to Communication Studies*] (S. İrvan, Trans.), Ankara: Bilim ve Sanat.

Gerrand, Craig, (2014), "Three Dimensional Printing: A Clinician's Experience", *The Bulletin of the Royal College of Surgeons of England*, Vol. 96, N. 7, pp. 230–231.

Google, (2020), https://www.google.com/glass/start/, (10.06.2020).

Güngör, Nazife, (2018), *İletişim: Kuramlar ve Yaklaşımlar* [*Communication: Theories and Approaches*], Ankara: Siyasal Bookstore

Harvey, Ailsa, (2020, March), "2020'nin Akıl Almaz Gelişmeleri" [Inconceivable Developments of 2020]. *How It Works*, Vol. 17, pp. 50.

Harvey, Ailsa, (2020, April), "Sanal Gerçeklik Ameliyathanesi" [Virtual Reality Operating Room]. *How It Works*, Vol. 18, pp. 54–57.

Ign, (2020), "Sex and Romance", https://www.ign.com/wikis/the-witcher-3-wild-hunt/Sex_and_Romance, (08.06.2020).

Kelly, Kevin, (2017), *Büyük Teknolojik Dönüşüm: Geleceğimizi Şekillendirecek 12 Teknolojik Kuvvet* [*Major Technological Transformation: 12 Technological Forces Shaping Our Future*] (Ü. Şensoy, Trans.), Istanbul: THY Press.

Laughey, Dan, (2007), *Key Themes in Media Theory*, New York: Open University Press.

Microsoft, (2020), https://www.microsoft.com/en-us/hololens/hardware, (10.06.2020).

Milgram, P., Takemura, H., Utsumi, A. and Kishino, F. (1994), "Augmented Reality: A Class of Displays on the Reality-Virtuality Continuum", *Telemanipulator and Telepresence Technologies*, Vol. 2351, pp. 282–292.

Miran, Leo and Franklin, Daniel, (Ed.), (2019), *MEGATECH: 2050'de Teknoloji* [*MEGATECH: Technology in 2050*] (B. Seyrek, Trans.), Istanbul: Siyah Kitap

Netflix, (2020), https://help.netflix.com/tr/node/100639, (10.06.2020).

Nreal, (2020), https://www.nreal.ai/, (10.06.2020).

Oskay, Ünsal, (1998), *Yıkanmak İstemeyen Çocuklar Olalım* [*Let's Be Children Who Don't Want to Wash*], İstanbul: Yapı Kredi Press.

Oskay, Ünsal, (2017), *İletişimin Abc'si* [*Abc of Communication*], İstanbul: İnkılap

Özgüneş, R.E. and Bozok, D. (2017), "Turizm Sektörünün Sanal Rakibi (Mi?): Arttırılmış Gerçeklik" [(Is it?) Virtual Competitor of Tourism Sector: Augmented Reality], *Uluslararası Türk Dünyası Turizm Araştırmaları Dergisi*, Vol. 2, N. 2, pp. 146–160.

Postman, Neil, (2014), *Televizyon: Öldüren Eğlence Gösteri Çağında Kamusal Söylem* [*Television: Public Discourse in the Age of Killing Entertainment Demonstration*] (O. Akınhay, Trans.), İstanbul: Ayrıntı Yayınları

Ryan, Johnny, (2019), *İnternetin Geçmişi ve Dijital Gelecek* [*Internet History and Digital Future*] (B. Keleş, Trans.), Ankara: Tübitak Yayınları.

Tdk, (2020), "Türk Dil Kurumu Sözlükleri", https://sozluk.gov.tr/, (13.06.2020).

Van Dijk, Jan, (2018), *Ağ Toplumu* [*Network Society*] (Ö. Sakin, Trans.), İstanbul: Kafka.

Williams, Raymond, (1985), *İkibin'e Doğru* [*Towards Two Thousand*], (E. Tarım, Trans.), Istanbul: Ayrıntı Press.

Yengin, Deniz and Bayrak, Tamer, (2017), *Sanal Gerçeklik VR* [*Virtual Reality VR*], İstanbul: Derin Press.

Eyüp ÖZ

Digital Technology and Sociopolitical Transformations

1 Introduction

We see that scientists have begun to spend more hours in developing new theories on the 'Digital Age', one of today's fashion concepts, and its social effects. In this chapter, which is written to make a small contribution to the studies on the subject, historical and current perspective evaluations are conducted concerning the limits, possibilities, and positive or negative aspects of digitalization without going into technical details.

When we look at history, we see that major technical transformations have forced humanity into deep-rooted sociopolitical, economic, and social transformations in short, medium and long terms. Once the transformation has begun, 'an entire intellectual environment ranging from astronomy to physics, from theology to social beliefs' is affected and changed (İnan, 2019: 63). Is digitalization process, which we all have witnessed, one of these deep-rooted transformers? Is digitalization imposing a new model on the industrial society? It is of course very difficult to answer these questions in advance. For now, however, we can examine the notable transformations in this first phase, which we can call the early period of digitalization.

Theorists working in this field have made several propositions. In Turkey, Kemal İnan, one of the key researchers working on the subject, in his study called *Teknolojik İşlevsizlik* (*Technological Nonfunctionality*) states that a new phase called 'information society' has been passed with the digitalization and automation. Futurist Ray Kurzweil (*The Singularity is Near*) suggests such an extreme theory as a new human species can be created with the peak of automation. Regardless of the point of view, the imaginations, expectations, or fears of digitalization are reminiscent of extreme optimism or pessimism toward the natural sciences and statistical data we see in European thinkers at the beginning of the Modern Age. It should not be forgotten that Saint-Simonism, Positivism, and similar movements, which were the children of the Age of Enlightenment (roughly between 1688 and 1789)[1]

1 It is based on a monthly magazine called *Nouvelles de la Republiques des Lettres*, which stood out with its human civilization and universal principles, published between 1684 and 1686 by Pierre Bayle, who was one of the Protestant philosophers going into exile with the Edict of Fontainebleau, which was declared in 1685 in the era of the French King Louis XIV, who ended the Edict of Nantes of 1598 that granted freedom for Protestants.

and sometimes carried themselves to a religious dimension, believed that science is capable of everything and promised a 'paradise' to humanity in this world in absolute conformity and harmony, or progress; Similarly, it should not be forgotten that the Enlightenment philosophers promised that they can extend human life up to a thousand years (Atalay, 2018: 20–21). The extreme optimism about the Digital Age is the belief that the scientific momentum caught today will continue forever without interruption. However, historical experience shows that such continuity is not possible and that scientific studies and civilizations, which are their life vessels, are born, flourish and then collapse for several reasons in a cycle of formation and deterioration (See, Martinez-Gros, 2014). For example, we can say that the atmosphere of threats and fear created by the COVID-19 pandemic that we are passing through has considerably weakened the globalization process used as a tool for the spread of neoliberal world order, which gained popularity in the 1975–1980s and supported digitalization to the utmost level and wild capitalism throughout the world. Therefore, this study aims to make a modest and indirect contribution to the discussions about the limits, threats, and possibilities of digitalization from the perspective of the past and the current.

2 'Digital Age': Historical Development Process and Its Basic Features

The concept of industry appeared in the thirteenth century with the scholastic thought, its meaning was expanded by Christian theologians, and the Italian Humanists reinterpreted the concept according to the conditions of the period. Industry was also used as trade competition, which was outside of the religious institutions' economic functioning (the Church's), occurring during civilian economic activities, and to name the skill in this area, thus to distinguish good or bad traders according to the criteria of service to the Christian community. In a broad sense, the word *indostrius* in Latin means diligence, industriousness, laborious work, expertise, competition, activism and experience (for the development process of the concept, see: Todeschini, 2017).

In fact, what we call the Industrial Revolution in Turkey is actually a process initiated by private entrepreneurs, now called the bourgeoisie, in the early eighteenth century, first in the cities of England, such as Manchester, and then Antwerp, Belgium, generally out of the central government control and initiative. However, we can say that the Industrial Revolution, which is stated to be at its fifth stage today, draws its strength from digitalization, automation systems, computers, and robots, which are formed by coding the numbers 1 and 0 in various combinations.

Every age has a fashion concept, or concepts. The motto of modernity is *progress*. While the Industrial Revolution has been seen as a process parallel to modernity, the word digital has moved beyond a motto and become, according to some researchers, the name of a new age: 'Digital Age'. It is suggested that the motto of this era is *transparency* (Han, 2017: 65). The word *digital*, which means *numerical* in

Turkish, is a system based on coding, and the coding language consists of different combinations of the numbers 1 and 0. The Cambridge Dictionary defines the digital age as "the present time, when most information is in a digital form." The active use of 'Digital Age' and its creations in daily life transforms the society and therefore the institutions depending on it (Işık, 2019: 55). Moreover, we see that digitalization has started to form its own social classes in a sense. Han expresses the commercial classification through a data company as follows:

> *(...) Axiom is a company trading in the personal data of about 300 million US citizens – in other words, practically all of them. By now, Axiom knows more about Americans than the FBI. The company divides people into seventy categories. In the catalogue, they are offered up like goods for sale. For any need, there is something to buy. People with a low economic value are designated as 'waste' – that is, 'trash'. Consumers with a higher market value are found in the group 'Shooting Star'. From ages thirty-six to forty-five they are dynamic, get up early to go jogging, have no children but are married, like to travel, and watch Seinfeld.* (Han, 2017: 72–73)

Zygmunt Bauman, on the other hand, argues that this process, which he describes as 'liquid surveillance', softens especially in the field of consumption chain and loosens the old status quo ties as the personal data extracted for one purpose can be easily deployed for any other purpose (Bauman and Lyon, 2013: 13). Therefore, the classification does not only remain in the commercial dimension. In addition to the use of commercial data for any other purpose, we see that the process brings new definitions such as digital alphabet, dictionary, symbol, digital identity, digital republic, digital democracy, digital poverty, and digital law. Furthermore, we see that the proletariat–bourgeoisie distinction, which is the main basis of Marxist ideology, has become meaningless and is replaced by the Robotariat-Big Brother dilemma, as the phrase is. But time will show us whether this status quo and class change corresponds to anything beyond evolution within the Industrial Society itself.

3 Techno-Social and Political Transformation in the Digitalization Process

Although it is difficult to answer the question above, considering that digital technology affects and encompasses the whole of life, namely individual, religious, cultural, legal, artistic, privacy, economic, social and political life, it is now an inevitable phenomenon to talk about a 'digital society'. Since we cannot think of civilized societies without a state, we can say that a new dialectic has developed, which considers the individual-society-state as well as the global dimension.

While the communication networks established by individuals through digital channels indicate a new type of sociology and congregation, they constitute destructively oppositional waves, which we can call 'cyber opposition', and are not completely controlled by the state in social media, and which mainly develop through disinformation, come from the bottom and deeply affect society. It is clearly seen in the example of the Islamic State of Iraq and Syria (ISIS) that especially the Internet and social media offer great new opportunities to radical religious and secular terrorist organizations. It was witnessed that the organization, which took over a UK-sized piece of land in a short period of one year, 'recruited militants' from almost all of the world via social media and the Internet.

ISIS has led to increase current concerns about the cyber world at the global level. In this respect, the threats posed by digital facilities on a global scale further have reinforced the necessity of comprehensive international cooperation and control. In addition to the threats arising out of the will of the states, it is inevitable that the process will cause great tensions in the future, concerning the claims that states themselves have pursued and developed policies such as intervening elections and firing social explosions through cyber operations implicitly conducted against other states.

Another finding about the threats of digitalization has been put forward by the research data carried out at Oxford University in 2013, which states that 47 % of today's professions are under threat with Industry 5.0, which is called unmanned technology (Frey and Osborne, 2017). 'For example, like radiology in which the cost is very high and this cost can be radically reduced by automation, high-paid jobs that are mentally qualified and have limited creativity and often require higher education are already under the threat of automation' (İnan, 2019: 233–235). This shows that industrially developed countries will face major unemployment threat in the coming years. In this sense, while the automation-based production process has brought the end of the classical working class in the Marxist sense in Europe, the shifting of 'wild capitalism' toward the less-developed countries, where labor is cheap, is one of the leading causes of the Yellow Vest protests that have recently started in France.

4 Surveillance: The State, Society, and Individual Triangle

In discussions about the threats and opportunities held in the magic of digitalization, states and societies are depicted as stray devices and masses designed by tireless and weariless scientists.

If we evaluate this approach from a social perspective, we see that when they could no longer bear the burden and deposits of the past, as we have seen before and after the modern period, they sometimes forced the power by rebellions or sometimes rebuilt their history from zero point with deep-rooted revolutions. Therefore, from the moment when neither states nor societies perceive digitalization to develop completely against themselves, it would be at least a mistake to think that they will remain spectators to this process.

From a state perspective, it should not be forgotten that the state is the engine of civil societies and scientific advances. The path of every science eventually leads to politics. The political sphere either supports scientific studies or prohibits what it considers to be dangerous. It can immediately adapt to processes it cannot control or tends to control them. One of the best examples of it is that the *British East India Company* was initially a semi-formal, semi-private initiative, and then the state nationalized the company and transformed it into one of the main means for colonization of India (see *Islamic Encyclopedia*, Vol. 22: 294–295).

It should be reminded that the tendency of digitalization to remain out of state control is in fact similar to the initial attitude of the Industrial Revolution. Because, when we look at history, we can see that the spark of the Industrial Revolution was ignited by private entrepreneurs who benefited from the weakness of the central authority of the European states. If we exclude this exceptional position of Europe, the traditional state capitals, where central control is extremely strong and therefore called 'eastern despotism', were generally the centers of organization and administration of scientific and technical advances. However, the places where the Industrial Revolution started in Europe were not capitals. But in these capitals as well, the states managed to take these processes under their control over time, as seen in the example above. This was carried out not only through the nationalization of private enterprises but also somehow transferring the economic rent here to the state vault. In the cases of failing to achieve these and resisting change, such as in France, revolutions leading to radical changes arose. Because the extremely strong central authority was the biggest obstacle to this rising new entrepreneurial spirit, and they would consider eliminating this resistance at all costs.

Nowadays, we are witnessing similar processes. We will witness that states will intervene areas such as cryptocurrencies, social media, and the development of new security-oriented software more and will accelerate this trend in the future. We can understand how quickly states adapt to this process from definitions such as *digital state*. Although some political leaders are now giving their messages on social media, we see that their relationship with these platforms may be tense from time to time. For example, US president Donald Trump, on the one hand, tries to steer and shape the international relations with his messages via Twitter, on the other hand, he threatens to block it.

As such tensions show, digitalization challenges the state, but it also allows the state to control its society, products, information, and capital in an unprecedented way. A recent example would be, especially after the terrorist attack at various places in America on September 11, 2001, 'travelers through airports everywhere started to be controlled with new digital devices such as rapidly increasing body scanners and biometric checks' (Bauman and Lyon, 2013: 11–12). In addition, the states under threat of immigration have begun to transform digital facilities into high-level security tools and use them more effectively.

'Axiom, an American Big Data company, promising clients a 360-degree customer view' (Han, 2017: 63), gives clues about the dimensions of modern digital control. The claims reflected in the media that daily collected data on commercial or social media platforms are transferred to the relevant security devices of the states, which always have a special interest in control, show how much the states focus on digital processes and use these media discreetly as a surveillance and data collection tool as much as possible. Likewise, we observe that an imaginary totalitarian regime surveillance depicted in George Orwell's famous novel *1984* is now strikingly realized in neoliberal and democratic regimes. The reason for this is that the traditions that the state phenomenon has created since its establishment have a steadfast continuity, which surpasses the different regime models and cyclical transformations in which they are governed. Therefore, it is evident that the Western democratic regimes that the bourgeoisie have created in an exceptional way cannot overcome the phenomenon of control and classification, which is one of the basic practices of the state. In this sense, Byung Chul Han describes the control model carried out in 'transparency' in neoliberal and democratic regimes as follows:

> *Orwell's surveillance state, with its telescreens and torture chambers, is fundamentally different from today's digital panopticon, with its internet, smartphones and Google Glass. Here there is no torture – just tweets and posts. Nor is there a mysterious 'Ministry of Truth'. Transparency and information have taken the place of truth. The new conception of power does not involve controlling the past, but steering the future psychopolitically.* (Han, 2019: 46)

As Bauman points out, Big Brother is the most commonly used analogy for Panopticon-like surveillance in the Western world. As a result of using the administrative device as a total control tool with its files and records, we are starting to talk about Big Brother when the state administration is gathered in the hands of a single person or party. Orwell's *1984* was intended as a post–World War II warning about the totalitarian potentiality of Western democracies, the state becoming pathologically absorbed with its own power and intimately involving in everyday control of its citizens' lives (Bauman and Lyon, 2019: 21–22). In this sense, it is obvious that Orwell is right in terms of the intensive use of digital surveillance facilities in Western democracies. Bauman also notes other aspects of modern surveillance as follows:

> *It is agreed that the panopticon was a key modern means of keeping control, by barring movement among inmates and promoting it among the watchers. But the watchers still had to be present sometimes. Of course the prison panopticon project was also expensive. It was designed to facilitate control through a semi-circular arrangement of cell blocks whose 'inspector' at the centre could see into any cell while remaining invisible to the inmates, behind a blind. It entailed the inspector taking some responsibility for the lives of inmates. Today's world, it is said, is post-panoptical; The inspectors can slip away, escaping to unreachable realms. Mutual engagement is over.* (Bauman & Lyon, 2013: 14–15)

As Bauman notes, the subjects of anonymity, confidentiality, and privacy cannot be ignored, but it should not be forgotten that it is also bound up with fairness and justice, civil liberties, and human rights. Because social classifications are primarily what today's surveillance achieves, for better or for worse (Bauman and Lyon, 2013: 24 25).

It can be said that since the new social classifications that digitalization has achieved and challenged in every field are not very visible for now, it has not yet fully influenced politics. As all the scientific advances intersect with, or will intersect, the path of politics, it can be stated that the transformations forced by digitalization will reckon with politics in a more commanding tone.

5 Digital Surveillance and Individual Freedom

Based on the concept of Jeremy Bentham's (1748–1832) panopticon (derives from the Greek words *pan* and *optikon* and means 'all-seeing place'), by overcoming the biopolitical imagination of Foucault's biopolitics, which explains how societies are controlled (and this has become a tool for power-building), we see that states no longer force but, in a sense, provide control and auditing through consent-based 'digital panopticon', which reminds us of a blind obedience like in Etienne de La Boétie's *Discourse on Voluntary Servitude* (1576).

While in the surveillance model envisioned by Bentham and Foucault, 'the blind spots in which the individual can experience their freedom remain, the surveillance perspective of the neoliberal order has expanded dramatically to condemn the psychology of individuals. Therefore, digital panopticon enables a 360 degree view' (Han, 2019: 63–64). In this regard, the societies' consent to state control due to security or concerns alike, the limits of individual freedom and their interaction with politics should be reformulated and rethought within the framework of digital codes (see, Han, 2019).

Another significant point about the relationship between digital surveillance and the new individual freedom is the expansion of public space and politics, including human psychology. This probably corresponds to a reality and future that humankind can direct by focusing on Hitler's idea of 'nationalization of people' (Hafner, 2019) or 'on panoptic discipline, or soul training, producing willing workers' (Bauman and Lyon, 2013: 22) that has a depth beyond Foucault's biopolicy, and going deep into psychology and mind. The memory of this person, which is psychologically expropriated, is full of symbolic words and virtual realities that increase expeditiously, and this type of individuals constitute the 'information society'.

However, 'in George Orwell's *1984*, every year, words decrease in number, and the space for conscious thought gradually diminishes. In a digital society, words are not destroyed, on the contrary, they multiply' (Han, 2019: 45). Ordinary things that people share for being transparent and for their freedom fall into a data store called Big Data, and this leads to a virtual self-confidence in the 'information society' that they know all the realities. As a result, the 'information society' and the 'information individual' will have to disappear in the multitude and probably have to walk hardly in the twilight of the blindness created by the information stack.

6 Conclusion

We see that the transition to computer system has considerably altered the way scientists write. With digitalization, new changes likely occur. Because networks, portals, Big Data, diversity of mass statistical tools, and a degree of knowledge that cannot be compared with any period of history require a new technique and intellectual agility[2]. Can today's political scientists say whether the phenomenon of digitalization can unite history, or if, as suggested, at the end of the process, humankind will lose control and become a slave to robots? Or can digitalization predict how far the state can force the *emic* and *etic* boundaries, or the biological, sociological, and psychological constants of humankind to transform?

The uncertainty about the future and the surprising fertility of time make it very difficult to give clear answers to these challenging questions. However, the global pandemic process called COVID-19, which appeared in Wuhan, China in December 2019, seems to have facilitated the answers partially. This process has shown that the states with strong digital infrastructure and the Internet are able to continue their activities. Despite negative scenarios, such as mass unemployment, digitalization and automation have played vital roles in overcoming the challenges posed by the global epidemic and prevented the deepening of unemployment in some sectors. For example, we witness communication opportunities offered by distance education during the protection of social distance and the productive role of factories based on an automation system in this period when human circulation is restricted. Therefore, digitalization has become more visible in this process we have been going through. In a world combined with the fear of COVID-19, while many aspects of politics and routine life are enclosed in parentheses even for a short time, digitalization and automation technologies are able to manifest themselves magnificently. Therefore, we need to reconsider digitalization and state, digitalization and society, and digitalization and individual interactions with their positive or negative aspects.

The intervention of nature with a deadly virus has changed the course of history, but it seems that it will affect the future of digitalization more positively. It seems certain that the states will focus more on this area to a greater extent. The concerns at the social level will likely decrease, and digital technologies will be perceived as realities that open new windows to life.

2 As Thomas Hobbes suggested fear of 'state of nature' in which every man against every man (Nemo, 2013: 147–148), as Jean Jacques Rousseau stated natural disasters such as floods or earthquakes, like in the case of COVID-19 (Nemo, 2013: 814), as Karl Marx insisted bourgeoisie united the history of humanity (Nemo, 2013: 935). That is to say, some powerful factors can unite history.

Bibliography

Akgül, Mehmet, (2017), 'Dijitalleşme ve Din', *Marife Dini Araştırmalar Dergisi*, Vol. 17, N. 2, pp. 191–207.

Atalay, Onur, (2018), *Türk'e Tapmak, Seküler Din ve İki Savaş Arası Kemalizm*, İstanbu: İletişim Yayınları.

Bauman, Zygmunt and Lyon, David, (2013), *Akışkan Gözetim*, İstanbul: Ayrıntı Yayınları.

Frey, Carl Benedikt and Osborne, Michael A., (2013), 'The Future of Employment: How Susceptible Are Jobs to Computerisation?', 'https://www.oxfordmartin.ox.ac.uk/downloads/academic/The_Future_of_Employmen.pdf, (30.04.2020).

Hafner, Sebastian, (2019), *Hitler Üzerine Notlar*, İstanbul: İletişim yayınları.

Han, Byung Chun, (2019), *Psikopolitika*, İstanbul: Metis Yayınları.

Işık, Alper, (2019), *Dijital Çağda Demokrasinin Dönüşümü*, (Unpublished Doctoral Dissertation), Marmara University.

İnan, Kemal, (2019), *Teknolojik İşlevsizlik*, İstanbul: İletişim Yayınları.

Kurzweil, Ray, (2006), *The Singularity is Near*, Londra: Penguin Books.

Martinez Gros, Gabriel, (2014), *Brev Histoire des Empires*, Paris, Éditions du Seuil.

Nemo, Philippe, (2013), *Histoire des Idées Politiques aux Temps Modernes et Contemporains*, Paris, Éditions PUF.

Todeschini, Giacomo, (2017), *Les Marchands et le Temple, La société chrétienne et le cercle vertueux de la richesse du Moyen Âge à l'Epoque moderne*, Éditions Albin Michel,

Yayla, Oğuz Turan, (2015), 'Dijitalleşme Çağında Eşitsizlik ve Ayrımcılık', *Liberal Düşünce Dergisi*, N. 73, pp. 43–53.

Esra Ebru MAVİ

Digitalization and the Changing Structure of the Government: An Evaluation on Turkish Public Administration

1 Introduction

In the post-1980 period, with the effect of economic policies called globalization and neoliberalism, there have been changes and transformations in many areas, from agriculture to energy, from defense to industry. The change in many sectors mainly refers to the change in the government. One of the areas most affected by the change is public administration, which represents the executive power of the government.

When the studies conducted in the public administration literature in recent years are analyzed, digitalization is shown as the actor of the change in management philosophy, practice and public policy formulation process and the implementer of the global system. However, the idea that digitalization is the main actor that influences the process of public administration and public policy formulation is open to discussion. Digitalization is biased in this process and helps to carry out projects that challenge the old management approach.

This study aims at understanding the effects of digitalization on the institutional structure of the changing Turkish public administration. Basic national/international policy texts on the subject are evaluated and a literature review is conducted. For this purpose, the transformed structure of the government is addressed; developments in this field are examined by focusing on what digitalization is and its dimensions. Then the institutional and structural transformation of digitalization in Turkish public administration is examined along with digitalization; it is discussed in the context of the change in the institutional structure and the provision of public services.

In June 2018, the parliamentary system model, which represents a strong bureaucratic structure, and was removed by claiming that it is not based on the will of the people and leads to dual executive (sharing the executive power between the president and the council of ministers), instability and slowdown, was replaced by the presidential government system model. With this model, the executive power has been gathered in the president and it has been justified that the president will conduct the administration from one source in a stable, effective and efficient way. In the new system, even though the first step to ensure that the public services are carried out quickly and efficiently with a single decision mechanism was executed through legal regulations, the implementation of the central power in the executive has been through institutional/structural transformation. Digitalization plays a medium role in the change in public administration. Digital-oriented projects and application examples, which are guides in ensuring the change proposed by the presidential government system model for the administration, have gained importance.

2 The Changing Structure of the Government and Public Administration

The specialization and standard business manners that stand out in the classic management approach, which represents the traditional management approach, have led to the routine-based production process. The clear definitions of the duty- and authority-based job descriptions of all employees from the manager to the lowest level and all the units that make up the links in the routine-based business processes created by the Weberian bureaucracy have facilitated the achievement of the goal by providing comfortable control of the management process.

POSDCORB (Planning, Organizing, Staffing, Directing, Coordinating, Reporting and Budgeting), which appeared in the early 1900s and explains the stages and functions of the classical and first known management process, has been the implementation guide for the administration. There was a dominance of the communication model influenced by the Weberian bureaucracy which prevails hierarchy, the strict chain of command, the strict division of labor, strict adherence to legislation and the professional perspective by keeping primary relationships in organizational communication in the background.

From the 1980s, globalization's criticisms toward traditional management approach has increased. The idea that the government is cumbersome and business processes that are burdened with unnecessary bureaucratic procedures lead to inefficiency has become popular. In this perspective, which is also referred to as the reinvention of management in the literature, there have been searches for an approach that enables the restructuring of public administration and its citizens to actively participate in the administrative process. It has been advocated that the government, which includes its citizens in the administrative process, will also become democratized. In the structure, the argument that an effective, efficient government and public administration approach will prevail has been consolidated.

2.1 Digitalization in Public Administration

Scholl (2014: 163–164) claims that, at the beginning of the twenty-first century, governments have faced three complex and intertwined and unique basic challenges:

(1) *Third Industrial Revolution: The simultaneous information revolution that facilitates the instant availability and dissemination of actionable information through computer-mediated networks in all areas of public and private life along*

with economic activity, which is the main part and driving force of the process called the Third Industrial Revolution

(2) *Rapid Change and Timely and Effective Lack of Intervention: Rapid change and lack of timely and effective intervention are said to lead to serious global and regional crises in the United governments, such as the 2008 financial collapse. Showing these crises as evidence that governments are insufficient to cope*

(3) *Wide Government Expenditures and High Public Debt Financing: Corroding the stability and long-term sustainability of all communities and nation governments*

The environment where the new public administration approach is carried out, which is offered as an alternative to the traditional public administration approach, has been implemented within the structure in which information and communication technologies (ICT) are used. Thus, the expectation that an effective and efficient public administration approach will be provided has been added to the agenda (Heeks, 2003: 2). Electronic government (e-Government) has become one of the frequently studied subjects in the field of information systems in recent years. The concept of e-Government was first mentioned in the USA in "National Performance Research" in September 1993 (Misra, 2007: 23). The concepts of digital government, electronic government and e-Government are often used interchangeably. The common features of these concepts are the development of ICTs in the provision of public service. Also, the terms that are referred to by different names today are generally used as e-Government (Wahedduzzaman, Miah, 2015: 1).

It is noteworthy that there is a lot of classification and naming concerning the methods of ICTs used by the government in the provision of services for citizens. Kamalov (2017: 189) suggests that the use of ICTs in public services for the last 30 years is a public governance model, but he also mentions that these technologies have three stages: the first stage is electronic government (e-Government), the second stage open government, the third stage smart government.

The combined use of the terms smart and government for the first time is based on the World Bank report on civil service reform. The term was also used without a formal definition in a report on the computerization of government operations in the government of Andhra Pradesh in India. More recently, former US president Bill Clinton used this term in the presentation of his views on the government's future role. Furthermore, the 2013 conference of the Digital Government Community, one of the main conferences on digitalization, was held under the slogan "From e-Government to Smart Government" (Scholl, 2014: 165).

Carter and Belanger (2005: 5–6) name all the uses of information and technologies in public services as e-Government services and divide the offered services into six categories according to their stakeholders:

1- *The government providing services to citizens.*
2- *The government and citizen relations within a democratic process: It exists as part of a democratic process, such as electronic voting.*
3- *The government and citizen-related business relationship: It is about companies' tax payments or filling out reports.*

4- *The government and business relationship in the market: It focuses on electronic purchases between the government and businesses.*
5- *From the government to employees.*
6- *From the government to the government.*

In the literature review (Carter, Belanger (2015), in national or international organizations, OECD (2003), DPT (2011), even though there is a general opinion that digitalization creates a transparent and accountable government structure and increases efficiency and productivity in the services provided by the government by safely accessing different services from different points from a single point and reporting that this will provide integrated service. It is noteworthy that Kamolov (2017: 185–186) makes a direct reference to the new ICTs in the Russian philosopher Nikolay Berdyaev's statement that the invention of the machine is stronger and more direct than the human dependence on nature.

Another issue about e-Government applications is related to users. Despite the belief that e-Government applications will provide efficiency, democratization, inclusion, participation and a common liberation, it is the idea that e-Government services are designed according to the notion of stereotype users and normal users (Wihlborg, Hedstrom, Larsson, 2017: 2549).

With the emphasis on transparency of information and data, named as an open book government, it is said that e-Government allows citizens to look at their own medical files, monitor their transactions or actively manage their own tax accounts and to utilize data storage. It is also stated that citizens are allowed to manage themselves more by abandoning the closed file system in the Weberian bureaucracy (Dunleavy, Patrick, Margetts, Helen, Bastow, Simon and Tinkler, Jane, 2006: 487). Instead of the traditional public administration approach, with the applications often described as e-Government and carried out with ICTs, the transfer of government-related transactions to the electronic environment has gained importance.

In the statements presented by the USA and China at the beginning of the century, it is reported that e-Government applications transform public activities and efficiency functions, and this brings organizational and process change in the commercial sector. Thus, attention is drawn to the need to investigate the destructive effects of ICTs in this environment. In New Zealand, the limited research on the impact of the change suggests that it has negative consequences; although e-Government has provided the illusion of change and has shown some operational benefits, it is argued that it basically does not bring any changes in institutional structures and its emphasis is mainly on technical aspects in e-Government research (Waller, Weerakkody, 2016: 6).

It is asserted that fully open systems are still not completely successful in the world, but there are suitable models for how they can work in smaller countries such as Sweden, and even there are models of "open book" companies in some regions of the private sector (Dunleavy, Patrick, Margetts, Helen, Bastow, Simon and Tinkler, Jane, 2006: 488). As it is pointed out about digitalization, the issue of drawing the boundaries of the services to be performed by the government rather than the technical dimension in performing the transactions and how to fill the content of the said datastores are open to discussion.

The governments have the authority to set up the rules of online activities and dictate them. It is acknowledged that the use of such force leads to a lack of control and a sense of helplessness of government practices on citizens. In order to participate in the e-Government process, citizens must adopt the process and intend to join the e-Government, send and receive information and documents and request e-Government (Warkentin, Gefen, Pavlou, Rose, 2002: 157,158,159). It is told that the markets are gradually becoming complicated and unpredictable; therefore, it is difficult to identify and correct the problems that arise. It is argued that e-Government applications can be the solution in such an environment. Less spending, limiting excess public expenditures and public debt financing are shown for the solution. At this point, it is stated that ICTs are beneficial in reducing costs and using resources. The given prescription is to focus on e-Government researches in order to understand the interdependencies well. There is an expectation that feasible options can be determined with these studies (Scholl, 2012: 326).

It is seen that the first applications on "e-Government" in the world are in local administrations. In the state–citizen relationship, since the sense of trust in the citizen's declaration is dominant, in the first e-Government applications, the security question has been designed with a focus only on unauthorized use or corruption of information (Türkiye Bilişim Derneği, 2001: 9).

Apart from the technical dimension, the discussions about digitalization, the definition of digitalization and what it encompasses in public administration management are very important within the scope of ICTs and the scope of services for those who receive them. Without creating a general public management theory for the networked governance model, it will be weak in assessing the implementation nature of the public services and programs and the strengths and weaknesses of the system (Stoker, 2006: 42). The importance given to the relationship between the technical and policy dimensions of digitalization in public administration becomes significant.

Scholl (2012: 326) mentions that e-Government studies have newly been carried out on the potential effects of electronic participation, social media and social networks on the government. He says, however, that the government has not yet fully understood the use of these technologies, and that extensive work is needed on smart government initiatives, in which all parties are involved.

The issue of digitalization has been included in the management processes for different purposes in the public and private sectors. It is stated that digitalization has changed the production methods, customer expectations and distribution channels and other practices in the private sector, and the private sector has achieved great gains in many areas, from the production and processing of information to decision-making and access to new markets. It is said that digitalization plays an important role in improving the performance of companies and reaching their goals and increasing their competitiveness (Türk Sanayicileri ve İş İnsanları

Derneği 2016: 5). It will not be misleading to say that the main motivation source of digitalization effort in the private sector is competition and profit maximization.

It is affirmed that telecommunications, insurance, finance and retail fields are the main owners of digitalization in the private sector, and these fields not only prioritize but also focus their investments on digitalization (Türk Sanayicileri ve İş İnsanları Derneği, 2016: 14,28). Considering the use of information and communication processes in digitalization and public service processes as a basic tool in the public sector is an effort to strengthen effectiveness, participation, transparency and accountability (National e-Government Strategy and Action Plan (2016–2019), 2016: 4). It can expressed that although the efforts to digitize in the private and public sectors are based on different grounds, the ultimate goal of both sectors is basically to focus on customers and citizens ,and one of the basic conditions of success in digitalization is to acquire high-level awareness (Türk Sanayicileri ve İş İnsanları Derneği, 2016:9, 53).

As developed and developing countries differ in the e-Government service processes due to mainly technological infrastructure and economic factors, the sociocultural structures and management differences of developed countries among themselves are also elements that may affect e-Government service processes.

Even though smart government applications exist in the USA in three powers (legislative, executive, judicial), or central or all levels of local governments, the government needs them more in terms of budget deficits and austerity measures. It is said that there is a need to discuss the issue of democratization in the smart government. It is exemplified that local governments like Seattle (population 660,000) and Bellevue (220,000) have overcome the financially difficult times with smart city administration (Scholl, 2012: 317, 324). In the USA, the Executive Office has given the necessary directives for making all government records unclassified in the units electronically and publicly available in order to provide transparency to the decision-making process of the government, improve accountability and promote cooperation and stakeholder participation. Furthermore, it has requested a detailed plan for cooperation and participation with other stakeholders, including businesses and citizens. The reason for this is to ensure direct participation in service delivery and to avoid returning to unclear government applications (Scholl, 2014: 165).

Based on the examples they give from the studies on the preparation process for e-Government applications in the world, Wahedduzzaman and Miah (2015: 4) emphasize that these studies mostly show a citizen-centered approach to preparation for e-Government and that they draw attention to the importance of sociocultural features in preparation for e-Government services. In addition to sociocultural factors, trust, networks and educational qualities are remarkable for researchers. It is also important that ideas of government systems, attitudes of officials, public administrators and authorities collaborate with each other in the preparation process of the use of electronic services.

Along with data collection and observation method to evaluate the holistic integration and readiness of local government and central government into the e-Government process, Wahedduzzaman and Miah (2015) conducted interviews with the local level ministry, senior executives and selected citizens in Bangladesh. In interviews with different stakeholders, it is seen that the local and central governments are not ready to harmonize citizens with the new management system and to include the public in the governance process in the management system of Bangladesh. The strict traditional bureaucratic structure in Bangladesh has caused failure to include the public in the government network system. In such an environment, it is stated that the authorities have not even received feedback from the citizens about the services provided. Although the public services are put into practice for the citizens, the citizens have not been integrated into this process, and their expectations cannot be met, and the atmosphere of the new working environment cannot be achieved. Therefore, governance systems such as e-Government have been offered as a solution, emphasizing the importance of involving citizens in the process. In the context of developing countries, it is explained that despite the deep-rooted policies on e-Government services, e-Government services are still in infancy, for example, Bangladesh ranks 138th out of 150 world countries on ICT. It is said that plans have been made to produce ICT policies nationwide and to establish collective e-centers, and regional training and administrative affairs have been facilitated by the use of computers and the Internet. It is also stated that internal automation works continue, and ICT support and an e-Government system will be important in monitoring the needs of users and activating the service delivery process.

Compared with the traditional government administration process, the belief that the e-Government has the potential to reduce costs and improve services by making its features accessible faster and easier and eliminate the risk of causing a large number of human errors due to the traditional management process requiring manual action is important for citizens to adopt e-Government (Warkentin, Gefen, Pavlou, Rose, 2002: 157, 158). However, the issue of easy access to e-Government depends on many factors, such as the infrastructure facilities in the region where the citizens are located, the status of having a device for access to the Internet and the opportunities for access to the Internet.

2.2 Change and Digitalization in Administrative/Institutional Structures

Factors affecting and stimulating change in the world and, accordingly, the approach and structures of management are addressed under four headings: (1) change in economic theory, (2) change in management theory, (3) competitive structure and progress of the private sector and (4) development of civil society with demand for social criticism and change (Prime Ministry, 2003: 21).

In the four classifications of Murali Chidurala (Sağsan, 2001: 3), the main reasons that the policies and applications developed by governments are based on ICTs are given as follows:

1. Increase in public and private sector partnerships
2. Union of private sector companies, official departments, civil associations and research institutes
3. Cooperation between companies gaining importance
4. The increasing importance of end users and the access of the information needed by the citizens without intermediaries

With the introduction of smart and open government administration smart governance has been mentioned to help effectively address three major challenges for social and individual welfare in the twenty-first century: (a) the core of the Third Industrial Revolution – the Information Revolution, (b) the pace of change and the lack of timely and effective government intervention and (c) broad government expenditure and exorbitant public debt financing. Although not considered a panacea, it has also been suggested that smart governance principles can more intelligently guide the relatively complex administrative arrangements of the smart and open government than traditional static and inelastic governance approaches can do (Scholl, 2104: 171).

When the problems and the suggestions on the issues of change and transformation in the government listed above are examined in general, the organizational structures are the subjects of the problem and the solution. In the criticism that the government is cumbersome and dysfunctional, the target is mostly institutional structures that form public administration. Improvements in institutional structures that form part of the government or the proposal of new ones instead of existing structures show that institutional/organizational structures are one of the main actors in the practices related to change in public administration.

It is claimed that adapting to new ICTs has different effects for managers, citizens or national and international stakeholders. Considering that governments are committed to knowledge-based economies today, it is assumed that public administrators promote innovation and social cohesion. This situation is thought to provide greater access to information for the citizen and the ability to gather and act together collectively (McNutt, 2014: 49). In governmental executions, users are said to be important actors, particularly during the implementation phase, but their roles and how they look are complex. Practice-based implementations are expressed to involve other marginalized stakeholder groups, apart from users, to change perception and strengthen changes (Wihlborg, Hedstrom, Larsson, 2017: 2552). Moreover, external stakeholders of public institutions are also told to be critical determinants of the regulation of public services. This situation is stated to emphasize the importance of the complex and interconnected global economy and the difficulty of connecting the government to a dynamic global environment (UNDP, 2015: 12).

The New Public Service model is active and citizen oriented, with a democratic perspective, and in such a perspective, citizens also consider the broader public interest beyond personal ones. Therefore, the role of public officials is explained to search for facilitating opportunities to strengthen citizen participation in finding solutions to social problems. It is significant that public administrators acquire skills to develop policies by negotiating complex problems with citizens. It is suggested that the government should be open, accessible, accountable and sensitive and operate to serve citizens in order to meet the public interest and social needs (UNDP, 2015: 10).

In the twenty-first century, public administration has undergone a dramatic change in developed and developing economies, and globalization and multiplication of service provision are the driving forces behind these changes. The policy challenges of the countries are becoming increasingly complex, imprecise and global rather than simple, linear and national. However, the dominant paradigms in which public sector reform is designed and implemented are relatively stable and do not fully cover the significance or consequences of these broader changes (UNDP, 2015: 4).

When studies on computer and communication technologies and e-Government are analyzed, it is seen that, contrary to the common view, the level of development and managerial expectations differ among countries in the world. While e-Government 3.0 discussions are on the agenda of the developed countries in the ICT and e-Government applications (Scholl, 2012), it is noteworthy that the need to use the e-Government at the administrative level is still on the agenda of the underdeveloped and developing countries.

The management systems of countries are the determinant of all public policies and establishing applications. It will not be misleading to assert that even the countries that seem to have the same management systems operate differently due to cultural and political structure differences. The management system of the USA is the Presidential System and the president, who holds the executive power, operates his power with the Presidential Office established in the Presidency. The Presidential Office consists of four units: The White House Office, The Office of Vice President, Specialist Offices and Advisory Boards, Councils and Committees (Ömürgönülşen, 2013: 299). These offices, which were created under the umbrella of the Presidency, can be said to be the administrative tool in America. Based on the example in the USA and the other management organizations that take this model as an example, the main purpose is to operate with the president without intermediaries in the executions to be carried out by the administration and thus to provide efficiency by making the operations to put into practice as quickly as possible in accordance with the decisions taken. Office-type management units expect simplification in management by removing units and business processes at intermediate levels.

Information and new technological systems are the biggest tools in the hands of a new public administration approach that has been stripped and redesigned from a traditional perspective. In other management models similar to the management system shown in the USA in the world or created to eliminate the weak, criticized features of the traditional management approach, digitalization has become the bridge that will connect the corporate structures with the sector they will serve. Governments must observe the rights of users whom they will deliver their services and also public services on equal and democratic and legitimate grounds (Wihlborg, Hedstrom, Larsson, 2017: 2551)

It is stated that leading countries in technology and economy have entered the competition to set standards in technology and economy in recent years; digital transformation should not be considered only as technological transformation, but the transformation covers a holistic transformation that institutions and organizations should put into practice in humans, business processes and technological elements by using digital technologies in accordance with the changing needs. Since technological developments are often prior to the standards to be created for digitalization, the need for a management approach with a visionary perspective that can predict this situation is emphasized (Türkiye Bilişim Derneği, 2019: 9).

e-Government can be defined as the uninterrupted and safe execution of the duties and services that the government is obliged to fulfill for its citizens, and in return for these, the execution of the duties and services of the citizens for the government in mutual electronic communication and operation environments. On the other hand, it can be said that e-Government is not a virtual government but works 24 hours by being constructed as a supported system in electronic environment in order to operate the real-life government more efficiently. (Türkiye Bilişim Derneği, 2001: 8).

Many social, political and economic problems cannot make clear implications about development and the result. They do not provide detailed estimates since change processes are inevitable and contain many unpredictable tendencies. The examples are developing and implementing strategies on possible effects inherent in climate data to reduce and adapt to the future effects of climate change on a country, as well as the uncertainty of potential reactions from farmers or other countries. For this reason, it is obligatory to work without expecting for very clear and stable information about the causes and effects of the programs put forward while providing services (Matei, Antonie, 2015: 347). This situation may help minimize surprises by keeping the A, B and C alternative plans aside according to the possible situations for the plans and programs prepared.

3 Digitalization and Turkish Public Administration

In line with the rapid change and development in ICTs, there have been changes in public administration policies and applications as well. The first examples of the mentality change in public management policy change in Turkey are seen in the reports prepared by the contributions of foreign experts. The "Preliminary Report on Administrative Reform and Reorganization" prepared in 1961 under the leadership of the State Planning Organization, the report with the short name "MEHTAP" (Central Government Organization Research Project) prepared in 1961 by TODAİE and the "Preliminary Report on the State Personnel Regime" prepared by the State Personnel Department in 1962 are the first official reports prepared to be the examples of reform studies in Turkish public administration. Following these reports, the Public Administration Research Report (KAYA Project) in 1991 is a continuation of improvement studies in public administration (Yılmazöz, 2009: 297–298).

By the end of the 1990s, it is seen that reform movements in public administration were mostly carried out through implementation. E-government applications based on ICTs in public administration have become prominent. The most concrete indicator of digitalization in the public is the Digital Transformation Office (DTO), one of the four offices that find their place in the presidential system. DTO is one of the basic structures in adapting ICTs with public administration.

3.1 e-Government Applications in Turkey

The process developed with the penetration of the Internet use in social life in Turkey since the mid-1990s has led to significant consequences in managerial operations. In parallel with the increase in diversity in public services and the acceleration of information flow with information technologies, there have been the processes that forced the changes in accessing and delivering services to citizens.

The first report on digitization in the public sector in Turkey prepared in conjunction with the World Bank in 1993 is the Informatics and Economic Modernization and Report (BilisimVeEkonomikModernizasyonRaporu.pdf, n.d). The second important report on digitization is the Turkish National Information Infrastructure Master Plan Final Report, published in 1999 by the Ministry of Transportation for preparing the Turkish National Information Infrastructure Master Plan (E-Turkey, 2014).

An e-Government Transition Action Plan was prepared in 1999 by establishing PublicNet Technical Board in 1998. The PublicNet Technical Board prepared the e-Government Transition Action Plan in 2002. With the Prime Ministry Circular No. 2003/12, all the projects, reporting and studies carried out until then related to e-transformation were combined as e-Turkey project under the responsibility of the State Planning Organization. In the Prime Ministry Circular No. 2003/12, the main goals of e-Turkey Project were demonstrated as the preparation of conditions that will create a government structure that has the principle of having participatory, transparent, effective and simple business operations in order to provide citizens with better quality and faster public services.

This project has aimed at the following:

1. *Revising the ICT policies and legislation within the framework of the European Union acquis and adapting the action plan envisaged for candidate countries within the scope of eEurope + to Our country*
2. *Developing mechanisms that will enable citizens to participate in decision-making processes in the public sphere with the help of ICTs*
3. *Contributing to making public administration transparent and accountable*
4. *Contributing to the implementation of the principles of good governance by making maximum use of ICTs in the provision of public services*

5. *Expanding the use of ICTs*
6. *Integrating, monitoring and evaluating the public's repeating or overlapping investment projects in order to reduce the waste of resources in the field of ICT and ensuring the necessary coordination among investor public institutions*
7. *Guiding private sector activities in the sector in the light of the above principles (TC. Ulaştırma Bakanlığı, 1999)*

Several e-Government action plans had been prepared by 2011. With the Decree-Law (DL) No. 655, which entered into force in 2011, the duties and responsibilities for e-Government policies were given to the Ministry of Transport, Maritime Affairs and Communications (UDHB). The E-Government Services Department was established to carry out e-Government studies under the umbrella of the Directorate-General of Communications within the Ministry. The Tenth Development Plan (2014–2018), which directs public policies at the highest level, was published in 2014. The Ministry of Transport, Maritime Affairs and Communications prepared the 2016–2019 National e-Government Strategy and Action Plan for shaping Turkey's e-Government policy with a holistic perspective of within the framework of information society policy. It is stated that the 2016–2019 National e-Government Strategy and Action Plan, which was prepared to follow a more participatory, transparent and holistic approach in the government's relations with other actors in the processes of determining, decision-making and implementing policies in public sector in Turkey, adopted a holistic and sustainable strategic perspective in shaping and implementing e-Government policy and was prepared to make e-Government more integrated, technological, participatory, innovative and qualified and the transition to the information society and locomotive of sustainable development (National e-Government Strategy and Action Plan (2016–2019)).

In Turkey, ranking studies–related e-Government applications are carried out by the Turkish Statistical Institute (TurkStat). TurkStat's measurements are institutional and aim at the citizens. Continuous statistical evaluation studies are performed by TurkStat, which is the national statistical institution. TurkStat currently measures the tendency of citizens and legal entities to use information technologies in line with the goal of becoming information society. For the first time in 2005, TurkStat conducted the Information Technology Usage Research in Enterprises and Household Information Technology Usage Research for Information Technology Usage Research. Another subject measured by TurkStat has been the satisfaction with public services provided electronically within the scope of the Life Satisfaction Survey.

In addition to TurkStat's studies, Information Society Statistics are prepared by the Department of Information Society in the Ministry of Development in order to follow the results of the strategies and action plans applied in the information society field (Index in Turkey, n.d).

According to the 2011 Information Society Statistics, the results of the Household Information Technologies Usage Survey and Information Technologies Usage Survey in Enterprises conducted by TurkStat provide information on the electronic access of citizens and businesses to public services and their use of these services. Accordingly, it is observed that the rate of using the Internet in the transactions of individuals and enterprises with the public has increased at a limited level over the years. On the other hand, it is understood that the enterprises have more intensive communication with the public in electronic environment compared to individuals. The Internet utilization rate of enterprises using the Internet in communication with public institutions was 72.8 % in 2009. This rate corresponds to 66.1 % of all enterprises. In the utilization rate, "insurance-related activities excluding financial and compulsory social security and pension funds" (89.8 %) and "repair of computers and communication equipment" (88.3 %) are in the first two places with the highest rates (e-Turkey, n.d).

In e-Government Development Index, which was carried out firstly by the United Nations in 2001 and since 2010 on a regular basis with a two-year period, Turkey has had a satisfactory performance in recent years and has raised its development level. In the index-calculating online service, telecommunication infrastructure and human resource components, Turkey, which was 68th in 2016, went up to the 53rd place in the survey/index conducted in 2018 (UN, 2018).

In the World Economic Forum's Network Readiness Index, Turkey was 48th out of 139 countries in 2016. It is expressed that these data show that Turkey, which has the target of attracting foreign investors to the country and prioritizes facilitating commercial activities, is not at the desired point in the network readiness index. In Waseda University International e-Government Index, Turkey is 41th out of 65 countries (The Presidency of the Republic of Turkey Digital Transformation Office – Public Chief Digital Transformation Officer, n.d.). According to the International Telecommunication Union ICT Development Index, Turkey was 70th out of 175 countries in 2016. In the Worldwide Web Foundation Open Data Barometer, Turkey was 47th out of 92 countries in 2016 (The Presidency of the Republic of Turkey Digital Transformation Office – Homepage, n.d.).

3.2 Institutionalization of Digitalization in the Public Sector: Digital Transformation Office

With the Law No. 6771 in 2017, amendments have been introduced to ensure the transition to a new government system. Among these amendments, the president is given the authority to issue a Presidential Decree (PD), which is stated to be the original regulatory authority. In addition, with the amendment made in Article 123 of the Constitution, it is possible to establish a public corporate entity with the PD. One of the first examples of the exercise of this authority is the establishment of the Presidential Offices founded with PD No. 1. With this decree, it is regulated that the offices, which are foreseen to assist the president in certain issues, such as following up some projects and reporting activities in the new system, are in the public corporate entity (Ocak, 2019: 189–190). The reason for the establishment of the DTO is explained on the official website of the Presidency as follows:

> *While the agenda of e-Government policies prior to the Presidential Government System was taking the institutional processes of public institutions and organizations to the electronic environment and updating the organizational structures. today's conditions have made it necessary to redefine the relations between all stakeholders that play a role in planning, decision-making and implementing processes and the government. With the transition to the Presidential System of Government, fast, transparent and effective management approach has been achieved and important steps, which are about assembling Digital Turkey (e-Government) and cyber security coordination under the same roof, have been taken. In accordance with developing technologies, social demands and reform tendencies in the public sector, aiming at bringing together digital transformation (e-Government), cyber security, national technologies, big data and artificial intelligence under the same roof, Presidential Digital Transformation Office was established within the scope of the Presidential Decree No. 1, which came into force after being published in the Official Gazette No. 30474 dated July 10, 2018."* (The Presidency of the Republic of Turkey Digital Transformation Office – About us, n.d.)

It is stated that with the transition to the presidential system from the parliamentary system in Turkey, a system in which policymaking, implementing and monitoring are carried out by a single person has been the goal by removing the guardianship mentality and similar formations that try to design politics and society in ways that are not based on the will of the people and the elimination of the dual execution (Sobacı, Miş, Köseoğlu, 2018: 1). The new presidential model specific to Turkey has been put into practice with the idea that it will eliminate dual execution and coalition governments seen as problems in executive (Bilir, 2017:4–5).

When country applications are analyzed, it is seen that these offices collect the data and information required for the formulation of public policies in their own policy expertise, produce the field information, develop appropriate policy alternatives accordingly and form policy and decision drafts and report these to the president. In the country applications, offices can sometimes be assigned to coordinate and evaluate implementation. Therefore, the offices are among the most important structures that play a role in improving the president authority to form and implement public policy. These offices are "Finance Office," "Human Resources Office," "Digital Transformation Office" and "Investment Office." In addition to being the closest staff to the president, these units are described as a "backyard" for the development of policies and projects. It is pointed out that these units manage the projects in accordance with the priorities of the president and have an executive qualification while working one-on-one with the president without any bureaucratic obstacle (Sobacı, Miş, Köseoğlu, 2018: 3).

The goal of DTO, with the slogan "Digital Turkey," is explained as carrying out digital transformation with the cooperation of public, private sector, universities and non-governmental organizations (The Presidency of the Republic of Turkey Digital Transformation Office – About us, n.d.). From this target of DTO, besides ensuring coordination among all public institutions, it can be said that the contact and cooperation with the private sector will accelerate. It is thought that the attitude toward cooperation with the private sector, which has been on the agenda since the 1990s, will also accelerate the transformation in the business practices in the public sector, thanks to the technological facilities brought by digitalization.

It is significant to note that even though digital transformation has a vital role in achieving Turkey's 2023 targets, only technology development is insufficient in digital transformation. In addition to this, it is important to have compliance with the standards to be created for this field, to obtain common sense, civil initiatives and awareness in the society in order to establish standards in digitalization (Türk Bilişim Derneği, 2019: 9–10).

It is stated that DTO, which was established with PD and has performed its executive function together with the president due to its office feature, provides the opportunity to work without intermediaries, is direct and fast in the transactions to be carried out and will not be involved in bureaucratic processes (Sobacı et al., 2018: 3). There are a few points to be mentioned at this point. First of all, the data needed by the operations to be performed by digitalization should be entered into the system. Since many institutions and organizations are involved in the stages of the service process, the lack of data entries will affect the whole system successively. Furthermore, there are disadvantages in itself that all relevant reports, information, documents, etc. will be kept in one hand. One of the most unique threats of the digital age is the problems that can be experienced in information security. Possible information security vulnerabilities or threats, information and documents carry the risk of being seized and destroyed as they will be gathered in one place.

The first precaution to be taken about information security is to establish the eligibility of the legal ground in order to intervene in this matter in the national or international arena when necessary. For this, a circular was issued under the name of Information Security Measures No. 2 in 2019. The aim of the circular was explained as "Moving information to digital environments, facilitating access to information, digitalizing infrastructures and using information management systems widely, brings serious security risks. For the purpose of reducing the security risks encountered, making them ineffective and ensuring the security of critical types of data that may threaten national security or lead to disruption of public order, especially when its privacy, integrity or accessibility are impaired, taking the following precautions are found appropriate."

The second measure to be taken about information security is related to the technical dimension. Operations such as storing and updating information require the intervention of technical personnel who are experts in this field. The issue of the personnel and hardware ensure information security necessitates the execution of a serious R&D activity in this subject. Apart from the technical personnel who will take part in ensuring the information security, it is very important that the personnel working in DTO receive user trainings on the security of information in the transactions performed.

There are eight projects carried out by the DTO. These projects are as follows:

1. *AçıkVeri (OpenData) Project: The slogan of the AçıkVeri Project is "Generating Value from Data for our Country with AçıkVeri." The aim of the project is to share public data by enabling transparency, accountability and the production of new value-added services within the framework of the privacy principles. Data are needed to develop artificial intelligence and innovative technologies. However, limited access to the required data hinders development. Within the scope of the AçıkVeri Project, it is aimed to generate value with artificial intelligence and innovative technologies by sharing the anonymized and confidential data.*
2. *Dijital Türkiye (Digital Turkey) Project: It is aimed to provide digital solutions that will ensure the delivery of high-quality, fast, uninterrupted and safe services to citizens. Within the framework of the task of preparing digital transformation roadmap and creating the digital ecosystem in the public sector, this project intends to provide integrated services in cooperation with the central and local governments and the private sector. In order to prevent publicly produced data from being requested as a document by another public institution, studies are conducted for zero-document policy. By reaching over 45 million users in Digital Turkey Platform, 5,170 services of 641 institutions are presented in Digital Turkey Portal. Twenty-one electricity supply companies and 72 natural gas distribution companies in 81 cities are integrated into Digital Turkey Platform. It is stated that electricity and natural gas subscription applications has decreased from at least one day to a couple of minutes on Digital Turkey; the dormitory registration procedures of students has reduced from 2 days to a few minutes via v1.0.*
3. *e-Yazışma (e-Correspondence) Project: The purpose of the project is to ensure that official correspondences between public institutions and organizations are made securely in an electronic environment. Within the scope of e-Yazışma Project, an encryption mechanism was established to assure that documents are transmitted securely between institutions by determining the official correspondence to be transmitted between public institutions and the package structure that will carry their metadata and electronic signatures. A software programming interface (API), which public institutions can integrate into their electronic document management systems, was developed, and its first version was released in December 2011.*
4. *Fikir Maratonu (Idea Marathon) Project: The aim of the project is to develop feasible, sustainable and technology-oriented ideas in the subjects related to Digital Turkey and e-Transformation and project proposals to design new services. It is intended to be a competition concept in which young people can comprehend the Digital Turkey (e-Government) service design processes and technical infrastructure and its participants to be students enrolled in undergraduate and graduate programs.*
5. *HackIstanbul Project: It is a cybersecurity talent competition that opens its doors to thousands of hackers worldwide. In the event that includes cyberattack scenarios that can be encountered in real life, contestants aim to prove that they are the best to raise awareness in this field by drawing attention to the issue of cybersecurity and to bring new talents to the sector.*
6. *KamuNet (PublicNet) Project: In order to ensure the transmission of the data shared on the KamuNet network at a security level compatible with the privacy, studies have*

been carried out to make the network end-to-end encrypted. Within the scope of the project, it is aimed to increase the security level to the highest level by using national encryption algorithms on local and national IP Crypto devices.

7. *KAYSİS Project: Electronic Public Information Management System (KAYSİS) is an information management system that will allow the integration of developed Digital Turkey (e-Government) applications from a single center defined electronically by identifying the elements in public administration with the legislative bases, from the description of the organizational structure of public institutions and organizations to the services provided, from documents used in services to communication and executive information of institutions.*
8. *TekDurak (OneStop) Project: It is aimed to provide digital services to the citizens having no possibility of access to Digital Turkey. Within the scope of the project, it is intended to increase the diversity by improving service delivery channels and to ensure that all citizens, especially disadvantaged groups, have access to public services. One-stop world examples are similar to e-Government Portal structures and physically have one-stop points. In addition to similar structures, the presence of physical one-stop points is noteworthy. It is aimed to ensure equal access by using two kinds of service deliver (The Presidency of the Republic of Turkey Digital Transformation Office – Projects, n.d.)*

It has been stated that the widespread use of e-Government applications to benefit from new ICTs will provide financial convenience for citizens and the business world. In this context, it is stated that it will be possible to make use of ICTs commonly via one-stop offices (physical or electronic), which are internationally recognized as well and among the abovementioned applications, with the simplification of processes, time limits in decision-making, studies to measure administrative burdens, administrative and legal simplicity and plainness (Prime Ministry, 2003: 158, 159).

Within the scope of digitalization in the public sector, apart from the projects carried out by the DTO and whose aims are described above, in "My Vehicles Concept," which can be expressed as the creation of a system in which all services are integrated with each other, it is aimed at accessing information from different public institutions such as the information on users' tax debt, fine and inspection related to their vehicles through a single service. e-Government Vehicles service aims to bring together multiple classic e-Government services to provide citizens with the information they need from a single point. With this service, instead of reaching the distributed services of each institution individually, citizens will be able to access all the information they want regarding their vehicles. The "My Vehicles" concept is an implementation example of the "increasing the integration between all management functions and tools" goal, which is the first among nine main goals[1] targeted in the effective e-Government goal.

---

1 Other subjects targeted in the strategy document are as follow: developing and delivering public services in a more qualified, fast and easily accessible way;

In a public statement done in July 2019 about studies related to the integrated service approach, it was stated that Digital Turkey Version 1.1's studies have been started, and with this, it will be focused on the delivery of the integrated service. Through this service, it was reported that citizens can access all information about their operations, such as insurance status, tax debt, fine, inspection status and rapid transit balance related to their vehicles from a single window to be created on Digital Turkey Portal (Dijital Türkiye'de yeni dönem hedefi 'bütünleşik hizmet sunumu, 2020).

In June 2020, many services of vehicles provided at the e-Government Gateway were gathered under the name of "My Vehicles" as a single service. The services offered by different institutions for citizens' vehicles are brought together in one place. Before "My Vehicles" service, it was required to operate on separate pages for 15 different services related to vehicles; however, the integrated service "My Vehicles" has provided to bring multiple e-Government services together. Operations that can be done about the citizens' vehicles via "My Vehicles" service are as follow: the citizens can see the fine written on the license plate of the vehicle registered on them, see and pay the motor vehicle tax, examine the car insurance and traffic policies, view the vehicle inspection and exhaust gas emission measurement reports and look at HGS and OGS transition, balance and infringement information (E-Devlet Kapısı'nın İlk Bütünleşik Hizmeti "Araçlarım" Kullanıma Açıldı, n.d.).

Conclusion

There have been many suggestions for many problems and solutions in the field of public administration from past to present. It is seen that the first suggestions were made mostly by foreign experts and that the relevant reports and was off the agenda before the implementation phase began. Although the studies in the following periods were prepared by teams of Turkish experts, it is seen that the source of ideas for the solution is the application examples of international organizations or other countries just as the previous reports.

Public organization is one of the first issues in the improvement or reform efforts in public administration. In the reform initiatives carried out on a sectoral basis in public services, the priority has been the unification of ministries or

strengthening public communication and democratic participation opportunities; responding to social expectations faster; increasing transparency and accountability in public services and corporate activities; determining public policies based on more data and with greater stakeholder engagement; obtaining and using public resources more effectively, economically and efficiently; creating a suitable institutional environment for adapting innovative approaches to public administration and ensuring sustainability in the e-Government ecosystem development (National e-Government Strategy and Action Plan (2016–2019), 2016: 26).

high-level institutions for services that are considered to be related for the purpose of reducing the number of public institutions and simplifying them. It is observed that the structural changes that have been brought to the agenda and implemented in the administrative system in every change of power have no serious effect on the improvement of the services.

The characteristics of the change in ICTs have become a tool for the principles aimed at achieving in structural change efforts on the public agenda. As exemplified above, it can be said that the merging of ministries to combine services that are considered to be similar or interrelated does not have a different purpose than the service integration purpose in the digital environment created by principles such as transparency, simplicity, speed and accountability, which are frequently repeated in almost all policy texts today, and the concept of "My Vehicles," which is the latest example of the integration of public services in terms of digitalization. It can be said that these efforts aim to simplify, speed up things by reducing paperwork, prevent duplication, etc. in the public. Although they pursue similar goals, the reason that the new management approach is in a more advantageous position in reaching their goals is that they take advantage of digitalization's features.

Although the idea of redesigning public services effectively, efficiently and accessible to all sounds good, it does not seem very possible that technological infrastructure services and vehicles can reach everywhere and every citizen. For example, it is seen that the difficulties in delivering distance education services to each student, which are produced as an idea during the days of the study and especially in the education sector, are tried to be overcome by using different technological tools (for example, options such as lecturing students who have a computer and Internet connection on different kinds of applications or preparing lecture videos and airing them on TV under the coordination of Ministry of National Education). While it is considered to use the same methods in order to provide equal and fair service to the needs of citizens with similar needs, it is thought that a little more time is needed to use the tools left to the arbitrary preferences of the people.

Making the methods developed under the influence of digitalization in public services sustainable, feasible and truly accessible to all citizens is one of the most important issues to be addressed in this regard. On the other hand, another point to be emphasized is the entries of the data that are essential for the services to be provided in digital environment. The issue of how much of the data will be entered into the relevant systems and the problems of updating the data in coordination with other relevant public institutions are very critical in terms of public transparency and reliability concerns.

In the change of approach in the public administration understanding, perhaps the most important among the objectives and recommendations listed is the human factor. Even though the physical presence of the manpower factor remains in the background while aiming to provide services through digital methods, it is important that public officials who understand the essence of the idea of digitalization and those who will receive this service demonstrate their awareness and desire to receive digital services. Even if it is imagined that all service sectors will be provided on digital platforms, it is important to educate and raise awareness of public officials and citizens in this sense.

Bibliography

Başbakanlık (2003), Kamu Yönetiminde Yeniden Yapılanma 1: Değişimin Yönetimi İçin Yönetimde Değişme, Ankara.

Bilir, Faruk, (2017), "Türkiye'ye Özgü Yeni Bir Hükümet Modeli: Cumhurbaşkanlığı Sistemi", *Yeni Türkiye*, N. 94, pp. 1–6.

Carter, L., Bélanger, F. (2005), "The Utilization of E-Government Services: Citizen Trust, Innovation and Acceptance Factors", *Information Systems Journal*, Vol. 15, N. 1, pp. 5–25.

Dijital Türkiye'de Yeni Dönem Hedefi 'Bütünleşik Hizmet Sunumu, (24.07.2019), "Dijital Türkiye'de.Yeni Dönem Hedefi "Bütünleşik Hizmet Sunumu", Aa.Com. Tr. https://www.aa.com.tr/tr/turkiye/dijital-turkiyede-yeni-donem-hedefi-butunlesik-hizmet-sunumu/1540451, (30.04.2020).

DPT, (2011), *Bilgi Toplumu İstatistikleri 2011* (Vol. 2826), Ankara.

Dunleavy, Patrick, Margetts, Helen, Bastow, Simon and Tinkler, Jane, (2006, July), "New Public Management Is Dead: Long Live Digital-Era Governance", *Journal of Public Administration Research and Theory: J-PART*, Vol. 16, N. 3, pp. 467–494.

e-Devlet Kapısı'nın İlk Bütünleşik Hizmeti "Araçlarım" Kullanıma Açıldı, (n.d.), Cbddo.Gov.Tr. https://cbddo.gov.tr/haberler/4346/e-devlet-kapisi-nin-ilk-butunlesik-hizmeti-araclarim-kullanima-acildi-, (19.06.2020).

e-Dönüşüm Türkiye, Bilgi ve İletişim Teknolojileri Dairesi Başkanlığı – T.C. Cumhurbaşkanlığı Strateji ve Bütçe Başkanlığı, http://www.bilgitoplumu.gov.tr/bilgi-toplumu/e-donusum-projesi/, (19.06.2020).

Endekslerde Türkiye, (n.d.), Dijital Akademi, https://www.dijitalakademi.gov.tr/olcumlerde-turkiye#BirlesmisMilletler, (19.06.2020).

Heeks, Richard, (2003), "Most E-Government for Development Projects Fail: How Can Risks be Reduce?", E-Government Working Paper Series, No. 14. University of Manchester, Harold Hankins Building, Precinct Centre, Manchester, UK.

Kamolov, Sergey, (2017), "Digital Public Governance: Trends and Risks", *Giornale di Storia Costituzionale*, Vol. 33, pp. 184–194.

Matei, Ani, Antonie, Catalina, (2015), "The Need for Positive Change: Adapting Management in Public Administration" *4th World Conference on Business, Economics and Management, WCBEM, Procedia Economics and Finance*, Vol. 26, pp. 345–350.

Misra, Dheeraj C., (2007), "Select Aspects of Conceptual Foundations of E-Government: Clearing the Fog for a Better Vision", *5th International Conference on E-Governance: Foundations of E-Government*, Hydebarad: ICEG, pp. 21–33.

McNutt, Kathleen, (2014), "Public Engagement in the Web 2.0 Era: Social Collaborative Technologies In A Public Sector Context", *Canadian Public Administration*, Vol. 57, N. 1, pp. 49–70.

National e-Government Strategy and Action Plan (2016–2019), Ulaştırma, Denizcilik ve Haberleşme Bakanlığı Yayınları, Ankara.

Ocak, Kasım, (2019), "Cumhurbaşkanliği Kararnamesi ile İdari Vesayet Yetkisi Düzenlenmesinin Bir Örneği Olarak Cumhurbaşkanliği Ofisleri", *Anayasa Yargısı*, N. 1, pp. 163–193.

OECD, (2003), The Case for E-Government: Excerpts from the OECD Report "The E-Government Imperative", *OECD Journal on Budgeting*, Vol. 3, N. 1, pp. 61–131.

Onuncu Kalkınma Planı, (2014–2018), Türkiye Cumhuriyeti Kalkınma Bakanlığı Yayınları, Ankara.

Ömürgönülşen, Uğur, (2013), "Amerika Birleşik Devletleri'nde Kamu Yönetimi", K. Karasu (Ed.), *Kamu Yönetimi Ülke İncelemeleri* içinde (313–412), Ankara: İmge Kitabevi.

Scholl, Hans, Jochen, (2012), "Five Trends That Matter: Challenges to 21st Century Electronic Government", *Information Polity*, Vol. 13, N. 3, pp. 317–327.

Scholl, Hans, Jochen, Scholl, Michael, C. (2014), "Smart Governance: A Roadmap for Research and Practice. iConference 2014 Proceedings", Available at: Smart Governance: A Roadmap for Research and Practice, (2014), IConference 2014 Proceedings. https://doi.org/10.9776/14060, (27.05.2020).

Sağsan, Mustafa, (2001), "e-Devlet: Toplumların Yeni Umut Işığı Mı?", *Stratejik Analiz Dergisi*, ASAM Yayınları, Vol. 19, pp. 1–11.

Sobacı, Mehmet Zahid, Miş, Nebi, Köseoğlu, Özer, (2018), "Türkiye'nin Yeni Yönetim Modeli ve Cumhurbaşkanlığı Teşkilatı", *SETA Perspektif*, Vol. 206, pp. 1–16.

Stoker, Gerry, (2006), "Public Value Management: A New Narrative for Networked Governance?", *The American Review of Public Administration*, Vol. 36, N. 41, pp. 41–57.

Türk Sanayicileri ve İş İnsanları Derneği (TÜSİAD), (2016), *Türkiye'deki Dijital Dönüşüme CEO Bakışı*.

Türkiye Bilişim Derneği, Türkiye Bilişim Derneği Raporu, (2001), Türkiye'de E-Devlet Nasıl Olmalı?, KAMU Bilgi İşlem Uygulamalarında Verimliliğin Arttırılması I–II–III Toplantıları Sonuç Raporu.

Türkiye Bilişim Derneği, (2019), *1. Dijital Dönüşüm ve Standartlar Zirvesi Sonuç Raporu,* Ankara.

Türkiye Cumhuriyeti Cumhurbaşkanlığı Dijital Dönüşüm Ofisi – Anasayfa. (n.d.). Cbddo.Gov.Tr., https://cbddo.gov.tr/kamu-dijital-donusum-lider, (19.06.2020).

Türkiye Cumhuriyeti Cumhurbaşkanlığı Dijital Dönüşüm Ofisi – Hakkımızda, (n.d.). Cbddo.Gov.Tr., from https://cbddo.gov.tr/hakkimizda/, (19.06.2020).

Türkiye Cumhuriyeti Cumhurbaşkanlığı Dijital Dönüşüm Ofisi – Kamu Dijital Dönüşüm Lideri. (n.d.). Cbddo.Gov.Tr., https://cbddo.gov.tr/kamu-dijital-donusum-lider/, (19.06.2020).

Türkiye Cumhuriyeti Cumhurbaşkanlığı Dijital Dönüşüm Ofisi – Projeler, (n.d.), Cbddo.Gov.Tr., https://cbddo.gov.tr/projeler/#3173, (19.06.2020).

Türkiye Cumhuriyeti Cumhurbaşkanlığı Strateji ve Bütçe Başkanlığı, http://www.bilgitoplumu.gov.tr/bilgi-toplumu/e-donusum-projesi/, (26.05.2020).

T.C. Ulaştırma Bakanlığı, (1999, October), Türkiye Ulusal Enformasyon Altyapısı Ana Planı Sonuç Raporu, http://www.bilgitoplumu.gov.tr/Documents/1/Yayinlar/991000_TuenaRapor.pdf, (25.05.2020).

United Nations, e-Government Survey, 2018, i Gearing e-Government to Support Transformation Towards Sustainable and Resilient Societies United Nations e-Government Survey 2018 Gearing e-Government to Support Transformation Towards Sustainable and Resilient Societies. (n.d.), From http://www.bilgitoplumu.gov.tr/wp-content/uploads/2018/08/un_e-gov_survey-2018.pdf, (16.04.2020).

UNDP, (2015), *From Old Public Administration to the New Public Service Implications for Public Sector Reform in Developing Countries,* Singapore.

Wahedduzzaman, W., Miah, S.J., (2015), "Readiness Assessment of e-Government: e-Developing Country Perspective", *Transforming Government People Process and Policy,* pp. 1–21.

Waller, Paul and Weerakkody, Vishanth, (2016*), Digital Government: Overcoming the Systemic Failure of Transformation – Digital Transformation Through Policy Design with ICT-Enhanced Instruments.* Working Paper-2, London: Brunel University.

Warkentin, M., Gefen, D., Pavlou, P. A., M. Rose, G. (2002), "Encouraging Citizen Adoption of e-Government by Building Trust", *Electronic Markets,* Vol. 12, N. 3, pp. 157–162.

Wihlborg, Elin, Hedstrom, Karin, Larsson, Hannu, (2017), "e-Government for All – Norm-Critical Perspectives and Public Values in Digitalization", *Proceedings of the 50th Hawaii International Conference on System Sciences,* pp. 2549–2558.

Yılmazöz, Mehmet, (2009), "Türkiye'de Kamu Personel Yönetimi Sorunu", *Mülkiye Dergisi,* Vol. 157, pp. 293–302.

Hatice OĞUZ and Metin ÖZKUL

A New Form of Violence in the Reality of Risk Society: Digital Violence

1 Introduction

The concept of risk has gained importance following the occurrence of unforeseeable changes regarding an individual's life during the process of transition from traditional society to modern society. Our era can be defined as the age of risk society and uncertainties in many respects. Among the significant characteristics of this era, we can include the increase of uncertainties in every facet of social life, the proliferation of the culture of fear, reflexive modernization, increased individualization, erosion of trust, decrease in trust towards authority, and fragmentation of cultural authority. Social change and transformation, which have become more observable in the last decade due to technological advancements, also cause significant change in cultural structures. One of the main objectives of technological advancements is to benefit society. However, certain negative characteristics and aspects of technology are unforeseeable and unintended emerge. This condition of being unforeseeable and unintentional is among the defining features of the concept of risk.

Violence, one of the undesirable components of social reality, has been in existence since ancient times. Despite the countless developments and progress, violence has remained as one of the most primitive aspects of humanity. It is also one of the important problems of the modern era that is outdated; yet it adapts to the era and diversifies with time.

Violence has assumed a new form under the influence of all these social changes and transformations. Types of violence such as physical, verbal, psychological, etc. generally or prevalently exist in social life both historically and at the present time. However, based on the increase of social relationships created by technology, we can also discuss digital violence now. This new type of violence manifests itself as intentional harming of another person through misuse and abuse of digital tools and applications.

Communication has been a basic need of humanity throughout history and the improving opportunities have shaped the ways of communication. Nowadays, humanity has numerous methods and tools to communicate, thanks to technological developments. With the opportunities provided by technology, information tools such as computers, mobile phones, the Internet, etc. have become the most vital parts of our lives, so to speak. This is because these tools provide the opportunity for simultaneous and sequential communication with numerous people without going away in the midst of the business of daily life, and they are preferred as sources of easy, cheap, and fast communication.

This study discusses how the current information technology provides the basis for the phenomenon of violence to take new forms in risk society and the repercussions of these new forms of violence that we are faced with in personal and social life. In the study, a theoretical analysis was performed by reviewing the literature review. Fundamental sociological theories about risk society and the phenomenon of violence are mentioned. In addition, scientific studies regarding digital violence were examined to concretize the characteristics of the new form of violence. In the conclusion of the study, the relationship between risk society and the phenomenon of digital violence was evaluated in terms of social reality by taking the results of the various empirical studies in the literature into consideration.

The industrial revolution that occurred during the transition from traditional society to modern society has opened the doors of a new era for the whole world. This new era brought with it great and radical changes, along with new risks. While risks such as natural disasters, epidemics, and wars were known in traditional societies, in modern societies, activities that humanity expected to be beneficial to social development have also caused latent risks alongside these risks. This situation has resulted in different types and effects of risks that are mostly unpredictable. Nuclear disasters, uncontrollable technological problems, disruption of ecological balance, biological weapons, and genetically modified foods are just a few of them.

When we pay attention to the daily use of the word risk, we can observe that it is employed to refer to meanings such as "danger," "harm," and "uncertainty." The word risk refers to a different concept in almost all branches of science in the social sciences literature. For example, in economics, it is rather emphasized as an "entrepreneurial virtue," the ability of rationally calculating the uncertainties that might fail the enterprise and facing it with courage. Although the word risk has differences regarding its conceptual meanings, today it is widely used as both a potential and phenomenal reality that should be measured and estimated in important aspects of life such as health, economics, management, law, and technology. In almost all of the mentioned fields, the concept of risk is used to refer to potential damages/hazards.

There are various approaches regarding the classification of the concept of risk in the literature. For example, while the cognitive approach proponents discuss the concept of risk more in terms of probability calculations in the fields of engineering, statistics, psychology, epistemology, and economics with a focus on hazards and damages and argue that risk is a phenomenon that can be scientifically calculated and managed. According to the advocates of sociocultural perspective, the phenomenon of risk cannot be understood without understanding the cultural and social context. On the other hand, social structuralism advocates tend to approach risk assessments relatively, which is a dynamic process that is constantly reconstructed according to time and place, and in fact emphasize the difficulty of determining the factors in defining the risk situation. According to them, risk cannot be evaluated objectively completely free from ethical values and belief systems (Lupton, 1999: 18–30).

Social scientists such as Ulrich Beck (risk society, reflexive modernization), Anthony Giddens (manufactured risk), Zygmunt Bauman (ambivalence and risk), and Mary Douglas (culture and risk, risk perception) have various views and explanations regarding the phenomenon of risk and the risk society. Moreover, there are also the ideas of theorists such as Sigmund Freud (theory of instinct), Albert Bandura (social learning theory), and Pierre Bourdieu (symbolic violence) which discuss the phenomenon of violence from various perspectives, such as both its causes and how it is realized.

Considered to be the inventor of the "risk society" discourse, Beck's approach to the phenomenon of risk is realized within the context of sociocultural perspective. Beck describes the transition from the industrial society to modern society by using the discourse of risk society. In this regard, risk society is a transition period definition. Modern society that emerged after the industrialization is faced with its own limits as a result of an understanding in which it does not think about the consequences of what it has generated and any means can be used for the sake of development (Beck, 2011: 49–55). In the risk society where relations and problems transcend national borders, conflict occurs not only among classes and groups but in a structure that also includes industrial relations. The future of the risk society is uncertain, and insecurity, which has an existential resource within this structure, has penetrated into every layer of social relations.

Giddens describes the phenomenon of risk as the dark side of modernity. According to him, risks should be discussed in terms of the consequences of modernity. Giddens, in his work called *Consequences of Modernity*, draws attention to the globalization of risk as the factor that creates risk's profile of modernity. He explains the globalization of risk through factors such as the risk intensity, expanding number of contingent events which may affect large numbers of people, ecological hazards as a result of the changing nature of the interaction between humans and nature, and new forms regarding the institutional dimensions of modernity (Giddens, 1998: 122–126).

In the light of the theoretical foundations about risk, if we think the phenomenon of violence in connection with a new form of violence, that is, digital violence, the following can be said:

Following the possibilities that we think of as modern society, which is sometimes referred to with terms such as "information society," "network society," and "global age" by various philosophers of the late twentieth century, humanity has made great progress in almost every field of individual and social life but also created new risks with these opportunities. While this structure, in which the phrases "risk society," "manufactured risk," "risk culture," and "uncertainty" are embodied, encompasses the developments that are considered to be paramount for humanity and nature, behind these developments, it also involves dangers that are sometimes visible, but mostly invisible. These risks arise from the damages humans cause to themselves, to the structures they are involved in or to the nature that is directly or indirectly caused by incalculable/lack of efforts in foresight or ignored consequences of the developments, innovations, etc. that happen in modern societies.

We can label the occurrence of digital violence as one of the invisible or ignored hazards of the risk society. As a matter of fact, individuals have been given ample opportunities with technological advancements, particularly in communication, besides many fields such as production, health, education, travel, etc. The use of these opportunities by individuals who can turn them into real means according to their interests by misusing or with malicious intent has started to uncover some risks. Especially, individuals' misuse of them for their self-interest and problematic use of digital applications that have produced faster information flow and easier communication have revealed various examples of digital violence. This condition is seen as the new form of violence becoming one of the phenomena that affect individuals quite negatively both in their private and social lives.

The phenomenon of violence, as a multidimensional, complex, and important social problem that exists in all areas of life, disrupts the harmony in human relations and the organization between social institutions, breaks the bond of solidarity, destroys hierarchies, negatively affects social order, and threatens the future of societies as a universal phenomenon. In a world surrounded by violence and aggression, we can be exposed to various violent behaviors such as abuse, harassment, physical or verbal attacks, etc. from foreigners, friends, partners, families, or relatives. Violent behaviors may occur in many different ways. The new form of violence that has emerged in the last quarter century of the history of violence is digital violence, which has been on the rise recently and has turned into a social phenomenon rather than being "isolated," as it was in the beginning.

There are various definitions of digital violence in the literature. The common ground in almost all of these is that violent behavior (intentional harm) is carried out by means of digital communication. Digital violence, which is also referred as "cyber violence," "cyberbullying," "online violence," "virtual harassment," "technological bullying," etc. in various sources can be defined as any activity that involves people threatening and harassing or blackmailing other people through technology-related ways such as e-mail and text messages (Steinhaver, 2008; Hazelwood & Koon-Magnin, 2013; as cited in Cebecioğlu & Altıparmak, 2017: 424). In this study, the concept of "digital violence" was preferred in an inclusive way to avoid confusion.

The most crucial difference between classical violent behaviors that are displayed in physical, verbal, psychological, etc. ways and digital violent behavior is that in cases of digital violence, the perpetrator and the victim are not face-to-face and the one who commits the violence is mostly unknown. To put it in a clearer way, the interaction between the perpetrator and the victim occurs over any digital communication platform, and in most cases, the perpetrator is unknown. In the presence of this uncertainty, avoiding violent behavior, finding a safe environment, and protecting against attempts of violence from other potential offenders, etc. can become difficult. Another difference is the possibility that the circumstances of violence will not be limited to the perpetrator and the victim and may spread to millions of people in a matter of seconds. For example, an image that can victimize an individual can be sent to numerous people online in seconds through any social media application. However, in cases of physical, verbal, psychological, etc. violence, this situation may remain limited to those who are involved/witnesses around, unless a special effort is made. In addition to these, we can say that other types of violence and digital violence share some aspects in terms of their origin. Such violent behavior can be exhibited in order to draw attention or prove oneself, for individual reasons such as revenge, or with the aim of establishing authority (Yüksel Özmen, 2018: 959–961).

Digital violence may lead individuals who are exposed to it or victimized by it into loneliness. As a result, it may cause the individual to be isolated from society. Besides the individual and familial factors, unawareness about tools like technology, the Internet, etc. also has a role as a reason for both resorting and being exposed to digital violence. There are also studies in the literature that demonstrate that those who spend a lot of time online exhibit more digital violent behavior compared to those who spend less time (Yüksel Özmen, 2018: 960). It can also be noted that the research on digital violence is conducted mostly on young people, while studies on adults and the elderly are limited. According to the data accessed, young people resort and are exposed to digital violence more than the middle-aged and elderly people; moreover, women are exposed to digital violence more than men (Russo & Pirlott, 2006; Staude-Müller et al., 2012; Baker et al., 2013; Devries et al., 2013; Woodlock 2013; Choja & Nelson, 2016; Lenhart et al., 2016; Lewis et al., 2016).

The improvement of communication technologies every passing day, new social platforms that make relations easier, and opportunities of sharing have caused social distances to disappear along with physical distances. Acceleration of establishing relationships and the purposes and rationale of establishing relationships changing and becoming easier, in a way, increase the efficiency of various businesses and popularize individuals' socialization activities. But despite that, on the flip side, communication and network technologies, which are produced for various purposes and are indispensable to our lives, have become a potential threat as a result of misuse. As a consequence, almost every individual of today's world, regardless of the society they belong to, can be exposed to potential threats of violence that may have effects in real-life environments and circumstances in terms of its possible consequences, even though it is limited to virtual environments at the beginning. Thus, today's people live not only in a society/world where technology makes our lives easier but also in a risk society/world where technology also makes life harder and even does not allow to experience the happiness to be obtained from the convenience.

2 Phenomenon of Risk and Risk Theories

According to the information in the sources of disciplines such as history, anthropology, sociology, etc., the idea of faith and fate is emphasized more in determining the private or social actions of the individual in traditional or feudal societies. In these societies, the risks were attributed mostly to nature or supernatural forces. For instance, individuals made sense of the epidemics, death, famine, war etc. as supernatural incidents or associated them with divine causes. The concept of risk, which was understood as a source of fear and anxiety in traditional societies, started to be interpreted as "to dare" and "heroism" in the period that started with the Enlightenment (Sennett, 2010: 84). Today, the word "risk," which has meanings such as "chance, danger, possibility of harm, taking a chance, putting something into danger," has been borrowed from French into Turkish and is defined as "hazard of being harmed" (Doğan, 1981; As cited in Soydemir, 2011: 170).

Many philosophers started to refer to social structures, which we have defined as modern society starting with the twentieth century and especially since the second half of this century, with different adjectives. For instance, we see that this grand transformation was called "postmodernity" by authors such as Bauman, Lyotard, and Harvey; "late modernism" by Giddens; and "global age" by Albrow. Knowledge age/society, information society, or network society among others can be added to these qualifications. We can say that the most critical of these concepts is "risk society," developed by Ulrich Beck and later recognized by Anthony Giddens and some other sociologists (Ritzer & Stepnisky, 2014).

2.1 Ulrich Beck: Risk Society and Reflexive Modernization

Beck analyzed what kind of risks and threats we are surrounded with within our world with the theory of risk society, which he developed in line with the relationship between sociology and modernity based on the concept of risk, which refers to the potential threats in the future.

Beck's conceptualization of risk society is the macro-sociological explanation of the transition from industrial society to risk society. The ground of discussion for this explanation consists of discussions of modernity. Risk is used as the transformative key concept of these discussions of modernity. As an intermediary concept, risk is the differentiation between the early stages of the industrial society and the definitions of the risk society. The new condition of risk is shaped by the fact that we are living in irreparable ecological consequences due to technological and social expectations (Wilkinson, 2001: 2).

Beck defines the new period we are living in with the metaphor of living in the "volcano of civilization" (Beck, 2011: 19–29). The social form of living in the volcano of civilization is the unforeseeable, unexpected structure the industrial society institutions take, which have been the source of rationalization and functional differentiation for today's world in line with the principle of evolution in the face of new transnationality of the living conditions. This condition underlies the theory of the risk society. In this kind of society, the risks, side effects, hazards, and overlooked matters directly influence our decisions. Beck places the discourse of "industrial society" in the center while defining how the concept of risk society emerged. According to this, the modern society that emerged after industrialization faces its own limits as a result of an understanding in which it does not consider the consequences of what it produces, and any means is acceptable for the sake of development.

According to Beck, the fundamental characteristics of the risk society are as follows. Firstly, in the risk society, problems and relations cross the national borders. Secondly, the risk society ensures that the political conflict is shaped in a context that includes the industrial relations, as opposed to only classes and groups. Finally, the risk society represents an uncertain future and the spread of ontological insecurity in social relations (Keyman, 1999: 28).

The concept of risk society in the way Beck defines it describes a world that is filled with globalized risks that may be political, economic, cultural, and environmental. According to Beck,

> *In modern society, risks induce systematic and often irreversible harm. Risks generally remain invisible, are based on causal interpretations, and thus initially only exist in terms of the (scientific or unscientific) knowledge about them. They can thus be changed, magnified, dramatized and minimized within knowledge.* (Beck, 2011: 22–23).

For Beck, the important issue is how the changes that occurred during the transition from the industrial society to risk society brought us to a condition of uncertainty and a stage that is defined by risks. According to him, analyzing our age or the age of reflexive modernity will provide the answer to this question (Esgin, 2013: 686). Reflexivity is concerned with the characteristic of modernity that critically values the fundamental logic and assumptions of the scientific and technological production, that is, the process of rationalization. This mechanism of criticism or rationalization manifests itself both in new social movements and in the questioning of scientific and technological progress (Medrano, 2004: 16–23). The society in the reflexive modernization process is described without a linear rationalization and with a keen distrust towards expertise systems. Beck calls this new condition of risk that is different from industrial society and its dynamics as "risk society."

According to Beck, the meaning of risk changes in different stages of the modernity. Whereas in premodern societies risks were not man-made but external, in modern societies risks and hazards are no more linked to God or fate. The characteristic of modern society is its aim to control risks. However, when risk society is achieved, risks have become incalculable, inexplicable, and limitless. In this sense, the stage of risk society has also become a stage defined by uncertainties. Because, risks that are potential threats for the future have been turned into problem-creating risks that do not have clear solutions by humanity through industry and science (Beck, 2011: 33–46).

Beck describes the risk society as follows: while the driving force in the industrial society is "I am hungry," in the risk society it is "I am scared." While the power that drives the individual into action in the industrial society is their needs, in the risk society it is the anxieties (Beck, 2005: 3).

According to Beck, the most significant difference between today's risks and the risks of the past is that the former is invisible and unpredictable. Therefore, they are difficult to take precautions against. The problems of uncertainty and uncontrollability have emerged in the most advanced stage of modernization. At the same time, the unpredictability of risks has brought about the idea of "noninsurability." Since atomic hazards and nuclear accidents do not fall into the classic "accident" category, they are not insurable. In today's world, the distribution of risks and the distribution of wealth have differentiated. Risks cross countries' borders and affect the whole world. They originate in a specific region and spread to the world in complex ways without being foreseen. The lack of authority and control results in hazardous aspects of risks becoming more widespread (Beck, 2005: 47).

Beck sees globalized risks affecting everyone in the world's risk society as "inequality." There is no difference between rich welfare states and poor countries in terms of their susceptibility to risk. However, there are significant differences in terms of distinguishing factors such as sharing of the resources and the ability to fight risk. As this situation progresses, the tension between those who are affected from the risk and those who benefit from it will increase. When the "invisible side effects" that are legitimized within industrial modernity suddenly become visible, a "boomerang effect" will occur. In other words, those who propagate risk and aspire to benefit from it will have to face hazards one day. According to Beck, sooner or later the perpetrator and the victim will be the same person (Beck, 2011: 22–27).

There are no class distinctions or social categories in Beck's theory of "reflexive modernization." There is no mention of a certain class becoming dominant or even gender discrimination. He defines the aim of reflexive modernization as abolishing the borders determined by nations, continents, and genders. This is because the risks that need to be struggled against do not recognize such distinctions. Beck emphasizes that modernization should not take place in an unconditional and unquestioned way, but it should rather occur with an awareness. According to him, modernization takes place in a continuity and cannot be opposed (Beck, 2005: 37–38).

Beck has a positive perspective towards the modernization process and sees facing risks and hazard and gaining awareness in the face of them as the best course of action to move in parallel with modernization. He states the necessity for everyone to have an inquiry, awareness, and control mechanism against the risks that arise in the risk society. This is because the dimensions of risks, threats, and hazards are much larger and unpredictable than they were in the industrial society (Beck, 2005: 37–38).

2.2 Anthony Giddens: Manufactured Uncertainty

For Giddens, as with Beck, the risk society emerges as a theoretical framework that we can use to establish a new form of modernity and examine its consequences. Unlike Beck, Giddens does not describe the concept of risk society as the last stage of modernity but considers it only as a part of the results of globalization which he describes as the expansion of modernity and its institutional dimensions. His risk discourse is the product of his analyses on modernity (Esgin, 2013: 688).

From Giddens' perspective, risks should be discussed in relation to the consequences of modernization since they are the dark side of modernization. Discussing the consequences of modernity also provides the accumulation needed to question the conceptual and factual phenomena of today's risk conceptualization. In this regard, describing the risk profile and classification criteria of modernity itself is necessary. According to Giddens, the risk profile and classification criteria of modernity are determined as follows (Giddens, 1998: 122–126):

Firstly, there is a globalization of risk in the sense of intensity. The phenomenon called risk intensity is the essential component of the threatening factor of the conditions we are living in today. The dangers such as a possible nuclear war, ecological destruction, unavoidable population explosion, and the destruction of global economic change relationships allude to a discouraging future with danger for everyone.

Secondly, the risk is globalized in terms of increase in the number of contingent events that may affect a large number of people. All displacement mechanisms take the events from the control of a particular individual or group. The more on a global scale these mechanisms are, the stronger their effects will be. In this approach where the concept of globalization is understood as the key, the sources or the services are not subject to a local control mechanism, and thus necessary adjustments in cases of unexpected disruptions, do not occur on a local scale.

Thirdly, there are risks stemming from the created environment or socialized nature entering the material environment in a more efficient and transformative way. The category of created environment or socialized nature category refers to the changing nature of the relationship between humans and the material environment. The variety of ecological hazards in this category stems from the transformation of nature by the information systems of humanity.

The fourth criterion is formed by the development of institutionalized risk environments that affect millions' life chances. For instance, the risks caused by the efficiency and prevalence of investment markets can be evaluated in this context.

The fifth criterion includes awareness of the risk as risk.

The sixth change criterion is formed by the tension created by the distribution of widespread awareness of risk information by large masses.

The seventh and final risk profile arises from the awareness of the limitations of expertise and no expert system being wholly expert about the consequences of the adoption of the expert principles. The belief that supports the trust in the expertise systems also prevents the ignorance that the people on the street exhibit when faced with expert knowledge; however, the existence of ignorance the experts encounter themselves as individual stimulant and in terms of all knowledge areas may weaken this belief for ordinary people or make it insignificant. Such a situation ensures the condition necessary to question the nature of expert knowledge.

While explaining the phenomenon of risk, Giddens starts with the phenomenon of "trust" in traditional and modern cultures. He approaches the phenomenon of trust in an ontological perspective, inspired by Eric Ericson. Ontological security is a very crucial form of feelings of security. This concept refers to the confidence that most people have in the continuity of their self-identity and the constancy of the surrounding social and material environments of action. Ontological security is related to existence or, in a phenomenological sense, being in the world (Giddens, 1998: 92).

The general context that provides the formation of an environment of trust in traditional and modern culture is also the determinant of the general risk environments of these societies. According to this, the formation of the risk environment in the traditional society is shaped by factors emanating from nature such as infectious diseases and climate and human-related threats such as marauding armies, local warlords, and robbers. The risk environment of modern society consists of its reflexivity, namely, the risks it creates itself, industrialized war machines, and the threat of meaninglessness caused by modernity.

Giddens' discussion of risk not only points out to a side of modernity that was shaped by the tension between modernity and traditionalism but also forms the ground for the concept of "reflexivity," which shapes today's analyses about risk. Accordingly, Giddens emphasizes that it is necessary to distinguish between simple and reflexive modernization. According to this emphasis, simple modernization is an old type, unilinear modernization. Reflexive modernization, by contrast, implies accepting the contradictions and limits of the modern order. According to him, reflexive modernization and globalization are intertwined, and this forms the basis of uncertainty (Giddens & Pierson, 2001: 228).

Social reflexivity refers to a society in which the conditions we live in are increasingly a product of our own actions, and on the opposite end, our actions are directed at challenging or managing the opportunities and risks we have created ourselves (Giddens & Pierson, 2001: 229).

Giddens firstly divides risks into two categories. The first is the "external risk," caused by the constancy of tradition or nature in industrial modernization. The second is the "manufactured risk," which generally defines environmental risks, born from the impact of the acquired knowledge on the world and without a historical experience. External risks are caused by events that affect individuals at an unexpected moment (from outside – bad harvest, floods, epidemics, hunger, etc.). These risks occur regularly and often enough to be predictable and to be insured across the population. External risks can be effectively calculated; insuring them can be determined by referring to time and risk charts. Manufactured risks are the most important indicators of the transition from industrial modernity to reflexive modernity. According to Giddens, manufactured risk was born out of the advancements during the development process of humanity, particularly science and technological advances (Giddens & Pierson, 2001: 222).

Apart from the concept of external and manufactured risk, Giddens draws attention to the "high consequence risks" that arise from the great impact of modernity and globalization. High consequence risks are those that bring along irreparable damage. Because the results are unknown and cannot be fully examined in the long run. Since there is a very deep, prevalent causality, well-known examinations do not help in analyzing and overcoming these types of risks (Giddens, 2010: 158–159). For instance, the long-term consequences of the accident that occurred at the Chernobyl nuclear power plant in 1996 are still not fully calculated. This also applies to cases such as the mad cow disease in the United Kingdom, the atomic bomb that was dropped on Japan, global warming, avian flu, COVID-19 in China, etc.

Giddens's risk society analysis is concerned with the effects of the spread of institutional dimensions of modernity on individual identities, as well as the issues of capitalism, industrialism, military power, and surveillance, which he describes as the institutional dimensions of modernity. According to Giddens, risk society is one of the inevitable consequences we face at this stage of modernity. He argues that "we live in a 'risk culture', which is to be explained by the radicalizing and generalizing of modernity" (Giddens & Pierson, 2001). Even a simple comparison between traditional society and modern society would lead us to the same conclusion. Indeed, it cannot be said the concept of risk existed in traditional cultures, because they were not in need of it. Risk refers to the hazards that are actively assessed in relation to future possibilities. Thus, it will become widely accepted only in a future-oriented society. Risk presumes a society that actively strives to break with the past as a prime characteristic of modern industrial civilization (Giddens, 2000: 37). Therefore, the concept of risk society should be considered in conjunction with modernization or the definition of radicalizing modernity.

According to Giddens, environments of trust and risk are now what is determining our lives. Environments of trust and risk differ in terms of premodern and modern cultures. Therefore, the negative aspects of risks and manufactured uncertainties regarding social life should also be evaluated based on different contexts. The preferred way of doing this is to develop a self-identity project that redefines the concept of ontological security in the face of existential anxiety. The concept of ontological security, which is the qualifier of the self-identity project, refers to the confidence that individuals have in the continuity of their self-identity and in the constancy of the surrounding social and material environments of action. Ontological security necessitates the existence of conditions that will ensure the continuity of individuals' routine actions in terms of the repetition of social practices. Individuals' activities in the social sphere are directly linked to their ontological sense of security. Since the ontological sense of security develops in a process starting from childhood, it coincides with the self-identity project that conditioned the reflexivity of modernity. In this sense, the opposite of ontological security is not insecurity but existential anxiety. Since radicalized modernity has expanded the environments of risk, ontological security has been replaced by existential anxieties. The effect of this on social life can be explained as the decontextualization of social relations. The global nature of modernity has transformed self-identity (Esgin, 2008: 465–466).

It can be said that Giddens' analysis of modernity aims to provide insights into how radicalized modernity shapes the new living conditions, manufactured uncertainties, and risks. Radicalized modernity is the result of modernity's self-renewal and developing a reflexive character as a result of its institutional dimensions emerging at a more advanced level. It also opposes to the claims of many thinkers that a postmodern period has started. With the new era, a period of radical modernity has been entered, in which the qualities of tradition and trust, which are the basis of knowledge, have changed. Radical modernity is an area of new motions where environments of trust and risk in which tradition is redefined take different forms (Esgin, 2013: 689).

According to Giddens, radicalized modernity reduces the general risks encountered in certain areas and lifestyles, while on the other hand creating new situations of risk defined in large scale and uncertainty in previous ages. These situations of risk involve high consequence risks that arise from the globalized nature of the social systems of the modern era and have not been encountered in previous periods. For instance, despite the importance attached to international treaties and control of armament, the risk of destructive warfare will continue to exist on a mass scale, as long as science and technology is concerned with the production of new weapons, where nuclear weapons and even the knowledge that produces them exist. The destruction that humans cause in nature due to their struggle to dominate has made the risks related with ecological disasters an inevitable part of our daily lives. Other high consequence risks, such as the crisis of global economic systems or the emergence of totalitarian superpowers, are an inevitable part of the risk society thesis (Giddens, 2010: 14–15).

2.3 Zygmunt Bauman: Uncertainty After Modernism

There is no direct emphasis on the concept of risk in Bauman's work. Rather, his discussions that aim to understand the age we are living in determine the framework of the concept of risk. He questions the factual problematization of modernity by defining the factual limits of the dominant tendencies of the age we are living in.

Bauman determines his approach to the phenomenon of risk through the concept of "ambivalence (uncertainty)." According to Bauman, the serious problem we are facing today is based on the fact that anyone can be redundant or replace anyone else in the new period, so that everyone is vulnerable to harm, and that no matter how elevated and powerful it may seem at the moment, any social position has the possibility to be precarious in the long run. Therefore, even the privileges in the new period are fragile and under threat. With such a definition, it tends to affect all kinds of social positions, regardless of what reference it is based on, whether the underdevelopment or the accumulation that arises from the discourses of modernization. In this context, the main characteristic of risk analysis to be carried out on modernity is shaped by the new trend of instability (Bauman, 2000: 181).

For Bauman, understanding uncertainty is closely tied to the concept of fear. Every historical period has its own fears that set it apart from others (Bauman, 2001: 181). Fears that feed the new sense of uncertainty and the positioning, regulating, and conforming spirit of modernity take over the reproduction of new social life from the collective social tools. This new production is largely privatized. However, the privatization in this case refers to not only the production of responsibility by stronger social powers and leaving social integration and systemic reproduction into the hands of private investments but also independent enterprises of the individual's self-investment by deinstitutionalizing. There is no longer a state-controlled and institutional service for those seeking escape from the uncertainty, ambiguity, and precariousness of existence (Bauman, 2001: 152). The interpretation of the individual responses to specialized fears leads not only to "the uncertainty of the defeat and victory in the never-ending struggle for self-creation with the way of thinking caused by the irrevocable criticism of the modernity process returns to individuals as the pain of inadequacy. And not being able to stand where people have chosen..." (Bauman, 2001: 153), but also the insecurity created by the postmodern sense of inability also lays the rich foundations of the new risk perception (Yılmaz, 2004: 37).

Unlike other theorists who use the concept of risk as a transformative subject in their conceptualization of risk society, Bauman considers risk as the result of an attack on the basis of epistemological premises of modernity rather than a conceptual focus. According to Bauman, beyond being meaningful and stimulating, "risk society" discourse is important as it makes it possible to discuss modernity. A discussion on modernity should be about ambivalence. According to him, ambivalence is a waste of modernity. At the same time, this waste also includes a reassessment of the problems caused out of striving to produce order and today's understanding of modernity (Bauman, 2003: 27).

According to Bauman, the effort of the modern era is based on building order, destroying and preaching to rebuild it. The strong actor of this period, the state, represented the project and regulating organizations of this rebuilding effort. These conditions are not as clear-cut in today's world in a way they can easily apply to everywhere. This new condition defines the framework of a life that is called "late modernity" by Giddens, "reflexive modernity" by Beck, "beyond modernity" by Blandier and "postmodern" by Bauman (2013: 31–32).

When Bauman notes that today's dominant feeling is "'uncertainty," he is pointing out that that this feeling is not related to the person's skill and luck but the future state of the world and the criteria of life suitable for this state. The uncertainty of the postmodern world is that it is not understood as a problem to be reduced or completely overcome by new efforts. The postmodern world is preparing itself for tomorrow under the condition of constant and irreducible uncertainty (Bauman, 2013: 35).

2.4 Mary Douglas: Cultural Construction of Risk

According to Douglas, although risk constantly exists as something belonging to the outside world, it is today to define it as a moral hazard. Douglas interprets risk not only as something of the modern world but as a true concept of all time. According to this interpretation, risk is rather defined within the context of hazard. Douglas does not use the concept of risk as a policy development tool, as in Beck and Giddens (Douglas, 1992: 38).

Douglas's definition of risk is significant as besides its focus on real-objective risks, it centers on risk anxiety and perception. Douglas highlights three general areas where risk anxiety exists. These are sociopolitical risks (Risks threatening the social structure, increase in crime rates, violence, ...), economic risks (threats to the economy or economic collapse risks), and natural risks (ecological threats to nature and human health, caused by technological risks) (Lash, 2000: 50).

People come up with numerous arguments in order to make their social relations determined and predictable against the atmosphere of uncertainty and anxiety. The main arguments are based on the tension between purity and danger. The sociocultural evaluation of the risk is built on concepts such as body, purity, and cleaning habits. Culture and its codes are used as a reference source for objects that individuals encounter through these concepts. Events that are evaluated by reference to cultural codes are seen as dangerous or risky. This evaluation should not disregard the phenomenon of relativism without making the mistake of thinking about culture as a whole. Hence, different groups in a culture can consider different actions to be risky (Douglas, 1992: 28–29).

Douglas argues that risk is also related to the perception of contamination caused by violations of the physical or cultural boundaries of the class or group. According to this, no object is essentially dirty. The location of the object also determines the framework in which it will be evaluated by the audience. For instance, while the shoe is not considered to be dirty when it is on its own, its presence on a dining table is what makes it considered dirty (Douglas, 1989: 2–3).

With an anthropological content, the tension between internal and external is used to coin definitions of risk. Accordingly, bodily control is also the determinant of social control. Thus, the development that caused the widespread use of the concept of risk is the discussions of globalization and the approaches of the world system. The act of defining a risk perception by an individual due to environmental pollution causes tension between the body and the outside (Douglas, 1992: 38–40). The uncovered tension, as a political tool, forms a strategy developed by Western societies about the danger and the other. The framework of the boundaries between the self and the other, which has been developed for the society, social groups and institutions can be interpreted as an outcome of using risk as a key concept.

Douglas specifically questions why individuals take actions that are labeled as risky. Here, the important point is to meaningfully define the behavioral choices that are not based on expert opinions. The unfolding of these actions can be understood through an in-depth analysis of traditions, expectations, and cultural categories rather than directions shaped by hereditary and spiritual models. Here, culture acts as the meaning of risk and the source of risky behavior retrospectively. Culture not only identifies risk but also serves as a reference source of experiences of living together as a whole. The richness of cultural relativism related to risk takes shape by different groups, values, inequalities, and ethical differentiation within the same culture (Douglas, 1992: 57–58).

Douglas has a different attitude compared to structuralists due to her emphasis on the importance of culture. The dangers were a source of fear in both the modern and premodern era. This discourse can be considered as a result of politicization rather than the actuality of dangers. This point is not emphasized sufficiently in other approaches. Cultural values and the meaning of integrity of the society present some facts to the members of the society by defining them as risks. In other words, Douglas defines risk as a tool of social structuring, rather than directly pointing to real dangers, as an implicit process which has been dressed in socio-cultural clothes (Yılmaz, 2004: 46).

All systems of cultural classification have vague deviations that may be placed in more than one category. Uncertainties and deviations may be understood as threatening, concerning facts that one should be aware of. Douglas emphasizes that discussions of globalization have an important place for the proliferation of risk idea. Thus, people have begun to realize that they are part of the world system. Depending upon this process, it is a fact that the state has to create new areas of protection (Douglas, 1992: 25–27).

2.5 Niklas Luhmann: An Age of Unmasked Anxiety

Luhmann is among the theorists who discuss social systems and function. He argues that social systems are present "apart from an environment ... anywhere people's actions are meaningful and interrelated." Thus, wherever there is any interaction between individuals, the social system emerges. According to Luhmann, there are three types of social systems: interaction systems (people's face-to-face interactions), organizational systems (where certain conditions are required for membership), and systems that concern large communities (all-inclusive social system, all societies) (Luhmann, 1982; as cited in Wallace & Wolf, 2015: 101).

He emphasizes the concept of "self-reference" when discussing social systems. According to Luhmann, self-reference is imperative for systems to adequately function. This means that the system is able to make observations on itself, think about itself and what it does, and make decisions accordingly. These systems (self-referencing) have the ability to "describe their own identity." They are able to identify themselves by drawing the boundaries on what they are and what they are not; in other words, the system has "structural autonomy" (Luhmann, 1982: 258).

According to Luhmann's perspective, self-reference applies to all subsystems such as politics, science, economics, family, education, and law. While stating that "scientific sub-system is thinking about itself in its fundamental theories and decisions whether to continue on the path of the traditions that history ascribed to it," he gives an example of a system's self-reference. "The systems that refer to themselves are not just those that organize or regulate themselves ... They exist as closed networks that reproduce themselves as a network by continuing to produce the units necessary to continue producing items" (Luhmann, 1985: 6). Claiming that a system self-references means assuming that the system has the ability to make decisions.

However, Luhmann is in the opinion that human or concrete social clusters should not be the center of social thought. According to Luhmann, social clusters are too complex to be handled in this way. They should not be thought of as consisting of people but rather as communication units. Thus, individuals are only a part of the social system environment (Luhmann, 1982: 232).

Luhmann argues that the fundamental problems of such a contradictory world can be solved or can be turned into smaller problems by religion or by some functional equivalents of religion such as art, love, absolute power, and money in the modern society (Luhmann, 1985: 39). The common aspects of these options are that they will, at least, provide some actors with standards of action that are accepted with faith and shared. These are complex interaction teams that will enable one to live in an otherwise chaotic and incomprehensible world.

Luhmann is essentially not optimistic about the future. According to him, the modern world is too complex to allow norms to be shared or even to generalize values. Luhmann states that what unites us is the fact that "systematic structural uncertainties have been accepted collectively." In his later works, Luhmann points to the negative aspects of modernity (broken functions). He sees society confronting the consequences of its own structural choices, such as ecological problems caused by its "rationality." Luhmann also points to increased awareness and concerns and the struggle to maintain the level of social welfare in the face of global dangers fed by modern environmental science issues. Luhmann defines it as "the era of unmasked anxiety" (Luhmann, 1988: 36).

In his book *Risk: A Sociological Theory*, Luhmann defines risk as the latent threat to an individual, based on a decision made by the said individual. This is the calculations about latent losses and gains over a period of time, such as whether to get rid of nuclear material or where to dispose of waste or whether to smoke a cigarette. In this analysis, Luhmann distinguishes the risk and hazard and identifies the latter as a latent loss that the individual has been passively exposed to that was not caused by decisions, such as a tornado, an earthquake, or a hurricane. Here, the hazard is "possible damage from external factors and attributed to the environment" (Luhmann, 1993: 11).

Luhmann also states that an individual's decision (risk) poses a danger to another individual (such as the effect of the smoker on the nonsmoker). Hence, the crucial difference between decision makers and those who are affected by this decision is that what is risk for one (the decision maker) poses a danger to the other (affected) (Wallace & Wolf, 2015: 104).

Returning to his thoughts on the communication units, Luhmann has focused on the question of how communication should seek to raise the level of risk awareness. He drew attention to the warnings against risks in product advertising and put "The multifarious efforts to influence sexual behavior in the face of the AIDS risk" under this heading (Luhmann, 1993: 4–5).

According to Luhmann, while people in primitive societies were actually under threat from many dangers, technological society makes decisions such as nuclear power generation that will affect the environment deeply. In the eyes of Luhmann, as we depend on decision makers in modern society, we are essentially under the threat of risks; thus, modern society is exposed to risk. Luhmann states, "Modern risk-orientated society is a product not only of the perception of the consequences of technological achievement. Its seed is contained in the expansion of research possibilities and in the knowledge itself." The carcinogenic effects of X-ray that became evident after the introduction of new technology are one of the consequences of the technological success that Luhmann mentioned (Luhmann, 1993: 28).

3 The Phenomenon of Violence

As the scope and effects of violence were analyzed above, its multidimensionality, significance as a social problem, and its destructive impact on all spheres of social relations such as the relations between individuals and institutions, etc. were pointed out. In addition to this, it should be noted that violence is observed both in social life and in intersocietal relations under many names such as war, torture, suicide, murder, and abuse. It can be argued that there is a common consensus that these forms of violence or cases of violence that mentioned names indicated have been experienced more or less frequently in almost every period of history. Therefore, violence, "as a phenomenon that we may encounter both individually and socially," appears in every society constantly but in different degrees in daily life (Kocacık, 2001: 1).

The concept of "violence" in Turkish is of Arabic origin and is used in the literature as the equivalent of the English word "violence." The origin of the word "violence" is based on the Latin word "violentia" and is derived from it. Vioelentia means tough, merciless, and strong personality in Turkish. The word "violare," which also derives from Latin, describes the act of using violence (Michaud, 1991: 7). In the dictionary of the Turkish Language Association, violence is defined as "the degree of a movement, a strength, intensity, hardness", "speed", "power from a movement," "using brute force to those who are in opposition," "brute force," and "excess in emotion or behavior" (TDK).

The word violence is one of the words used in the social sciences literature by many scientists, which is often used closely in terms of meaning, but has also many definitions. For example, one of these definitions is "Violence is an act of punishment through psychic and / or physical damage to a person or persons" (Çelebi, 2010: 12). Another example is the definition of violence by the World

Health Organization (WHO). According to this definition, violence is defined as "the intentional use of physical force or power, threatened or actual, against oneself, another person, or against a group or community, that either results in or has a high likelihood of resulting in injury, death, and psychological harm in the person exposed to violence" (WHO, 2019). As these two examples are compared, it can be seen that they overlap regarding their similarity in meaning, but they differ in terms of their inclusiveness in terms of the areas where violence may emerge. The sociological meaning of the word violence is also quite extensive. In this regard, violence, in general, whether face-to-face or through indirect ways/tools, can be defined as "doing something to someone or making someone do something against their will by using force or pressure, act of violence, the natural tendency to express the feelings in an ill-mannered way, the irresistible power of something, the rudimentary structure of the action, bullying" (Kesici, 2007: 128). In the world report on "Violence and Health," violence is also defined as "the intentional use of physical force or power, threatened or actual, against oneself, another person, or against a group or community, that either results in or has a high likelihood of resulting in injury, death, psychological harm, maldevelopment or deprivation" (Krug et al., 2002: 5).

Considering the source of violence, there are approaches arguing that violence is a sociological problem and it is the result of environmental factors as well as approaches relating it to individual factors. Family, school, social institutions, economic groups, etc. are the social factors that affect the emergence of violence (Tezcan, 1996: 107). On the other hand, in his study "Where Does Violence Begin?" (1996: 24), Mosses states that institutions such as church and the army can put pressure on the individuals who will use violence by affecting the personal structures and can also justify violence. Moses exemplifies this condition with the ability of a person to perform homicide, which would not be welcome in social life, with the title of a soldier. Ergil (2011: 40–41) evaluates violence from another perspective and he states that the reason for violence is motivation. Motivations such as social pressures on individuals in social life, expectations about life standards, honor killings, and blood feuds can also cause violence. At the same time, he has argued that violence can be considered as a concept that is caused by conflicting social interests which lead individuals to resort to violence due to psychological feelings of emptiness, devaluation, and anger. Likewise, Polat (2016) stated that as a concept that is considered as one of the two fundamental impulses of the human structure along with sexuality regarding human psychology, violence is an instinctive concept and strongly related with the relationship between mother, father, child, and family.

Violence tendency is biologically related with the individual's survival instinct. Furthermore, the violence is also known to occur with the low secretion of the hormone of happiness. Besides hormonal and biological factors, psychobiological and neurological processes play an important role in the emergence of violence (Kılınç & Tuncer, 2013: 4).

Along with these definitions, legal definitions on violence also comes into prominence. In this context, apart from the general definitions of violence, the most common analysis of violence is the definition of the concept from a legal point of view (Kocacık, 2001: 2). The most essential reason for the existence of the modern state is described as protecting the state against violence. This violence in the legal sense appears in two ways: the protection of the state from the violence between individuals and protection against the violence that might originate from other states (Boyar, 2016: 55). From the perspective of legal experts, violence is defined as an act that includes aggression that may harm more or less other people in the social life as the person himself (Polat, 2016: 16).

Violence, which has manifested itself in different forms as a social problem since the beginning of human history, has been redefined and reproduced constantly as a result of historical processes and the development of humanity. As violence becomes widespread, it becomes difficult to understand and classify it. In the process where the form of violence is changed, reproduced, and reshaped with its physical, psychological, sexual, and verbal components, the traditional definitions of violence, which treat violence as an attack against bodily integrity, are insufficient for the current conditions and do not fully meet the connotations of the word (Benjamin, 2010: 20–21). Describing violence through traditional definitions paints a picture of violence limited to physical actions, the results of which can be seen in legal regulations, such as killing, injury, beating, and rape. However, the concept of violence should be examined by considering changing social conditions, power relations, and hegemony.

One of the important factors in defining violence as a complex phenomenon recently can be claimed to be the expansion of its content. Due to this expansion, sensitivity to the subject has increased. Now, the prejudice that the visible damage of physical violence is heavier than emotional-verbal violence has left its place to the thought that hurt feelings sometimes lead to more severe and irreparable consequences. As a result of this condition, violence, which has been addressed in many aspects of life from economic and social violence to workplace violence, have created new concepts such as school violence, youth violence, violence against women, and digital violence (Özgür Sayar, 2006: 114).

More than one million people die each year because of self-directed violence, interpersonal violence, or societal (collective) violence, and many more face non-lethal violence practices. When the statistical data are analyzed, it can be seen that in parallel with the population growth, there has been a higher increase in the violence rates than the rate of population growth in some years. In addition to the numerical increase in the rates of violence, the content of violence has been expanding and the age of resorting to violence decreases (Kocadaş et al., 2010: 16).

There are various different categorizations regarding the types of violence. The major reason for the diversity in these categorizations, as we have already observed while discussing the definitions of violence, is the complex and comprehensive characteristics of the phenomenon of violence, which has many different types.

We can categorize the types of violence, which will be explained by considering its form and content below, under the headings of physical, psychological (emotional-verbal), sexual, economic, political, and societal (collective) violence. As a result of the recent developments, digital violence (cyberbullying) has been added to the literature. Besides, it has been observed in some sources that violence is also categorized based on the offender–victim relationship (self-directed violence, interpersonal violence, and collective violence) (Violence Prevention Alliance [VPA], 2018) and the purpose of violence. It is possible to encounter one or more of these types of violence in any dimension of our daily life, including living spaces such as home, work, school, or street.

Physical violence is the form that is based on power and often leaves heavy and visible marks on the opposite side (victim) and even comes to mind when violence is mentioned. The emergence of visible results such as scratch, break, bruising, fracturing, burning, demolition, etc. explains the measurability of physical violence. "Power" is at the core of physical violence, and we observe this type of violence as power transforms into action. Power is generally used for the purposes of domination, suppression, and establishing superiority (Neugebauer, 2000:35).

Psychological violence is the use of voluntary force in various ways and methods against another person or group that negatively affects their physical, mental, spiritual, moral, or social development. Verbal abuse includes bullying, harassment, and threats. Psychological violence, which is also called "verbal violence" or "emotional violence" in the literature, manifests in the form of very heavy insults and words uttered periodically in order to shake the person's confidence in the subjects valued by the person and to hurt the person (Gökkaya, 2009: 173).

Sexual violence is defined as encountering sexual unsolicited statements, suggestion, action, or attack by the person or persons that the victim knows or does not know, in any environment (street, house, workplace, etc.). Sexual violence also includes various behaviors such as rape, forced prostitution, sexual harassment, circumcision, incest, catcalling, inappropriate gestures, sexual objectification, and excessive jealousy (Yıldız et al., 2011: 3).

According to the WHO (2002), deprivation/neglect-related violence practices can target psychological or physiological needs, as well as taking the form of economic violence, which includes depriving the person of the economic opportunities necessary for survival. Economic violence, especially in patriarchal societies, emerges as striving to have power and control over women encouraged by unequal power relations between women and the roles shaped by social life after raising children in certain ways (Gökkaya, 2009: 129).

In general, the victims of all these forms of violence are the elderly, women, and children, who are the vulnerable individuals of social life. It is obvious that the classification of violence under four categories (physical, psychological, sexual, and economic) is not sufficient to explain the phenomenon in a sociological sense. A complete understanding of the phenomenon of violence can only be possible by revealing its political and social dimensions. This is because violence based on political, collective, religious, ethnic or social classes, and war, at the top, severely destroys social tissues, affects large masses, and results in socioeconomic outcomes that leave scars in the societies of the world.

The WHO has addressed the social aspect of violence under the headings of both interpersonal violence (violence among youth, incidents in schools, workplaces, prisons, nursing homes, etc.) and collective violence. Collective violence affects large groups and is examined and analyzed under three categories as social, economic, and political violence (Yıldız et al., 2011: 1–3).

While collective violence takes place in an organized and orderly manner, there may also be unplanned, spontaneous forms of collective violence; for instance, a protest action may turn into destroying, burning, and plundering. While violence is sometimes a means for a specific purpose, sometimes it is an action or a series of actions that occur randomly and do not have a purpose beyond providing individual satisfaction to its perpetrators. According to the classification adopted by the International Police Organization (Interpol), collective violence includes the relationship between citizens and power (state, power, and authority). "Violence perpetrated by the individual or the group may be directed against someone else or the opposing group; in terms of targets and victims, it can be directed against the state as well as being directed by the state against its citizens." The presence of collective violence, in which the actor assumes a collective identity rather than merely being an individual, is known in history and remains an important issue of today's societies (Ünsal, 2005: 32).

According to Boston University's definition, although interpersonal violence is a form of physical violence, it can also occur emotionally in the form of threats, use, intimidation, bullying, and oppressing and controlling any person. The target of interpersonal violence may belong to any group of society, such as women, men, lesbian-gay-bisexual-transsexual-queer (LGBTQ) communities, individuals with disabilities, and the elderly (www.bu.edu).

In the type of violence that is directed at the self, the person who inflicts the violence and the one who is exposed to violence are the same individual. Intentional injuring of oneself or the attempt to do so manifests as self-harm or suicide (VPA, 2018). Self-directed violence, despite the victim and the offender being the same person, can affect larger groups such as the family and friends of the victim who passed away or who was injured due to the tragic consequences (Crosby et al., 2011: 418).

After all of these explanations, we can define the phenomenon of violence as an act of harming the physical and psychological integrity of the person, depending on his/her feelings of grudge and hate, in various ways other than his/her demands by applying force and pressure to himself or the people s/he is in contact with or the society s/he lives in. These actions take place in a wide range from coercion, assault, oppression, torture, hitting, injury, and murder to socioeconomic problems.

3.1 Theoretical Explanations About the Phenomenon of Violence

Since violent behaviors involve complex processes, this phenomenon is examined through different perspectives by various science disciplines. For instance, different disciplines such as sociology, psychology, biology, psychiatry, etc. examine the causes and consequences of the violence phenomenon from different perspectives.

The biological approach explains the causes of violent or aggressive behavior mostly in relation to the neurological structure of the brain and hormonal balance. On the other hand, the psychological and psychiatric approach explains the phenomenon of violence in relation to the mental process of the individual, personality traits such as psychopathological syndromes, being instinctual, egocentrism, hyperactivity, intelligence level, mental disorders, personality disorders, and irritability. The sociological approach analyzes the act of violence on the basis of values, social and institutional structure, socialization process, and interpersonal relationships. When the criminology literature is investigated, it can be observed that violent behavior is analyzed in relation to various variables such as family, poverty, unemployment, inequality, culture or the subculture of violence, socio-economic status, drug and alcohol use, social capital, and mass media (Kızmaz, 2006: 248).

According to the instinct theory that was pioneered by Freud, aggression is another instinct that needs to be satisfied similar to hunger and sexuality. According to Freud, humans cannot be purified from these emotions, but it must be ensured that they use this emotion in nondestructive ways. Thus, this approach anticipates that aggression may be transformed into acceptable models of behavior within socialization processes such as sporting activities, all kinds of competitions, professions that involve physical strength or surgery, etc.. The aim is to drain or elevate or express the energy caused by the aggression drive by channeling it towards less harmful targets in transformed ways such as creating a backlash (Okumuş, 2014: 11–12).

Social learning theory states that aggression occurs because of environmental factors. It draws attention to the fact that human behaviors are learned and focuses on how human behaviors can change with events and experiences. In other words, social learning theory argues that behaviors are shaped by experiences in addition to being affected by observations (Kırhan, 2007: 36). In this process where the learning process develops successfully and quickly when the individuals follow the examples suitable to their own social lives (Bandura, 1971: 3), behaviors are learned by finding social clues, and this approach becomes an indispensable part of learning. Social learning theory argues that the individual's emotions, thoughts, and subconscious accumulations that were acquired during the childhood, and consequent impulses, lead the individual to violence. Especially children raised in abusive households tend to observe and learn that people with more power in the relationships can use aggression to control less powerful people. It was found that males who have witnessed the father resorting to violence or who have experienced harmful behavior in childhood were drawn to violence (Okumuş, 2014: 11–12).

According to Goleman, it is possible for people who experienced domestic violence in their childhood to turn their family life into a home of aggression in their adult lives. These troubled individuals repeat the arbitrary and merciless discipline used by their parents in their childhood when they become parents themselves. It does not matter whether the mother or the father was the extremely aggressive one during childhood. Girls and boys who were angry when they were young turn into mothers and fathers who act in an arbitrary and harsh manner. At the same time, this model of aggression introduced by the parents is carried by their children to school and playground, and then other facets of life. Such parents are not necessarily mean-spirited people or not wanting the well-being of their children. They merely use the model they learned from their parents on their children (Goleman, 1998: 84).

Siegel (1989: 251) explains violent behavior with social factors such as family breakdown, relative poverty, being involved with deviant peers, academic failure, and weakened adherence to traditional groups and norms. Croall (1998: 190) argues that violence is also a behavior associated with low self-confidence in a low-class cultural structure. Moreover, he points out that conflicts and tensions that develop in the family due to unemployment and poverty can trigger violent acts. In his study in which he explains the relationship between socioeconomic disadvantage and violent behavior, Markowitz (2003) also states that individuals from low socioeconomic backgrounds are more likely to use physical violence as a form of punishment towards their spouses and children. Blau and Blau (1982) also argue that economic inequality or income inequality is the reason for violent behavior rather than lower class values. This is because according to these theories, income inequality leads to feelings of injustice, frustration, and anger in individuals. As opposed to the theorists who explain the phenomenon of violence in relation to inequality, poverty, and unemployment, some researchers argue that socioeconomic factors cannot be separately discussed from the cultural factors while explaining the violence behavior. Cohen, who conducted the first research on gangs, and some recent studies on young subcultures associate socioeconomic and cultural causes with each other. Heimer, for instance, who states that the socioeconomic backgrounds of families have an impact on individuals' inclination to violence, argues that violent behavior develops during the learning process. According to him, parents generally have an oppressive authority in families with a low socioeconomic background, which is one of the causes of children's tendency to violence. To sum up, Heimer argues that the phenomenon of violence cannot be discussed separately from the social economic status, cultural issues, or cultural contexts in which these definitions are acquired (Kızmaz, 2006: 259).

Cultural theories understand violent behavior as a consequence of prevalent values/behaviors learned and shared in society. This approach helps us understand how the cultural structure in some societies encourages acts of violence. On the other hand, structural theories explain violence through hindered positions and relationships caused by poverty, unemployment, and lack of opportunities. In a sense, it is assumed by structural theories that violent behavior stems from the hindered/blocked state of the individual in terms of his social position. From this point of view, being hindered leads the individual to unlawful acts. The interactive theory also understands the phenomenon of violence as the consequence of the dynamic relationships between individuals and their social status. The interaction theory is based on the premise that the nature of individuals' social status and the attributes of the social and physical environment have an important impact on the direction and consequence of social interaction. In this theory, social and material factors such as family and peer relationships, presence of alcohol and drugs, and easy access to weapons are considered to be important criteria that impact/determine the tendency to violence. To sum up, the cultural factors approach aims to connect the weakness of ethical awareness and family values with the phenomenon of violence, while the structural approach explains the phenomenon with economic, educational, and other socioeconomic disadvantages that impact the individual negatively. The interactive approach also explains it with the individuals' constant exposure to negative experiences and factors in family, school, and community settings (Durant, 1999: 270).

Debarbieux emphasizes the fact that violence is a relative phenomenon. Such that it may be a phenomenon that changes according to a certain period, a social environment, distinct conditions in addition to legal, and political codes in different periods and places where it makes sense. Even the individual's fatigue level at different times of the day is an effective factor on this relativity (Nair, 2015: 11).

When the advancing technology is considered, it becomes imperative in the twenty-first century, when human history is most susceptible to interaction, to think beyond defining violence only through the harm that was caused. Galtung (1969) defines violence as the cause of the difference between what could have been and what is, the factor that increases the distance between the potential and the actual, and that which stops the distance from decreasing. He argues that if a society has the knowledge and technology to eliminate the deaths caused by tuberculosis and yet there are still people who die from tuberculosis, structural violence is present in that society. He states that that violence is impeding humans from meeting their needs in an avoidable way, to put it more broadly, damaging human life through a lesser realization of the ability to meet human needs than what is possible. According to Galtung, the risk of being subjected to violence is violence. If we consider the present day through this definition, violence is also present in the structure, which causes people to feel unsafe in a time when private space is becoming increasingly blurred, and even that chooses not to provide this sense of security when it is possible to do so because of certain power relations. Not removing the solvable and/or avoidable situations on purpose and the existence of a group of people who are forced to live with the constant risk of being subjected to violence are also a part of violent behavior (Memiş, 2015: 179–182).

Another component that needs to be examined while analyzing the sociocultural reasons of acts of violence is the patriarchal structure or perception of masculinity. Feminist theorists see patriarchal or masculine structure as an important source of violence, suggesting that male violence is constructed structurally and culturally, and its source lies in the form of male image creation (Messerschmidt, 1999; Kersten, 1993; Anderson & Umberson, 2001; Totten, 2003; as cited in Memiş, 2015: 185–190).

Dissimilar to these, Bourdieu applies the concept of "symbolic violence" while explaining the phenomenon of violence. Symbolic violence refers to a new type that has evolved in line with the new forms of bullying and oppression used by the power. This new form, where violence is rendered imperceptible with gentleness, makes the domain of power attractive and renders oppression and bullying invisible. The concept of symbolic violence is significant in terms of giving nonphysical violence an actuality (Bourdieu & Passeron, 2019).

The only force that can answer questions regarding "symbolic violence," which can also be called invisible violence, that is, what is violence or what is legal/illegal, is the power that defines it. Written rules have a significant role in terms of preventing violence to occur uncontrollably. This approach, which once more confirms the fact that violence exists in any case, authorizes the state to keep it under control. The possibility of war and thus the authority to make restrictions on humans' lives is thus completely given to the state. In this sense, the state has the freedom to exercise this power, but it predominantly manages to rule over the majority without needing to use violence. To comment on Bourdieu's theoretical framework, the state prefers to turn its citizens compliant rather than making them revolt against the state because of violence. Thus, the state demonstrates its power "in a certain order and sameness and in accordance with the known rules" (Özsöz, 2009: 31).

3.2 Digital Violence

After the general explanation about the phenomenon of violence, we can return to the main objective of this study. The main purpose of this study is to describe the digital violence phenomenon that we face as a new form of violence, the foundations of which are laid by today's risk society and the new conditions and opportunities.

Aside from the fast access to information and many other benefits provided to the users by the Internet, the digital world has a dark side that leads to consequences that can be considered within the scope of the concept of violence (Campbell, 2005: 2). The fact that information shared on digital communication platforms, which are also called new media, is easy to hide and is quick to spread, has led to the emergence of digital violence (Turan & Dilmen, 2016: 26). This type of violence, which is called "cyberbullying" in English, in other words, digital violence, first emerged with the increasing use of networked computers and mobile phones among young people (Smith et al., 2008: 376).

Monitoring an individual on a digital network, requesting evidence of where the person is or what the person is doing, obtaining a secret record or spreading a record obtained with permission, or using these records to threat, blackmailing with personal information obtained through these media or exploitation, controlling and inquiring private accounts or threatening to do so, and insulting, humiliating and hateful actions towards an individual on the Internet are digital violence behaviors (http://erktolia.org/).

It can be said that digital violence, which is defined with different concepts such as cyberharassment, cyberbullying, and online bullying in the foreign literature, has all the characteristics and outcomes of the traditional form of violence other than physical harm and injury. Considering that all types of violence involve deliberate and willful harm, it can be said that the components of digital violence are the same (Mishna et al., 2012: 64).

Both traditional violence and digital violence are concerned with relationships, power, and control (Privitera & Campbell, 2009). Overall, digital violence can be explained as harming another person who is vulnerable through the use of electronic media (social networking sites, e-mail, chat, SMS, MMS, etc.) (Palermiti et al., 2017). According to definitions by Privitera and Campbell (2009) and Piotrowski (2012), digital violence refers to inappropriate, unsolicited, negative, and aggressive behaviors carried out by the perpetrator against a target(s) and by using online or wireless information and communication technologies (D'Cruz & Noronha, 2013). Digital violence may be inflicted by any means used to communicate electronically (Watts et al., 2017).

The main distinction of digital violence from traditional violence is that it involves technology. In addition, while explaining the differences between traditional violence and digital violence, Lawrence (2016) stated that traditional violence takes place in a face-to-face situation, it is possible to find a safe place or escape, audience is limited, the perpetrator can be identified, and it possible to receive gestures or bodily reactions from the victim or the witnesses. On the other hand, digital violence can be encountered any day and any time, there is no safe place and it is difficult to escape, it is shared by a wide audience, it can virally spread in seconds, the perpetrator may remain unknown, it is difficult to understand the feelings of the victim, there is no geographical limit, and the victim can easily become the perpetrator (Kanbur & Kanbur, 2018: 243). It should also be added that there is no time limit on the effects of digital violence in terms of how long they last. The ability to preserve the formations and interactions in the digital environment also makes it possible to use them over and over again.

In the report prepared for the Internet Governance Forum (IGF, 2015) addressing the topic of "Gender Based Violence and Online Abuse" by the United Nations that aimed to understand the scope of digital violence, behaviors of gender-based digital/cyber/online violence are listed as follows: " infringements of privacy," "surveillance and monitoring," "damaging reputation or credibility," "harassment," "direct threats or violence," and " blackmail and attacks against communities." These behaviors include activities such as accessing private data without an individual's permission (hijacking personal accounts, stealing passwords, using/stealing identities, accessing another user's computer when logging into a user's accounts, etc.), monitoring or surveillance of online and offline activities, use of spyware or keyboard recorders without the user's consent use, deleting or sending emails or content without consent, creating and sharing false personal data (such as online accounts, advertisements, or social media accounts) in order to harm reputation, fake photo or video editing, spamming messages, "digital violence" of unsolicited messages or contact, repeated harassment, threats of sexual or physical violence, abusive comments, unsolicited sending or receiving of sexually explicit material, encouraging physical violence, hate speech messages or e-mails targeting gender or sexual identity, money or property theft, hijacking websites, social media accounts and/or e-mail accounts of certain organizations and communities, threats of direct violence to community members, shelter addresses, etc., and disclosure of confidential information.

Although these behaviors are discussed in relation to gender and women in the report, they are digital violence behaviors that all members of the society are/can be exposed to or can/will exhibit regardless of demographic characteristics.

Digital violence differs from traditional violence in the following ways: there is a high rate of anonymity and a lack of supervision, increasing number of victims, high rate of reflecting on the violence and maximizing the harm, and a potentially wider audience to witness the violence (Boulton et al., 2014: 145). Another crucial point is that digital violence is a widespread problem among the youth (Peker, 2012). Studies show that digital violence among the youth is a serious public health problem. Especially in the last decade, research on the prevention of this process, in which young people can be both the aggressor and the victim, has entered the academic literature (Modecki et al., 2014). Additionally, studies about peer bullying also lead the studies on digital violence behavior, which is more common among youth. The findings of the study conducted by Ekşi and Yaman (2010) on adolescent violence, for instance, have significance in this sense. Ekşi and Yaman made observations based on the survey data of 3,468 people who were receiving education and 891 who were not between the ages of 11 and 18 about the tendency towards violence and its causes. Findings about why young people think violence occurs are significant. According to young people, causes of violence include problems about girlfriend/boyfriend, economic reasons, a lack of purpose in life, using violence for pleasure/to blow off steam, influence of friends, seeking to prove oneself or get revenge, discipline problems in school, and lack of legal punishment. It can be said that many of the reasons given for violence are also causes of digital violence.

Additionally, in various studies on digital violence (Bayar & Uçanok, 2012; Eroğlu & Peker, 2015), referring to social learning theory, it is emphasized that children determine their behaviors by imitating the behaviors of their peers during social interactions.

In order to understand the online behavior of users, firstly understanding the nature of human behavior is imperative. Every individual's behavior is unique, which makes answering questions about when and how exactly they behave challenging. However, for the sake of determining a general framework, it is possible to state that needs and motivations are the most important factors that determine human behavior. Like other living beings, humans need to meet their minimum basic needs in order to sustain their organic life as living beings themselves. Among the basic needs, in terms of our subject, the most relevant ones are defense/aggression and sexuality. As among the needs and drives that are necessary to survive, particularly sexuality and aggression have been able to find a place within computer and Internet technologies (Arıcak, 2015: 21–22). The fact that the needs and motives that are present in our lives and are often repressed are mostly uncontrollable in the Internet environment, which is perceived as a nonsanctioned area, enable individuals to see their behaviors in a way that justify them. Exposure to online content that has violence and sexuality, especially at a young age, in other words, Internet misuse, can lead to numerous negative outcomes such as Internet addiction, game addiction, and related negative user behavior, online hate speech, and digital violence.

Although it is thought that aggression emerged to protect our biological existence, fear, anger, rage, and hatred are also feelings that can lead to aggression. Although aggression exists innately, it is defined and shaped under the influence of the environment, particularly in childhood. If violence is an acceptable behavior in the environment or if the child is exposed to violence, the child is expected to exhibit aggressive behavior. Similarly, manifestation of these behaviors in real life is easier if the family is not bothered by violent videos or games due to not considering them as real or if the games are rewarded (Arıcak, 2015: 22–23). It is expected for people to consider the images they are exposed to for a long time to be legitimate after a certain period of time. Particularly with the present-day graphic designs (augmented reality glasses and virtual realities should be considered within the scope of graphic design) visually minimizing the difference between real and virtual can be expected to expedite the aforementioned process of legitimization of violence.

Binark and Bayraktutan (2013) in their study, which was conducted within the scope of the new media and ethics violation, classified ethical violations on the Internet and social media as political, economic, and cultural. Among the political reasons, there is the struggle for power/authority. In order not to lose their power/authority, they attempt to censor or control the social media according to their interests and cause ethical violations. Capitalist initiatives that monopolize the new media technologies constitute the economic causes of ethical violations in social media. Factors such as intense exposure of users to advertisements, disappearance of the separation between news and advertisements cause ethical violations. Lack of awareness and education on many issues such as violation of personal rights, privacy, etc. is also among the cultural causes of ethical violations.

The fact that digital technologies get gradually more involved with our lives brings with it an increase in the related issues. This increase naturally caused a rise in multidisciplinary curiosity on the part of scientists. In recent years, some of this curiosity has started to be concerned with digital violence. The findings of some studies that discuss the phenomenon of digital violence can be summarized as follows:

Although definitions for digital violence are similar to each other, some issues are currently being discussed by researchers regarding the definition. For instance, some definitions state that a conscious repetition is mandatory for digital violence, while others argue that a single post on social networks can also be considered as repetitive behavior, since by the nature of social networking sites, a single post is seen and shared again by hundreds of people (Yaman et al., 2011: 167). In their study on digital violence, Dredge et al. (2014) found that as a result of a semi-structured interview with 25 social network users between the ages of 15 and 24 who had negative experiences, the participants especially emphasized the negative impact of digital violence on the victim. Once again, it is important to discuss and draw attention to the experiences of the victim in terms of emphasizing the difference between the violence that takes place in the material environment and the violence in the digital environment rather than the topics that have been mentioned more frequently while defining digital violence, such as intent or power relation between the perpetrator and the victim. The high number of people who witness the victim in the digital sphere, the continuation of the victimization because of sharing, the fact that the perpetrator may remain anonymous, and the ability of the perpetrator to engage in worse behavior due to witnessing the effects of his/her actions in person cause the victims of violence in the digital sphere to suffer more severe consequences and impacts.

Staude-Müller, Hansen, and Voss (2012) emphasize that the online harassment, online mobbing, and online victimization or digital violence we are encountering after the development of the Internet are not only the problems of the youth but also threaten adults. In the survey study conducted on 9,000 people in Germany, it was found that women and girls between the ages of 10 and 50 were exposed to different forms of digital violence such as persistent follow-up and cyberharassment. In the outcomes of the study, digital violence is defined to be a continuation of the causes of conventional violence (offline violence). Gender-based violence due to online abuse can be directed at men or women. Likewise, men and children can be victims of online abuse and violence. Nevertheless, it is important to note that online abuse and gender-based violence are caused by the same existing structural inequalities and discrimination as other forms of gender-based violence (Staude-Müller et al., 2012; IGF, 2015).

In one of the comprehensive studies named "Cyberbullying Behavior Among University Students" (Bayram & Saylı, 2013) on digital violence, it was aimed to reveal the extent of digital violence behavior among university youth. The sample of the study consisted of 612 university students. In the study, it was revealed that 30.6 % of the students were exposed to digital violence at least once with the purpose of threatening or humiliation. The most prevalent method of digital violence was e-mail, followed by activities in chat rooms/message boards and comments/social networking sites. In the study, it was found that male students were victims of digital violence/bullying more than female students (Bayram & Saylı, 2013: 107–116).

There are many studies that show women are subjected to digital violence behavior more than men (for a few case studies, see: Baker, Campbell, & Barreto, 2013; Devries et al., 2013; Woodlock, 2013; Choja & Nelson, 2016; Lenhart, Ybarra, Zickuhr, & Price-Feeney, 2016; Lewis et al., 2016). Russo and Pirlott (2006) argue that gender-based violence, which is the framework of digital violence against women, is based on patriarchal social structures, takes multiple forms, and is reinforced by the cultural roles and media images of men and women. In addition, a survey of more than 9,000 German Internet users aged between 10 and 50 years showed that women were more likely to be victims of online sexual harassment and cyber surveillance than men, and the impact of this type of violence was more traumatic for victims (Staude-Müller et al., 2012). This finding has also been confirmed by research conducted by the Pew Research Center (PEW, 2014) in the United States; in that study, it was found that men were more likely to be subjected to partial "mild" forms of digital violence (such as name-calling, taunting, and shaming) whereas women (especially young women aged between 18 and 24 years) were subjected to disproportionate digital violence, name tracking, and online sexual harassment.

Although some research on the gender of digital violence victims suggests that women are more in numbers, some research states that there is no definite tendency in terms of gender. Again, in addition to the studies indicating that there is a relationship between age and victimization, there are also studies that prove that there is no such relationship (Dredge et al., 2014: 17).

Digital bullying can be defined as intentional, repetitive, and harmful behavior by individuals and groups by misusing the Internet or other communication tools (Patchin & Hinduja, 2006; Juvonen & Gross, 2008; Slonje & Smith, 2008; Willard, 2008; Li, 2007; Arıcak, 2011). Digital victimization, on the other hand, is the situation in which individuals or groups are subjected to behaviors that are technically or personally harmful and suffer economically or psychologically because of these behaviors (Smith et al., 2008; Arıcak, Kınay & Tanrıkulu, 2012). Recent research shows that there is an increase in numbers of perpetrators and victims of digital violence. In a study conducted by Calvete et al. (2010) with adolescents, it was found that 44 % of the respondents inflicted digital violence. The study of Spears et al. (2015) revealed that 27 % of the participants were victims and 18 % were both perpetrators and victims. In addition, in Serin's (2012) study, 26 % of the participants were found to be involved in digital violence. In the study by Peker (2013), it was determined that 46 % of the students were victims of digital violence at least once and 65 % inflicted digital violence at least once. Digital violence behavior and victimization are particularly more common during university years, which coincides with the period when individuals transition from adolescence to young adulthood. To illustrate, in the study conducted by Dilmaç (2009) with university students, 22.5 % of students stated that they had resorted to digital violence at least once, while 55.3 % had been exposed to digital violence at least once in their lives, and 20 % stated that they both resorted to and was exposed to digital violence.

The idea of not being able to be reached because of the opportunity to hide his/her identity (Lenhart, 2016), getting the approval of his friends, attracting their attention and earning their respect (Dilmaç, 2009; Varjas et al., 2010), having the opportunity in online platforms to express digital violence offender's ideas that cannot be expressed during face-to-face communication because of the gender, ethnicity or social status of the victim, the desire to take revenge on those with whom the offender has problems (Kowalski et al., 2012), lack of warm family environments in family relations, insecure attachment between the parents and the child, and bullying behaviors of parents (Li, 2007) are considered as the reasons for digital violence. Besides, it was found that those who were exposed to digital violence had problems in friendship and family relationships, had high levels of social anxiety, low self-esteem, depression, anger and frustration, some social problems, conflicts with their parents, and tended to display aggressive behavior and experience social problems (Ybarra et al., 2004; Nishina et al., 2005; Patchin & Hinduja, 2006; Wolak, Mitchell & Finkelhor, 2007; Campfield, 2008; Topçu et al., 2010; Yaman et al., 2012).

In Peterson and Densley (2017)'s research, characteristics of individuals who engage in cyber aggression and violence, which include both traditional criminal risk factors (i.e., externalizing traits) and potentially new risk factors (i.e., internalizing traits), were examined. In the study, internalizing traits and externalizing traits such as low self-control, impulsivity, psychopathy, and lack of empathy also were strong risk factors for individuals who engaged in traditional (i.e., in person) forms of aggression and violence. However, unlike traditional forms of aggression where prominent attackers are invisible to victims and often not anonymous, online aggression can affect individuals with different internalizing features, such as depression or shyness.

We can assume that the social and psychological factors mentioned affect the individual's being a perpetrator or victim of digital violence and that the Internet is becoming a crucial part of human life today. The integration of the Internet and our social life has reshaped the way we communicate, and (Castells, 2002; ac cited in İğrek, 2009), social lives, communication styles, social relations, and social norms have been deeply affected by this new communication technology. Social support in relation to social harmony, family relationships, relationships with the community, perception of the individual about the community they live in, positive and negative emotions, compliance with social norms, and anti-social tendencies may also affect whether one becomes a perpetrator or victim of digital violence.

Whether the victim knows the identity of the aggressor or not is one of the variables of digital violence. Wolak et al. (2007) found that the amount of harm caused by someone the victim knows is more than people with whom one communicates online only. In a study of over 1,500 young people, they also found that approximately 50 % of the known attackers sent messages that everyone could see and gossip and that this created a situation that was hard to stop. In the research, they state that the behavior of digital violence occurs more than once if the harmed person is known and the attack is usually one-time only when there is only online communication.

In the study of Yaman et al. (2011), it was stated that there is a connection between digital violence and traditional violence, in which a person who is a victim in the physical environment can turn into a person who inflicts harm by changing roles in the digital environment. A person who is weak in the physical environment can be stronger in the Internet environment and the situation may cause a change in roles. There is no need to be physically strong on the Internet. One can hide personal information on the Internet; there is no time and space limitation to harm (you can see the harm done to you while you sleep during the night only when you wake up in the morning or in a later period of time); again by hiding their own credentials, the perpetrator can make more people to witness compared to physical environment. These and similar situations distinguish digital violence from traditional violence. In addition, another important point is that the perpetrator does not fully comprehend the harm caused to the victim since s/he does not see the victim face-to-face. This may lead to further harm (Aslan & Önay Doğan, 2017: 104).

Digital violence behavior may also differ based on different social media platforms. Social media platforms contain different behavior patterns based on their design. For example, Facebook is a space where personal representation is foregrounded, people come to be seen, followed, to watch, and share. In their research conducted on 316 people aged between 15 and 24 years that examined the relationship between Facebook use and digital violence, Dredge et al. (2014) found that 51 % of those who participated in the study had been exposed to digital violence more than once in the last six months. This shows that the increase in the number of Facebook friends, similarly the frequency of profile posts and the high number of negative wall posts, increase the risk of exposure to digital violence. Based on the findings, it can be stated that the increase in watching and sharing on Facebook, that is, the intensity of use, also increases the risk of being victimized, at least for the aforementioned research.

The most important instance of digital violence today is "hate speech" on social media. Users' comments that involve swearing and insults under the posts on each other's profile demonstrate the digital violence perpetrated by social media users through hate speech (Özer, 2019: 17).

4 Conclusion

Rapid changes and advances that occurred in science and technology in the late twentieth century paved the way for the new era to be called the "information age" and the societies of this era "information society." As every era has its own unique dynamics of transformation, key dynamics of information age that shape the social, cultural, economic ,and environmental structure are accessing information, developments in technology, and the tools and resources that emerged as a result of these developments. Changes that occurred through technology and science in information societies called attention to the importance of topics such as computers, Internet, and information. Tools such as the Internet, computers, and smartphones, which have become widespread due to technology, have facilitated the lives of individuals and met many needs in the beginning. However, the circumstances turned negative in the long run both for the individuals and the societies.

Nearly infinite resources that have been provided by technology to individuals in almost all areas, from education to health, communication to economy, etc., have brought with it risk, fear, and anxiety about safety both in individual and social spheres.

Concepts such as modernization and globalization are particularly drawn upon in sociology for explaining said changes. In this study, fundamental theories that were addressed regarding social change also discuss modernity in terms of its negative aspects. Namely descriptions such as "Risk society," "manufactured uncertainties," "ambivalence," "risk culture," "age of anxiety," etc. are used to explain the changes in the social sphere along with the modernization process.

Factors such as technological developments, advancements in science, differentiation of personal, and social opportunities in fact have laid the foundations of modernity. Having said that, globalization is another crucial concept that should not be overlooked. Along with globalization, the process of modernization has changed the meanings of conventional concepts of space/time. All kinds of positive/negative relations have "intensified" and become "prevalent" in terms of resources that have been achieved with modernization.

Scientists who explain risk society essentially point out to the increase, differentiation, and prevalence of personal and social resources as well as factors such as risk, uncertainty, and fear that came with modernization and globalization. More importantly, these factors have become uncontrollable. Various disasters, outbreaks, conflicts, climate changes, etc. that occur every so often throughout the world are actually the consequences of the risks, dangers, and uncertainties embodied in the developments in modern societies.

Humanity has faced a great number of risks, dangers, and threats since the beginnings of its existence. These risks, dangers, and threats that have been affecting humanity from past to present have not disappeared in time, but their nature has changed. As emphasized earlier, while the risks in the early periods of human history were directed at basic needs and security of life, today's risks are also directed at our needs for socialization in various areas.

The aim of this study was to draw attention to the digital violence that has been threatening our safety in the digital sphere. That is to say, digital safety is among the areas where the fear, which is created by risk, hazard, and threats that have been in existence since the past, resurfaces in today's context.

The Internet, computer, and smart mobile devices, which have become important tools for our lives since the twenty-first century is the age of technology, despite providing many resources for storing, reserving, and spreading information, are also sources of risk, fear, and danger environments. Photo and video sharing sites and electronic communication tools are used increasingly more by individuals, which results in a new type of violence to increase and take its place in our lives. Particularly with the widespread use of the Internet, social media platforms YouTube, Twitter, Facebook, Myspace, Instagram, Swarm, etc. and communication platforms WhatsApp, WeChat, Skype, etc. have emerged. Because users share private and social information in such applications, such as photos, videos, addresses, personal information, etc., it enables individuals who they know or do not know to use this information maliciously. Additionally, while using social media or e-mail, individuals may also encounter dangers such as harmful applications, fake links, and connections and suffer from technical problems. Individuals whose personal accounts and information are stolen may be subjected to harassment, abuse, and digital violence.

What causes digital violence is essentially the misuse and abuse of network technology that was produced for communication or information sharing purposes. The reasons that lead individuals to commit digital violence may be personal as well as social. The fact that there are too many activities that can be carried out with information technologies and the lack of regulations and bans regarding the digital sphere and its use encourage the spread of digital violence behavior.

Digital violence includes a wide variety of behaviors such as online sexual harassment, insulting or threatening others through social media tools, sending messages with viruses or harmful materials, posting others' photos or videos online without consent, and using others' photos in social media or dating sites to chat while disguising their own identity.

The Internet, which the humanity met in the 1970s as a result of the influence of modernity and became widespread around the world in the 1990s, thanks to globalization and following the direction of economic and commercial circles, has taken its place in our lives as a force that provides science and technology, communication, knowledge, chat, and education quickly and at very low cost. And in today's world, as a result of the rapid developments in communication and information technologies and the conditions we are living in, the use of these technologies is increasing daily in most aspects of our lives. This technology, which was originally created to facilitate our lives, has also brought with it hazards such as digital violence due to misuse.

Finally, although this study tried to encapsulate "risk society" through the phenomenon of "digital violence," which is one of its novel outcomes, we should not overlook the fact that, alongside the "opportunities" we create in both our personal and social lives, there might also be "risks" due to various social changes and developments in different periods.

Bibliography

Anderson, Kristin and Umberson, Debra, (2001), "Gendering Violence: Masculinity and Power in Men's Accounts of Domestic Violence", *Gender and Society*, Vol. 15, N. 3, pp. 358–380.

Arıcak, Osman Tolga, (2011), "Siber Zorbalık: Gençlerimizi Bekleyen Yeni Tehlike", *Kariyer Penceresi*, Vol. 2, N. 6, pp. 10–12.

Arıcak, Osman Tolga, (2015), *Siber Alemin Avatar Çocukları*, İstanbul: Remzi Kitabevi.

Arıcak, Osman Tolga, Tanrıkulu, Taşkın and Kınay, Hüseyin, (2012), "Siber Mağduriyet Ölçeğinin İlk Psikometrik Bulguları", *Akdeniz Eğitim Araştırmaları Dergisi*, Vol. 9, N. 1, pp. 1–6.

Aslan, Alev and Önay Doğan, Betül, (2017), "Çevrimiçi Şiddet: Bir Siber Zorbalık Alanı Olarak "Potinss" Örneği", *Marmara İletişim Dergisi*, Vol. 27, pp. 95–119.

Bandura, Albert, (1971), *Social Learning Theory*, New York: General Learning Press.

Baron, Robert A. and Byrne, Done R. (2000), "Agression, Social Psychology", *USA: Allyn and Bacon*, Vol. 11, pp. 438–476.

Bauman, Zygmunt, (1997), *Modernite ve Holocaust*, (S. Sertabiboğlu, Trans.), İstanbul: Sarmal Yayınları.

Bauman, Zygmunt, (2000), *Siyaset Arayışı*, (T. Birkan, Trans.), İstanbul: Metis Yayınları.

Bauman, Zygmunt, (2001), *Parçalanmış Hayat-Postmodern Ahlak Denemeleri*, (İ. Türkmen, Trans.), İstanbul: Ayrıntı Yayınları.

Bauman, Zygmunt, (2003), *Modernlik ve Müphemlik*, (İ. Türkmen, Trans.), İstanbul: Ayrıntı Yayınları.

Bauman, Zygmunt, (2013), *Postmodernlik ve Hoşnutsuzlukları*, (İ. Türkmen, Trans.), İstanbul: Ayrıntı Yayınları.

Bayar, Yusuf and Uçanok, Zehra, (2012), "Okul Sosyal İklimi ile Geleneksel ve Sanal Zorbalık Arasındaki İlişkiler: Genelenmiş Akran Algısının Aracı Rolü", *Türk Psikoloji Dergisi*, Vol. 27, N. 70, pp. 101–118.

Bayram, Nuran and Saylı, Müslüm, (2013), "Üniversite Öğrencileri Arasında Siber Zorbalık", *İstanbul Üniversitesi Hukuk Fakültesi Mecmuası*, Vol. 71, N. 1, pp. 107–116.

Beck, Ulrich, (2005), *Siyasallığın İcadı*, (N. Üner, Trans.), İstanbul: İletişim Yayınları.

Beck, Ulrich, (2011), *Risk Toplumu, Başka Bir Modernliğe Doğru*, (K. Özdoğan and B. Doğan, Trans.), İstanbul: İthaki Yayınları.

Beck, Ulrich, Giddens, Anthony and Lash, Scott, (1994), *Reflexive Modernization: Politics, Traditional and Aesthetics in the Modern Social Order*, Cambridge: Polity.

Benjamin, Walter, (2010), "Şiddetin Eleştirisi Üzerine", (Aykut Çelebi, Ed.), *Şiddetin Eleştirisi Üzerine*, İstanbul: Metis Yayınları, pp. 19–43.

Binark, Mutlu and Bayraktutan Sütcü, Günseli, (2013), *Ayın Karanlık Yüzü: Yeni Medya ve Etik*, İstanbul: Kalkedon Yayınları.

Blau, Judith R. and Peter M. Blau, (1982) "The Cost of Inequality: Metropolitan Structure and Violent Crime", *American Sociological Review*, Vol. 47, pp. 114–29.

Boulton, Michael, J., Hardcastle Katryna, Down, James, Fowles, John and Simmonds, Jennifer, A., (2014), "A Comparison of Preservice Teachers' Responses to Cyber Versus Traditional Bullying Scenarios: Similarities and Differences and Implications for Practice", *Journal of Teacher Education*, Vol. 65, N. 2, pp. 145–155.

Bourdieu, Pierre and Passeron, Jean-Claude, (2019), *Yeniden Üretim: Eğitim Sistemine İlişkin Bir Teorinin İlkeleri*, (A. Sümer, L. Ünsaldı and Ö. Akkaya, Trans.), Ankara:Heretik Yayınları.

Boyar, Oya, (2016), "Anayasa Hukuku ve Siyaseti", *Marmara Üniversitesi Hukuk Fakültesi Hukuk Araştırmaları Dergisi*, Vol. 22, N. 1, pp. 53–80.

Calvete, Esther, Orue, Izaskun, Estévez, Ana, Villardón, Lourdes and Padilla, Patricia, (2010), "Cyberbullying in Adolescents: Modalities and Aggressors' Profile", *Computers in Human Behavior*, Vol. 26, N. 5, pp. 1128–1135.

Campbell, Marilyn, A., (2005), "Cyberbullying: An Old Problem in a New Guise?", *Australian Journal of Guidance and Counselling*, Vol. 15, N. 1, pp. 68–76.

Campfield, D. Carroll, (2008), *Cyberbullying and Victimization: Psychosocial Characteristics of Bullies, Victims and Bully/Victims* (Master's Thesis), University of Montana, Missoula.

Cebecioğlu, Gülçin and Altıparmak, İpek Beyza, (2017), "Dijital Şiddet: Sosyal Paylaşım Ağları Üzerine Bir Araştırma", *Sakarya UniversityJournal of Education*, Vol. 7, N. 2, pp. 423–431.

Cohen, Dov (1998), "Culture, Social organization and Patterns of Violence", *Journal of Personality and Social Psychology*, Vol. 75, No. 2, pp. 408–419.

Croall, Hazel, (1998), *Crime and Society in Britain*, London: Longman.

Crosby, E. Alex, Buckner, V. Ayanna and Taylor, D. Beverly, (2011), "Addressing Self-Directed Violence Prevention for Preventive Medicine Practitioners", *American Journal of Lifestyle Medicine*, Vol. 5, N. 5, pp. 418–427.

Çelebi, Aykut, (2010), "Sunuş: Bir parıltı ... Sonra Gece", (Z. Direk and E. Göztepe), *Şiddetin Eleştirisi Üzerine, Metis Defterleri*, İstanbul: Metis Yayınları.

D'Cruz, Premilla and Noronha, Ernesto, (2013), "Navigating the Extended Reach: Target Experiences of Cyberbullying at Work", *Information and Organization*, Vol. 23, pp. 324–343.

Dilmaç, Bülent, (2009), "Sanal Zorbalığı Yordayan Psikolojik İhtiyaçlar: Lisans Öğrencileri İçin Bir Ön Çalışma", *Kuram ve Uygulamada Eğitim Bilimleri*, Vol. 9, N. 3, pp. 1291–1325.

Doğan, Mehmet, (2011), *Büyük Türkçe Sözlük*, İstanbul: Gerçek Hayat.

Douglas, Mary, (1989), *Purity and Danger: An Analysis of the Concept of Pollution and Taboo*, New York: Ark Paperbacks.

Douglas, Mary, (1992), *Risk and Blame: Essay in Cultural Theory*, London: Routledge.

Dredge Rebecca, Gleeson, John and Garcia Xochitl de la Piedad, (2014), "Cyberbullying in Social Networking Sites: An Adolescent Victim's Perspective", *Computers in Human Behavior*, Vol. 36, pp. 13–20.

Dredge Rebecca, Gleeson, John and Garcia Xochitl de la Piedad, (2014), "Presentation on Facebook and Risk of Cyberbullying Victimisation", *Computers in Human Behavior*, Vol. 40, pp. 16–22.

Durant, J. Thomas, (1999), "Violence as A Public Health Problem: Toward an Integrated Paradigm", *Sociological Spectrum*, Vol. 19, N. 3, pp. 267–280.

Ekşi, Halil and Yaman, Erkan, (2010), *Çocuk ve Ergende Şiddet-Karma Metodolojiye Dayalı Bir Alan Çalışması*, İstanbul: Kaknüs Yayınları.

Ergil, Doğu, (2011), "Şiddetin Kültürel Kökenleri", *Bilim ve Teknik*, Vol. 399, pp. 40–41.

Eroğlu, Yüksel and Peker, Adem, (2015), "Ergenlerde Akran İlişkileri İle Siber Zorbalık Statüleri Arasındaki İlişkinin İncelenmesi", *Electronic Turkish Studies*, Vol. 10, N. 11. 593–606.

Esgin, Ali, (2008), *Anthony Giddens Sosyolojisi*, Ankara: Anı Yayıncılık.

Esgin, Ali, (2013), "İmal Edilmiş Belirsizlikler Çağının Sosyolojik Yönelimi: Ulrich Beck ve Anthony Giddens Kaynaklı Risk Toplumu Tartışmaları", *Gaziantep University Journal of Social Science*, Vol. 12, N. 3, pp. 683–696.

Giddens, Anthony, (1998), *Modernliğin Sonuçları*, (E. Kuşdil, Trans.), İstanbul: Ayrıntı Yayınları.

Giddens, Anthony, (2000), *Elimizden Kaçıp Giden Dünya*, (O. Akınhay, Trans.), İstanbul: Alfa Basım Yayım Dağıtım.

Giddens, Anthony, (2010), *Modernite ve Bireysel Kimlik: Geç Modern Çağda Benlik ve Toplum*, (Ü. Tatlıcan, Trans.), İstanbul: Say Yayınları.

Giddens, Anthony and Pierson, Christopher, (2001), *Modernliği Anlamlandırmak-Anthony Giddens'la Söyleşiler*, (S. Uyurkulak and M. Sağlam, Trans.), İstanbul: Alfa Yayınları.

Goleman, Daniel, (1998), *Duygusal Zeka*, (B. Seçkin Yüksel, Trans.), İstanbul: Varlık Yayınları.

Gökkaya, V. Bilican, (2009), "Türkiye'de Şiddetin Kadın Sağlığına Etkileri", *Cumhuriyet Üniversitesi İktisadi İdari Bilimler Dergisi*, Vol. 10, N. 2, pp. 167–179.

Hazelwood, D. Steven and Koon-Magnin, Sarah, (2013), "Cyber Stalking and Cyber Harassment Legislation in The United States: A Qualitative Analysis", *International Journal of Cyber Criminology*, Vol. 7, N. 2, pp. 155–168.

Heimer, Karen, (1997), "Socioeconomic Status, Subcultural Definitions, and Violent Delinquency", *Social Forces*, Vol. 75, N. 3, pp. 35–79.

IGF, (2015), *Internet Governance Forum-Best Practice Forum on Online Abuse and Gender-Based Violence Against Women*, http://www.intgovforum.org/cms/documents/best-practice-forums/623-bpfonline-abuse-and-gbv-against-women/file, (28.10.2019).

İğrek, A. Şeyda, (2009), *Yeni Bir Sosyal Ağ Oluşumu: İnternet'in İnsan İlişkileri Üzerindeki Etkileri* (Master's Thesis), Uludağ Üniversitesi, Bursa.

Juvonen, Jaana and Gross, F. Elisheva, (2008), "Extending the School Grounds? – Bullying Experiences in Cyberspace", *Journal of School Health*, Vol. 78, N. 9, pp. 496–505.

Kanbur, Aysun and Kanbur, Engin, (2018), "İşyerinde Siber Zorbalık Ölçeği'nin Türkçeye Uyarlanması: Geçerlik ve Güvenirlik Sınama Çalışması", *Psikoloji Çalışmaları-Studies in Psychology*, Vol. 38, N. 2, pp. 237–258.

Kersten, Joachim, (1993), "Crime and Masculinities in Australia, Germany and Japan", *International Sociology*, Vol. 8, N. 4, pp. 461–478.

Kesici, M. Rauf, (2007), "Yoksulluk Şiddet Döngüsünün Sosyal Politika Açısından Analizi", *Çalışma Ve Toplum Dergisi*, Vol. 2, N. 13, pp. 121–158.

Keyman, E. Fuat, (1999), *Türkiye ve Radikal Demokrasi*, İstanbul: Bağlam Yayınları.

Kılınç, Hakan and Tuncer, Deniz, (2013), *Türkiye'de Fiziksel Şiddet Eğiliminin Sebepleri ve Analizi*, Retrieved from, http://myweb.sabanciuniv.edu/bac/files/2013/10/T%C3%BCrkiyede-Fiziksel-%C5%9Eiddet-E%C4%9Filiminin-Sebepleri-ve-Analizi_final.pdf (09.07.2019).

Kırhan, Aylin, (2007), *Üniversite Öğrencilerinin Tematik Televizyon Kanal Tercihleri Kullanımlar Doyumlar Kuramı-Sosyal Öğrenme Kuramı Çerçevesinde Maltepe Üniversitesi'nde Bir Çalışma* (Master's Thesis), Maltepe Üniversitesi, İstanbul.

Kızmaz, Zahir, (2006), "Şiddetin Sosyo-kültürel Kaynakları Üzerine Sosyolojik Bir Yaklaşım", *Fırat Üniversitesi Sosyal Bilimler Dergisi*, Vol. 16, N. 2, pp. 247–267.

Kocacık, Faruk, (2001), "Şiddet Olgusu Üzerine", *Cumhuriyet Üniversitesi İktisadi İdari Bilimler Fakültesi Dergisi*, Vol. 2, N. 1, pp. 1–7.

Kocadaş, Bekir, Özgür, Özge and Özbulut, Mahmut, (2010), *Gençlik ve Şiddet, Toplumsal Alanda Şiddetin Yükselişi*, Ankara: Atatürk Kültür Merkezi Yayını.

Kowalski, M. Robin, Limber, P. Susan and Agatson, W. Patricia, (2012), *Cyberbullying: Bullying in the Digital Age*, New Jersey: Wiley-Blackwell.

Krug, Etienne G., Dahlberg, Linda L., Mercy, James A., Zwi, Anthony B. and Lozano, Rafael, (2002), *World Report on Violence and Health*, Retrieved from, https://apps.who.int/iris/bitstream/handle/10665/42495/9241545615_eng.pdf;jsessionid=F8BCBE221710FD86335F288252870D50?sequence=1 (04.11.2019).

Lash, Scott, (2000), "Risk Culture", (Joost Loon, Ulrich Beck, Barbara Adam, Eds.), *The Risk Society and Beyond- Critical Issues for Social Theory*, London: SAGE Publications.

Lenhart, Amanda, (2007), *Cyberbullying And Online Teens*, Retrieved from, https://www.pewresearch.org/internet/2007/06/27/cyberbullying/, (11.01.2020).

Li, Qing, (2007), "Bullying in The New Playground: Research into Cyberbullying and Cyber Victimisation", *Australasian Journal of Educational Technology*, Vol. 23, N. 4, pp. 435–454.

Luhmann, Niklas, (1982), *The Differentiation of Society*, New York: Columbia University Press.

Luhmann, Niklas, (1985), "Society, Meaning, Religion-Based on Self-Reference", *Sociological Analysis*, Vol. 46, pp. 5–20.

Luhmann, Niklas, (1988), "Tautology and Paradox in the Self-Descriptions of Modern Society" *Sociology Theory*, Vol. 6, N. 1 pp. 21–37.

Luhmann, Niklas, (1993), *Risk: A Sociological Theory*, New York: Routledge.

Lupton, Deborah, (1999), *Risk and Sociocultural Theory: New Directions and Perspectives*, New York: Cambridge University Press.

Markowıtz, Fred E., (2003), "Socioeconomic Disadvantage and Violence: Recent Research on Culture and Neighborhood Control as Explanatory Mechanism", *Aggression and Violent Behavior*, Vol. 8, pp. 145–154.

Marohn, C. Richard, (1982), "Adolescent Violence: Causes and Treatment", *Journal of the American Academy of Child Psychiatry*, Vol. 21, N. 4, pp. 354–360.

Medrano, J. Diez, (2004), "Questioning Modernity: A Test of Giddens's, Beck's, and Inglehart's Theories, Internet", Retrieved from, http://www.sscnet.ucla.edu/soc/groups/ccsa/medrano.pdf (30.12.2019).

Memiş, Emel, (2015), "Şiddet ve Ekonomi", (Betül Yarar, Ed.), *Şiddetin Cinsiyetli Yüzleri*, İstanbul: Bilgi Üniversitesi Yayınları, pp. 175–191.

Michaud, Yves, (1991), *Şiddet*, (Cem Muhtaroğlu, Trans.), İstanbul: İletişim Yayınları.

Mishna, Faye, Khoury-Kassabri, Mona, Gadalla, Tahanya and Daciuk, Joanne, (2012), "Risk Factors for Involvement in Cyber Bullying: Victims, Bullies and Bully–Victims", *Children and Youth Services Review*, Vol. 34, N. 1, pp. 63–70.

Modecki, Kathryn, L., Minchin, Jeannie, Harbaugh, Allen, G. and Runions, Kevin, (2014), "Bullying Prevalence Across Contexts: A Meta-Analysis Measuring Cyber and Traditional Bullying", *Journal of Adolescent Health*, Vol. 55, N. 5, pp. 602–611.

Moses, Rafael, (1996), "Şiddet Nerede Başlıyor?", (Ayşe Kul, Trans.), *Cogito Dergisi*, Winter-Spring Period, Vol. 6, N. 7, pp. 23–28.

Nair, Güney, (2015), *Gençlik Şiddetinin Toplumsal Nedenleri* (Doctoral Dissertation), Cumhuriyet Üniversitesi, Sivas.

Neugebauer, Richard, (2000), "Research on Intergenerational Transmission of Violence: The Next Generation", *The Lancet*, Vol. 355, N. 9210, pp. 1116–1117.

Okumuş, Eylem, (2014), *Doğamızdaki Şiddet*, İstanbul: Siyah Beyaz Kitap.

Özer, Asuman, (2019), *Reklamlarda Erkeklere Yönelik Simgesel Şiddet: Banka Reklamları Üzerine Bir Analiz* (Master's Thesis), Yaşar Üniversitesi, İzmir.

Özgür Sayar, Özge, (2006), "Yetiştirme Yurtlarında Akran Şiddeti- Ankara'da İki Yetiştirme Yurdu Örneği", *Toplum ve Sosyal Hizmet Dergisi*, Vol. 17, N. 1, pp. 113–132.

Özsöz, Cihad, (2009), *Pierre Bourdieu Sosyolojisi ve Simgesel Şiddet* (Master's Thesis), Hacettepe Üniversitesi, Ankara.

Palermiti, Anna, Lisa, Servidio, Rocco, Bartolo, Maria Giuseppina and Costabile, Angela, (2017), "Cyberbullying and Self-esteem: An Italian Study", *Computers in Human Behavior*, Vol. 69, pp. 136–141.

Peker, Adem, (2013), *İnsani Değerler Yönelimli Psiko-eğitim Programının Problemli İnternet Kullanımı ve Siber Zorbalık Üzerindeki Etkisi* (Doctoral Dissertation), Sakarya Üniversitesi, Sakarya.

Peker, Adem, (2015), "Ergenlerin Saldırganlık ve Siber Zorbalık Davranışları Arasındaki İlişkilerin İncelenmesi", *Ekev Akademi Dergisi*, Vol. 19, N. 61, pp. 323–336.

Peterson, Jillian and Densley, James, (2017), "Cyber Violence: What Do We Know and Where Do We Go from Here?", *Aggression and Violent Behavior*, Vol. 34, pp. 193–200.

Pew Research Center (PEW), (2014), "Online Harassment", Retrieved from http://www.pewinternet.org/2014/10/22/online-harassment/, (10.11.2019).

Piotrowski, Chris, (2012), "From Workplace Bullying to Cyberbullying: The Enigma of E-Harassment in Modern Organizations", *Organization Development Journal*, Vol. 30, N. 4, pp. 44–53.

Privitera, Carmel and Campbell, Marilyn, A., (2009), "Cyberbullying: The New Face of Workplace Bullying?", *Cyber Psychology & Behavior*, Vol. 12, N. 4, pp. 395–400.

Polat, Oğuz, (2016), "Şiddet", *Marmara Üniversitesi Hukuk Fakültesi Hukuk Araştırmaları Dergisi*, Vol. 22, N. 1, pp. 15–34.

Ritzer, George and Stepnisky, Jeffery, (2014), *Sosyoloji Kuramları*, (H. Hülür, Trans.), Ankara: De Ki Basım Yayım.

Russo, Nancy Felipe and Pirlott, Angela, (2006), "Gender-Based Violence, Methods, and Findings", *Annals of the New York Academy of Sciences*, Vol. 1087, N. 1, pp. 178–205.

Sennett, Richard, (2010), *Karakter Aşınması*, (B. Yıldırım, Trans.), İstanbul: Ayrıntı Yayınları.

Serin, Hüseyin, (2012), *Ergenlerde Siber Zorbalık/Siber Mağduriyet Yaşantıları ve Bu Davranışlara İlişkin Öğretmen Ve Eğitim Yöneticilerinin Görüşleri* (Doctoral Dissertation), İstanbul Üniversitesi, İstanbul.

Siegel, J. Larry, (1989), *Criminology*, St. Paul: West Pub. Company.

Slonje, Robert and Smith, Peter, K., (2008), "Cyberbullying: Another Main Type of Bullying", *Scandinavian Journal of Psychology*, Vol. 49, N. 2, pp. 147–154.

Smith, Peter, K., Mahdavi, Jess, Carvalho, Manuel, Fisher, Sonja, Russell, Shanette and Tippett, Neil, (2008), "Cyberbullying: Its Nature and Impact in Secondary School Pupils", *Journal of Child Psychology and Psychiatry*, Vol. 49, N. 4, pp. 376–385.

Soydemir, Suat. (2011), "Modernizmin Karanlık Yüzü: Risk Toplumu", *Sosyal ve Beşerî Bilimler Dergisi*, Vol. 3, N. 2, pp. 169–178.

Spears, Barbara, A., Taddeo, Carmel, M., Daly, Anthony, L., Stretton, Alexander and Karklins, Larisa, T., (2015), "Cyberbullying, Help-Seeking and Mental Health in Young Australians: Implications for Public Health", *International Journal of Public Health*, Vol. 60, N. 2, pp. 219–226.

Staude-Müller, Frithjof, Hansen, Britta and Voss, Melanie, (2012), "How Stressful Is Online Victimization? Effects of Victim's Personality and Properties of The Incident", *European Journal of Developmental Psychology*, Vol. 9, N. 2, pp. 260–274.

Tezcan, Mahmut, (1996), "Bir Şiddet Ortamı Olarak Okul", *Cogito Dergisi*, Winter–Spring Period, Vol. 6, N. 7, pp. 105–108.

Topcu, Cigdem, Erdur-Baker, Özgür and Yıldırım, Ali, (2010), "Turkish High School Students Perception of Cyber Bullying: A Qualitative Study", Retrieved from, https://eera-ecer.de/ecer-programmes/conference/3/contribution/4393/ (03.02.2020).

Turan, G. Seda and Dilmen, E. Necmi, (2016), "Bilişim Okuryazarlığı Bağlamında Siber Şiddetin Meşrulaştırılması", *Yeni Symposium Psikiyatri, Nöroloji ve Davranış Bilimleri Dergisi*, Vol. 54, N. 3, pp. 25–29.

Türk Dil Kurumu Sözlüğü, Retrieved from, https://sozluk.gov.tr/ (02.03.2019).

Ünsal, Artun, (2005), "Genişletilmiş Bir Şiddet Tipolojisi", *Cogito Dergisi*, Winter-Spring period, Vol. 96, N. 6–7, pp. 29–36.

Wallace, Ruth A. and Wolf, Alison, (2015), *Çağdaş Sosyoloji Kuramları Klasik Geleneğin Genişletilmesi*, (L. Elburuz and M. R. Ayas, Trans.), Ankara: Doğu Batı Yayınları.

Watts, Lynette K., Wagner, Jessyca, Velasquez, Benito and Behrens, Phyllis, I., (2017), "Cyberbullying in Higher Education: A Literature Review", *Computers in Human Behavior*, Vol. 69, pp. 268–274.

Wilkinson, Iain, (2001), "Social Theories of Risk Perceptions: At Once Indispensable and Insufficient", *Current Sociology*, Vol. 49, N. 1, pp. 1–22.

Willard, Nancy, (2008), "The Authority and Responsibility of School Officials in Responding to Cyberbullying", *Journal of Adolescent Health*, Vol. 41, N. 6, pp. 64–65.

Wolak, Janis, Mitchell, Kimberly, J. and Finkelhor, David, (2007), "Does Online Harassment Constitute Bullying? An Exploration of Online Harassment by Known Peers and Online-Only Contacts", *Journal of Adolescent Health*, Vol. 41, N. 6, pp. 51–58.

Yaman Erkan, Eroğlu Yüksel and Peker Adem, (2011), *Başa Çıkma Stratejileriyle Okul Zorbalığı ve Siber Zorbalık*, İstanbul: Kaknüs Yayınları.

Yaman, Erkan and Peker, Adem, (2012), "Ergenlerin Siber Zorbalık ve Siber Mağduriyete İlişkin Algıları", *Gaziantep University Journal of Social Sciences*, Vol. 11, N. 3, pp. 819–833.

Yalçın, Nagehan, (2015), *Sosyal Medyada Simgesel Şiddet: Ekşi Sözlük Örneği* (Master's Thesis), Pamukkale Üniversitesi, Denizli.

Ybarra, Michele, L. and Mitchell, J. Kimberly. (2004), "Online Aggressor/Targets, Aggressors and Targets: A Comparison of Associated Youth Characteristics", *Journal of Child Psychology and Psychiatry*, Vol. 45, N. 7, pp. 1308–1316.

Yıldız, Ali Naci, Kaya, Mehmet, Bilir, Nazmi, (2011), *İşyerinde Şiddet*, Ankara: Türkiye İşçi Sendikaları Konfederasyonu.

Yılmaz, Cevdet, (2010), *Risk Kapıyı Kırınca Kentlerde Yoksulluk, Dayanışma, Güven ve Güvenlik*, İstanbul: Libra Yayınları.

Yüksel Özmen, Şule, (2018), "Dijital Şiddet, Siber Zorbalık ve Yeni Medya Okuryazarlığı Üzerine Bir Değerlendirme", *Uluslararası Sosyal Araştırmalar Dergisi*, Vol. 11, N. 61, pp. 958–966.

Esra DOĞAN

The Problematic of Digitalization as the Social Dimension of Fiscal Transparency

1 Introduction

The term digital, which corresponds to the term "numerical" in Turkish in the Turkish Language Institution dictionary, is also described as "electronically displaying data on a screen." Accordingly, it has a meaning content in the form of digitalization, digitization, and data in electronic environment. While digitalization requires the presence of technological development on the one hand, it also requires that the nondigital elements have the quality to be digitalized. From a scientific perspective, while technological development is already a targeted situation, digitization of nonnumerical elements brings along a narrowing of meaning and reductionism. This narrowing of meaning and reductionism is a necessity and corresponds to the nature of the digitization logic. However, narrowing of meaning and reductionism contradicts the logic of scientific approach in social sciences. Indeed, Popper, one of the thinkers who dealt with reductionism theoretically, did not reject it as a whole but evaluated it by dividing it into two. According to this distinction made in the form of scientific and philosophical reductionism, while philosophical reductionism is considered to be a completely wrong path (Popper, 1994: 85), considering scientific reduction as a goal, it is thought that "nothing can be counted as great a success as reduction in science" (Popper, 1994: 48).

When digitalization, which embodies the problematic of reductionism, which is required by the digitalization of data in a scientific sense, is considered in the context of factual reality, it is seen that it is the result of the information revolution in the twentieth century, following the agricultural revolution and the industrial revolution. At the point reached by the information revolution, while being limited to data access and data processing, integration of data beyond data processing has been made possible with digitalization. This integration brought about by digitalization emerges not only in a certain area but also as a whole, including all of the relations of production, consumption, and distribution. This holistic structure has become evident with the twenty-first-century project, which is ultimately referred to as Industry 4.0. With this European-centered project, which envisages the combination of the production process connected with digital networks in the industry with the analysis of data and the integration of production with smart factories, it is proposed to integrate not only the processes involved in manufacturing but also the whole process that connects from manufacturing to suppliers, from design to customer relations, and to collect and process the data of the production process (Narin, 2018: 233). With this project, in which the form of production itself is shaped, the consumption and distribution processes and relationships that are defined in the form of production and arise as a result of the production relations within this scope are also determined. Accordingly, it seems likely that different contrasts brought about by digitalization will emerge in the social context, just as capitalist society relations based on the contrast between labor and capital replaced feudal society relations based on the contrast between serf and master. However, this situation does not bring any change in the form of production at the moment, but it brings with it a differentiation in the current capitalist mode of production in the form of the understanding of industrial society being replaced by the understanding of information society just like it happened during the transition from the industrial revolution to the information revolution. However, the difference between those who are in direct contact with production network and those who are in indirect relation leads the society to be divided into different segments within itself. This situation has emerged in the urbanization phenomenon brought by the industrial revolution in the form of ruralization and urbanization of the society in the first phase and as suburbanization within the city in the secondary phase. With the formation of the information society, this situation has moved beyond a spatial-based distinction to a certain degree. The new distinction emerging on the basis of access to and use of information on a social scale has revealed the importance of access to information. Accordingly, action and strategy plans have been prepared in the form of developing information society strategies on the basis of states. Access to digital media, which is one of the basic requirements of digitalization in the social context, has also been included among the determined strategic goals. Therefore, in the social structure where digitalization is included, rather than access to information, through what the information is accessed and how the accessed information is formed gain importance. It can be said that the social structure, which is able to access the information in question through digitally produced information and digital resources within a holistic structure associated with each other, is the social equivalent of digitalization.

On the other hand, in the social structure that comes into contact with the information produced in certain contexts through reduced data, there are public political actions with basic political content that are required by being a society. One of these basic actions is the feasibility of the social participation mechanism. Although there are many different requirements of the social participation mechanism, it can be said that its basic requirement is the accessibility to the said participation and the quality of participation afterwards (Doğan, 2019). When these requirements are considered in the context of digitalization, it is seen that in the first place, there is the problematic of accessibility of digital media as mentioned above. Access to digital media socially requires broadband Internet subscription and the presence of electronic devices where this subscription can be used. Secondly, there is the problematic of having the quality of being able to use the information accessed through digital media. This situation requires a sufficient level of digital literacy on one hand, and it necessitates technical information literacy beyond the level of education on the other hand. Indeed, as stated above, the use of reduced data, which enables digitization, requires that nonnumerical attributes be also digitizable. However, as stated by Popper, reductionism in social sciences brings wrong or incomplete results. What is important at this point is not only to understand the reduced data, but also to be able to evaluate it with the missing aspect beyond understanding it. This situation, which requires full expertise, includes a contradictory structure as a result of the technicalization problem that is necessarily embodied in digitalization. It seems inevitable that similar to the situation that arises in the context of the process of production in the form of alienation of labor as a result of mechanization, the contradictory structure in question brings alienation to what is political by means of digitalization on a social scale.

Digitalization can be addressed on the basis of alienation on a social scale in the context of different problematics. It can be said that one of the points where the problematic feature in the financial sense comes to the fore is fiscal transparency, which is the financial reflection of the social scale. Fiscal transparency is the fiscal management principle that has emerged as a result of neoliberal policies, especially financial discipline concern, which has gained prominence in the twenty-first century. However, in terms of public financial management, unlike accountability, fiscal transparency is basically not a tool for control but for participation. In this context, it is ensured that financial processes are shared with the public as a result of the quality of being political. However, the main problem in this context is that the financial processes are reduced to a number of criteria, and as a result, their political content becomes a technical one. This situation, which has emerged as the technicalization of the financial, has also gained depth with digitalization. The public sharing of the financial processes that are digitized and converted into digital data being made in digital environment within the framework of fiscal transparency criteria is the problematic of evaluation that firstly requires access to what is digital and expertise afterwards, as mentioned above.

In this context, the social dimension of fiscal transparency will be evaluated in the context of digitalization within the scope of this study. In this context, digitalization will be handled in two different contexts, and the evaluability of both digital access from a social perspective and the digitalized data will be examined. In this respect province-based digital accessibility in Turkey will be evaluated empirically. Within the scope of this evaluation, spatial statistical analysis is planned by using Turkish Statistical Institute (TUIK) statistics through mapping method using GeoDa program. As a result of the review and analysis, the differentiation level of digitalization, which forms the context of fiscal transparency, according to the provinces in Turkey, will be revealed. Accordingly, an empirical and social-based critical study proposal regarding the reductionism considering the increasing impact on social sciences.

2 Overview of Fiscal Transparency

Fiscal transparency refers to the information available to the public about the government's fiscal policymaking process. It refers to the clarity, reliability, frequency, timeliness, and relevance of public fiscal reporting and the openness of such information (IMF, 2018: 1). Within this definition, fiscal transparency has a content that is shaped on the basis of the relationship between the two parties. On the one hand, there is the administration responsible for ensuring transparency, and on the other hand, there is the citizen to whom this transparency is presented. Accordingly, the reason for the existence of transparency is that the public and political context whose legislative, executive, and judicial processes are realized by the administration covers the citizen as well. Indeed, this is the main point which the fiscal transparency diverges from the principle of accountability, with which it is closely related. While the principle of accountability basically takes control of the use of public resources from the public quality of the sources in question related to the partner, the fiscal transparency principle includes the publicity of the partner in addition to the public quality related to the partner. According to accountability, the extent of scope that fiscal transparency has stems from the fact that the public regulation and activities carried out by the administration have a political context beyond the publicity quality.

Within the bilateral structure of fiscal transparency, the administration is responsible for ensuring transparency and making the necessary documentation and statistical data accessible, while the citizen must be able to access and process the data provided to him/her. Accordingly, fiscal transparency can be expressed as the establishment of a transparent structure by the administration in a way to enable participation of the citizen in a political context. Considering participation that constitutes the distinctive part of the scope of fiscal transparency, it is seen that public participation is described as "public participation refers to the variety of ways in which citizens and the general public, including civil society organizations and other nonstate actors, interact directly with public authorities with respect to the design, implementation and review of public policies by any form of communication. Participation ranges from one-off consultations to on-going and institutionalized relationships that leave records subject to access to information." Participation in the financial context is mainly ensured through fiscal policy. Accordingly, there are four main domains in which direct public participation should be sought in the design and implementation of fiscal policies. The first domain is the annual budget cycle from fiscal strategy and the preparation of the annual budget proposal by the executive, through presentation of the budget and its adoption by the legislature, to budget implementation, in-year reporting on and amendment of the budget, and end-of-year reporting, auditing, and review. The second domain is new policy initiatives, plans, or reviews on revenues, expenditures, financing, assets, and liabilities. These are fiscal policy initiatives that may have been subject to public engagement outside of the annual budget cycle or over a longer period than the window for preparation of the annual budget. The third domain is the design, production, and delivery of public goods and services from service delivery planning and the setting of service standards, engagement during service delivery, through feedback from service recipients, independent review mechanisms, and monitoring and evaluation. The fourth domain is the planning, appraisal, and implementation of public investment projects from national and sector planning through project preparation, appraisal, and selection to project implementation, audit, and review.

Public authorities should endeavor to ensure that citizens and other nonstate actors have effective opportunities to participate directly in public debate and discussion with respect to the design, implementation, and review of fiscal policies, observing the following interdependent principles as "accessibility, openness, inclusiveness, respect for self-expression, timeliness, depth, proportionality, sustainability, complementarity, reciprocity" (The Global Initiative for Fiscal Transparency, 2020). Moreover, the principles of public participation in fiscal policies are intended to apply to all public authorities, that is to executive entities in all levels of the government as well as more generally to legislatures and supreme audit institutions. They are a set of interdependent codes intended to be applicable to all country contexts, although the specific institutions, processes, and programs for public participation reflect diverse country circumstances. The approach aims to ensure that official authorities encourage and facilitate the open, responsible, and constructive engagement of nonstate actors, civil society organizations, business organizations, citizens, individuals, and the general public in government fiscal policies and budget-making. Considering these principles of participation, which constitute the main distinguishing aspect of fiscal transparency, innovations in information and communications technology have made public participation more feasible and easier by dramatically lowering the costs and time requirements of such interaction and dialogue. They have also enabled the creation of new virtual spaces which governments. Citizens and nonstate actors are increasingly dialoguing on social, environmental, and economic challenges and opportunities. While public participation involves costs both for official entities and for civil society and needs to be tailored appropriately in each case, direct public participation as a policy tool is an increasingly cost-effective means of developing and implementing public policies.

3 Fiscal Transparency in the Context of Digitization

The establishment of organized structures for public and political facts and providing data to citizens in this context are the basic conditions of feasibility of fiscal transparency. The fact that this condition has digital content facilitates the feasibility of fiscal transparency if the necessary infrastructure and access opportunity are available. Digitalization for the organizational structure that forms the administrative side of fiscal transparency is provided primarily through the production, use, and conservation of all kinds of data infrastructure owned by administrative structures in digital environment. Following the establishment of a digital data network on the basis of administrative structure, the administrative context of digitalization is completed by providing data integration in a digital context between different administrative structures. Within this context, while the transactions carried out by the administration are direct and digital, the processes before and after the execution process also have a digital content.

The relationship of the administrative structure with a digital quality with the citizen who forms the other side of fiscal transparency can occur in the context of public service provision in the execution phase of public services as well as in the form of information presentation regarding the processes before and after the execution. In the context of the relationship between the administration and the citizen, it can be said that the factual equivalent of this structure, which is digitalized, is e-government. E-government is defined by the Organization for Economic Co-operation and Development (OECD) as "the use of information and communications technologies (ICTs), and particularly the Internet, to achieve better government" (from OECD The e-Government Imperative, 2003). This definition focuses attention on why countries are implementing e-government rather than on the ICT tools themselves. Faced with the pressure of increasing government performance while being responsive to citizens' needs, OECD countries have realized that e-government goes beyond the simple exercise of putting information and services online and can be used as a powerful instrument to transform the structures, process, and culture of the government and make it more efficient, user oriented, and transparent. (OECD, 2006). However, beyond this general definition of e-government, digitalization has occurred mainly with the transformation of e-government. Accordingly, the transformation from e-government to digital government has occurred as a result of a four-step process. The first stage is referred to as e-Government 1.0, characterized by the provision of online services and setting up of government websites and IT systems within public administrations in the late 1990s and early 2000s. In the second half of the 2000s the discourse shifted to e-Government 2.0, also sometimes referred to as "open" government, also covered under the concept of "eGovernance." It went hand in hand with Web 2.0 collaborative technologies and aimed at creating an open-source platform in which government, citizens, and innovative companies could interact. Since the mid-2010s e-Government 3.0 has proceeded as the "smart" or "intelligent" government, powered by innovations related to open and big data, administrative and business process management, Internet of Things (IoT), and blockchain. The final stage in the late 2010s is the concept of e-Government 4.0 as a transformed and citizen-driven government, which adapts itself to the needs and expectations of citizens, businesses, nonprofit organizations, and other stakeholders and creates relations and exchanges that are personalized, interactive, and easy to access (Barcevičius et al., 2019: 10). While each of these stages is used in the literature as the counterpart of the digital government, digital government requires innovations in all areas: internal processes, governance, service delivery, and policies. This way of understanding the digital government is in line with the definition provided by the OECD, according to which digital government refers to the use of digital technologies, as an integrated part of the governments' modernization strategies, to create public value. It relies on a digital government ecosystem comprised of government actors, nongovernmental organizations, businesses, citizens' associations, and individuals that support the production of and access to data, services, and content through interactions with the government. The digital transformation can be defined as both, transformation of internal processes and as a transformation of the relationships between governments and other social and political actors (institutional transformation), tailored to specific policy areas and needs (Barcevičius et al., 2019:18). The technology based areas required for the transformation to digital government are artificial intelligence (in the broad sense), predictive analytics, robotics and automation, IoT, geo-spatial data, blockchain, and open government data (OGD).

Unlike others, OGD, one of the innovation-based technological areas required for digital transformation, is directly included in the government's digitalization in public financial management. OGD is a "philosophy" and set of policies promoting transparency, accountability, and value creation by making government data available to all. By opening the huge quantities of data and information collected by public organizations and encouraging their use, governments can promote business creation and innovative citizen-centric services (Barcevičius et al., 2019: 27). Accordingly, fiscal transparency gains a digital dimension with the access and participation of citizens. However, there are some requirements for fiscal transparency to be achieved within this context, which basically depends on the relationship between the administration and the citizen. The absence of these requirements constitutes the main barriers to digitalization. Accordingly, obstacles due to these requirements are grouped as technological (IT infrastructures, interoperability, access to data, data storage), organizational (digital strategy, workforce skills, data sharing, alignment of agencies), legal (new legal and regulatory frameworks, privacy, cybersecurity), ethical (citizens' trust, accountability of new technologies, human rights), sociocultural (citizens' adoption, conservative habits, and cultural barriers), and economic (lack of financial resources to implement and scale innovations) factors (Barcevičius and diğerleri, 2019: 57). Among these factors, especially legal and technological ones constitute the basis for digitalization to be achieved. The applicability, development, and sustainability of the formed digital context are provided through organizational, ethical, sociocultural, and economic factors that complement the legal and technological qualities.

When digitalization is considered especially in the context of fiscal transparency, the existence of a legal basis that enables the digital formation of the financial documentation system takes priority. As a matter of fact, the provision of fiscal transparency depends mainly on public administration and the public financial management approach shaped in this direction. In this regard, when public financial management issue is considered within the context of Turkey, it is seen that there basically exists structural transformation in the form of a new public administration which emerges in the direction of public management and governance approach and in the form of the reflections of public administration reform on financial management as a result of this approach. These reflections on public financial management are seen in the form of efficiency, effectiveness, and productivity, which are the main criteria of the performance evaluation requirement as a result of the logic of public management. As a result of the same logic, the traditional state has been replaced by the e-government understanding by accepting the issues that are the subject of management as counting, classification, association, planning, and reducing them to being categorizable (Dikmen, 2003). The formation of the e-government that has emerged in this process is basically a development indicator with the introduction of technological context to the management, but it becomes the mediator of performance-based management within the scope of the new public administration approach. This situation, in turn, involves a two-step problem when the administration–citizen relationship of the transformation that has occurred in the public administration management and the public financial management approach is considered in the context of the citizen. The first stage is the technicalization problematic that arises when the phenomenon of public financial management is carried to the logic of performance-based public management. Accordingly, what is political is reduced to what is technical by transforming the processes of political nature, which are within the scope of technocratic management and public financial management, which have replaced the bureaucratic management approach into technical expertise. This situation turns the citizen into a passive actor of the public financial management area, by transforming the financial issues into a technical specialty, ignoring the "public" in the content of the public financial management where the citizen is included. The second stage is the state of making citizens passive through technicalization in the field of public financial management gaining depth with e-government and digitalization. Although there is a general acceptance to ensure fiscal transparency regarding digitalization, it can be said that the citizen, whose presence has already been passive in the field of public financial management, has become even more passive with the legitimacy of fiscal transparency. As a matter of fact, with this transformation that requires citizens to access technological opportunities, there is also the problem of having information about using technology beyond the access problem. Therefore, public financial management has become a technical expertise, access to technological infrastructure, and ability to use technology.

4 Fiscal Transparency in the Context of Digitalization on the Scale of Turkey

When fiscal transparency is considered in the context of digitalization on the scale of Turkey, it is seen that there is an administrative context shaped within the scope of the Law on Public Financial Management and Control No. 5018. Within the scope of the Law on Public Financial Management and Control No. 5018, fiscal transparency is described as "The public is informed in a timely manner in order to provide control in obtaining and using all kinds of public resources". For this purpose, it is mandatory: a) to clearly define duties, powers and responsibilities; b) to ensure transparency and public's access to the reports regarding government policies, development plans, annual programs, preparation of strategic plans and budgets, their negotiation in authorized bodies, and implementation and implementation results; c) to publicize the incentives and supports provided by public administrations within the scope of general administration at certain periods no more than one year; and d) to form public accounts in compliance with a standard accounting system and accounting order complying with generally recognized accounting principles. Accordingly, with the enforcement of the law, most of the documentation that has been included in the budget process since 2006 and which must be prepared and presented to the public within the scope of budget transparency has been regulated and made accessible to the public.

On the other hand, in accordance with the relevant law, arrangements have been made to create city councils, beyond the documentation system, both to facilitate the contact of the citizen with the related documentation process and also to support fiscal transparency on a local scale. Of the arrangements, the basic one is the Municipal Law No. 5393 dated March 3, 2005, which allows the creation of the city councils in the places where municipalities are located. Pursuant to Article 76 of the Law, the content of the city councils is as follows: "The city council tries to implement the principles of developing the vision of the city and citizenship awareness, protecting the rights and the legal order of the city, sustainable development, environmental awareness, social assistance and solidarity, transparency, accountability, participation and decentralization in the urban life. Municipalities provide assistance and support in the effective and efficient execution of the activities of the city council, which is formed by the participation of representatives of public professional organizations, industrial unions, notaries, universities (if available), nongovernmental organizations, political parties, public institutions and organizations, and neighborhood mukhtars and other related persons."

As mentioned in the previous section, legal and technological infrastructure is required in the formation of digitalization. In this respect, when the legal structure is viewed in the context of fiscal transparency, thus, public financial management in Turkey, there is a legal ground for city councils, but there are also eight basic documents regarding the public financial process. The aforementioned documents within the scope of law numbered 5018 and their accessibility level are shown in the table below.

Documents	2006	2008	2010	2012	2015	2017	2019
Pre-Budget Statement	◉	◉	◉	○	○	○	○
Executive's Budget Proposal	◉	◉	◉	◉	◉	◉	◉
Enacted Budget	◉	◉	◉	◉	◉	◉	◉
Citizens Budget	●	●	●	●	●	◉	◉
In-Year Reports	◉	◉	◉	◉	◉	◉	◉
Mid-Year Review	●	●	○	○	◉	◉	○
Year-End Report	◉	◉	◉	◉	◉	◉	◉
Audit Report	◉	◉	◉	◉	◉	◉	◉

● Not produced; ◉ Available to the public
○ Published late, or not published online, or produced for internal use only

Tab. 1: Documents for budget transparency index

As can be seen from the table, relating to the budget period, the government budget proposal, the central government budget, the annual report, the end-of-year report, and the audit report have been arranged regularly since 2006, allowing public access. However, while the pre-budget report is not available to the public since 2010, as of 2017, the half-year report and the citizen budget are open to public access. Accordingly, it is seen that the legal infrastructure regarding the digital context of fiscal transparency has been completed to a certain extent, but it does not have a long history.

It can be said that the legal and technological dimension of digitalization has been completed to a certain extent by the transfer of infrastructure within the legal context to digital context via e-government. However, the contact of the citizen with the legal and technological context in question is possible with the existence of a technological infrastructure that enables access to such information beyond having knowledge that requires technical expertise. In this context, it can be said that the availability of Internet access for citizens is the basic indicator of the said technological infrastructure.

In this context, digitalization is handled within the context of the social dimension of fiscal transparency. Accordingly, the Internet access level of the citizens is examined firstly in order to reveal the possibility of contact with the digital. Secondly, the level of participation of the citizens to the social and the political is examined in order to reveal the participation dimension beyond the accountability of the fiscal transparency. Finally, digitalization level of the city councils, which allows social and political participation as the solution to the technicalization problem related to the public financial management, is examined empirically.

4.1 Empirical Evaluation

With the province-based aim of questioning the digitalization in the scope of social perspective of fiscal transparency, an empirical evaluation has been performed to make it possible to reveal the real circumstances in Turkey.

4.1.1 Objective and Scope of the Research

The purpose of this study is to reveal the opportunity of the public to benefit from the said transparency in a structure where fiscal transparency is provided through digital possibilities. Accordingly, the scope of this research is to create a province-based data for all 81 provinces in Turkey. In this context, in order to make an assessment appropriate for the purpose of the research, the provincial broadband Internet subscription published by Information Technologies and Communication Authority and the election statistics published by the Turkish Statistical Institute were used.

4.1.2 Methodology

While the use of data in the analysis process appears in four stages and the first two stages are independent from each other, the last two stages are formed as the result of the first two stages. Accordingly, the possibility of digitization on the basis of provinces in the first stage, the level of political participation on the provincial basis in the second stage, the possibility of digitalization on the basis of the provinces on the third stage, and the possibility of digitalization of the political participation on the basis of the provinces on the fourth stage are examined. Following the introduction of digitalization on a provincial basis comparatively with broadband Internet subscription rate, the level of political participation is also examined comparatively through the participation rate in local elections. Based on the evaluations made in the first two stages, in the third stage, a categorization is made for the comparative distribution of the political participation level of the local people and the possibility of digitalization. With this categorization, 81 provinces are divided into 4 groups, as summarized in the table below.

Tab. 2: Categorization in the third stage

	Broadband access rate	**Participation rate in local elections**
First category	High	High
Second category	High	Low
Third category	Low	High
Fourth category	Low	Low

In making the categorization, the provinces were ranked according to the distribution of broadband Internet subscription rate evaluated in the first and second stages and the distribution of participation rates in local elections. The distributions are based on percentiles and the division of these slices into six groups. Accordingly, in the ranking made for each distribution, the tax broadband Internet subscription rate and the rate of participation in local elections are low for the provinces in the first, second, and third groups within six groups and considered high for the provinces in the fourth, fifth, and sixth groups. Secondly, the combination of each ratio was made with the other ratio in the form of high and low. Based on this combination, 81 provinces are divided into four groups.

In the fourth stage, the possibility of digitalization of political participation on a provincial basis is revealed through the categorization made in the context of the availability of the city councils' web pages and the coverage of these pages. When the web pages of the city councils are analyzed, it is seen that there is information such as "information about the structure of the relevant city council, legislation related to the city councils, events and announcements, relevant councils and working groups, study and activity reports." However, there are web pages containing some of the information in question, as well as some containing all of them. Accordingly, as a result of the examination made on the web pages of the city councils belonging to the municipalities in 81 city centers, the provinces that do not have a web page are given zero points, while the provinces with a web page are categorized by scoring the five information categories listed above. With this categorization, 81 provinces are divided into 6 groups.

The spatial data analysis (SDA) method was used in the formation of the above mentioned distributions in this four-step analysis process. Spatial data analysis is data analysis in which the interactions, structures, and processes of space data are explained in the context of possible relations with other spatial events (Bailey and Gatrell, 1995). It includes techniques that reveal the spatial distribution and the spatial density level of this distribution by examining the relation of the space-based data with the data in other spaces and thus tests the level of spatial clustering. With the method of spatial distributions by determining spatial differences, the spatial relation models and spatial heterogeneity level are allowed to be defined and revealed (Anselin, 1998). In this context, firstly, the distribution of relevant data on the provinces based on maps is evaluated through GeoDa program.

4.1.3 Results

The findings of this research are presented in accordance with the four-stage process specified in the method.

First Stage: Digitalization Opportunities by Province

At this stage, the distribution of broadband Internet and broadband mobile subscription rates over the years is analyzed in 81 provinces. First, broadband Internet and broadband mobile subscription rates are evaluated through the figure below for 2011–2018.

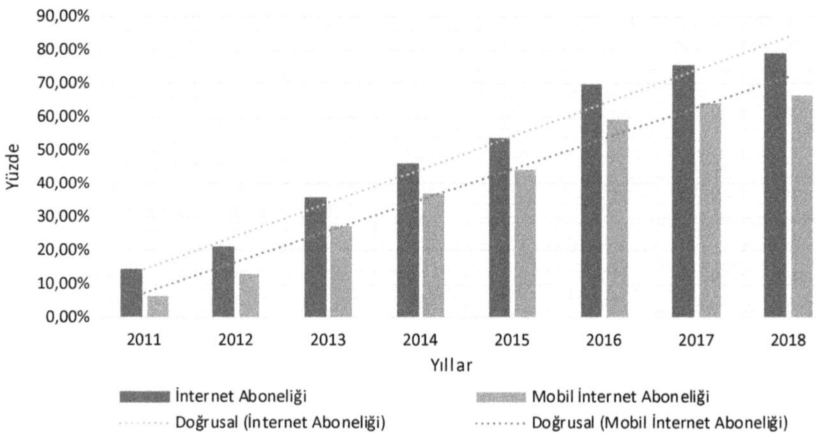

Fig. 1 : Broadband internet subscription rates (2011–2018). Source: Information Technologies and Communication Authority's Annual Provincial Statistics have been used

Looking at the rate of broadband Internet subscription in 81 provinces, it is seen that it increased from 14 % to 79 %. An important part of this rate, which is the sum of fixed and mobile broadband Internet subscription rates, is the mobile Internet subscription rate. As a matter of fact, the broadband mobile Internet subscription rate has increased from 6 % to 66 %.

In this context, broadband Internet subscription rate is an indication of the overall presence of the digitization opportunity. This opportunity, which also showed a significant improvement in the historical process and the scale of Turkey, is said to be quite high at this point. However, the fact that the development of digitalization was realized through mobile subscription instead of fixed subscription is due to the rapid increase in the number of subscribers with the introduction of 3G in 2009 and 4.5G in 2016. This situation is considered as a development of the mobile sector as well as the broadband sector within the scope of the annual reports published by the Information and Communication Technologies Authority. Accordingly, the fact that the digitalization facility in question also has mobile technology content can be considered as a high level of digitalization.

On the other hand, the results corresponding to the average 81 Turkey provinces may differ in distribution. Accordingly, the following figures have been prepared in order to make a comparative evaluation between provinces.

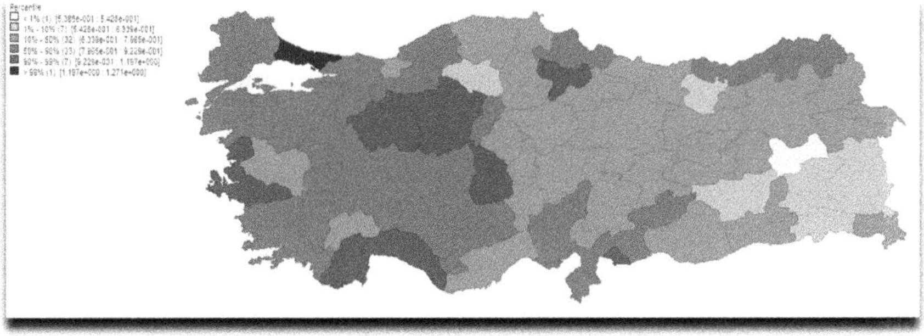

Fig. 2: Broadband internet subscription (2018)

The broadband Internet subscription rate increases as you move from light to dark. Accordingly, the provinces where broadband Internet subscription is high are located in the western and southern regions, while the provinces with the lowest are located in the eastern and northern regions. However, when the provinces are examined in comparison with the neighboring provinces, it can be said that the provinces are in different groups than at least one neighbor. Therefore, despite the concentration at the regional level, there is diversity on a provincial basis.

Beyond the general structure of the distribution, the province with the highest subscription rate is Istanbul and the province with the lowest is Muş. Gümüşhane,

Bitlis, Şırnak, Diyarbakır, Van, Siirt, and Çankırı are in the category of provinces with low subscription rates after Muş. Ankara, Amasya, Aksaray, İzmir, Kilis, Antalya, and Eskişehir are in the category of provinces with high subscription rates after Istanbul.

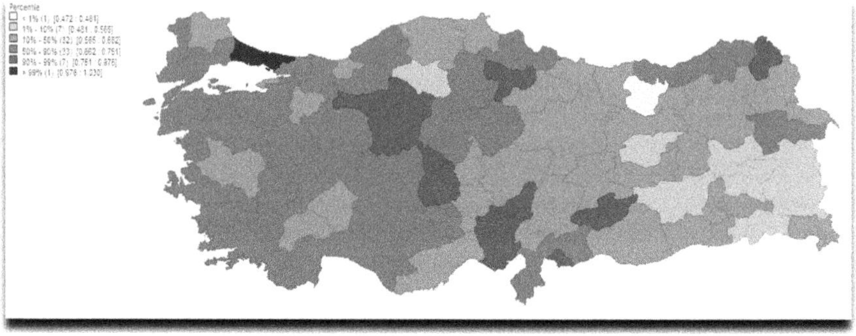

Fig. 3: Broadband internet subscription (2018)

The broadband Internet subscription rate increases as you move from light to dark. As assessed within the scope of Fig. 1, the general structure of the distribution is similar to the scale of Turkey because the broadband Internet subscribers rate majorly consists of the mobile broadband subscribers. Accordingly, as in broadband Internet subscription (Fig. 2), the provinces with high mobile broadband Internet subscription are located in the western and southern regions, while the lowest provinces are located in the eastern and northern regions. However, when the provinces are examined in comparison with the neighboring provinces, it can be said that the provinces are in different groups than at least one neighbor. Therefore, despite the concentration at the regional level, there is diversity on a provincial basis.

Within the scope of the distribution, when the provinces are evaluated specifically, the province with the highest subscription rate is Istanbul and the lowest is Gümüşhane. The provinces with low subscription rates after Gümüşhane are Muş, Bitlis Diyarbakır, Çankırı, Şırnak, Van, and Tunceli. Kilis, Amasya, Aksaray, Ankara, Adıyaman, Adana, and Ardahan are among the provinces with high subscription rates after Istanbul.

Figs. 2 and 3, which show the distribution of digitalization possibilities on the basis of 81 provinces as a whole, show that there isn't a significant difference between the provinces in the same distribution category, while there is a significant difference between the regions. Accordingly, it can be said that the broadband and mobile sector is higher in the provinces, especially in the western part of the country, and therefore the province-based distribution of the country's digitalization possibility differs significantly.

Second Stage: Political Participation Level by Province

At this stage, in order to reveal the level of political participation of the local people, the participation rate in the local elections is divided into 81 provinces and shown and evaluated through the map.

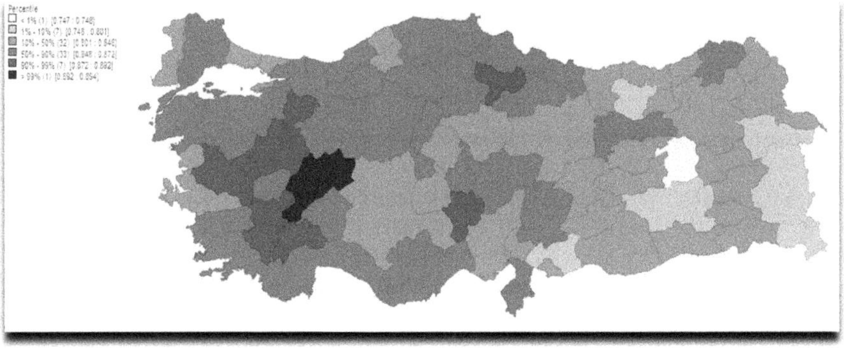

Fig. 4: Participation rate in local elections (2019)

As we move from light to dark, the participation rate in local elections increases. Accordingly, the provinces with high participation rates in the local elections are located in the western and southern regions, while the provinces with the lowest are located in the eastern and northern regions. However, when the provinces are examined in comparison with the neighboring provinces, it can be said that the provinces are in different groups than at least one neighbor. Therefore, despite the concentration at the regional level, there is diversity on a provincial basis.

When the province was evaluated within the scope of the distribution, the province with the highest participation rate was Afyonkarahisar (89 %) and the lowest was Bingöl (75 %). The average of 81 provinces is 84 % and the level of participation is quite high. In this context, other provinces with relatively low participation rates are Ağrı, Gümüşhane, Diyarbakır, Van, Gaziantep, Hakkari, and Batman. In the category of provinces with high participation rates after Afyonkarahisar, Manisa, Bilecik, Burdur, Denizli, Amasya, Kütahya, and Niğde are located.

Third Stage: Categorization of Digitalization Opportunities by Province and Political Participation Level

At this stage, all 81 provinces are divided into four categories according to the digitization opportunity and the level of political participation and are shown and evaluated through the map.

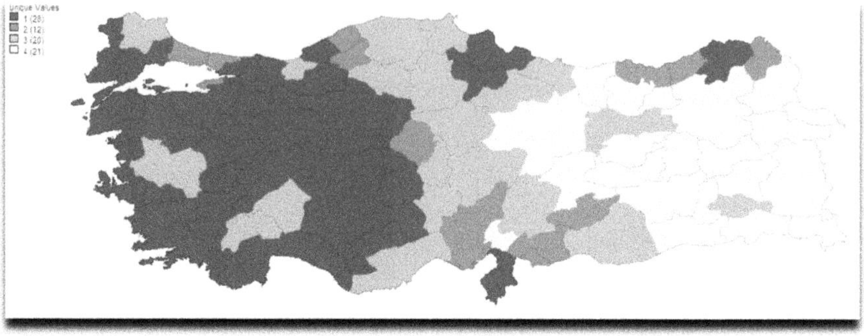

Fig. 5: Grouping of Internet access and elections' participation rate

The four categories presented through the map were created with a combination of situations where the rate of participation in local elections is low and high for each situation where the Internet access rate is low and high. As we move from light to dark, the lightest color indicates Internet access and participation rates in the elections are low, light color indicates high participation rate and low Internet access rate, dark color indicates low participation rate and high Internet access rate, and the darkest color indicates the provinces with high Internet access and participation rates.

In the provinces in the first category (darkest color), Internet access and elections participation rates are high. There are 28 provinces in this category, and all of these provinces are located in the west of the country except Hatay, Artvin, Samsun, and Amasya. Therefore, it can be said that there is a regional concentration in the western part of the country in terms of both Internet access and high political participation rates.

In the provinces in the second category (dark color), the Internet access rate is high; however, the rate of participation in the elections is low. There are 12 provinces in this category, and the neighborhood relationship between these provinces is low and they are located in different regions. Therefore, it cannot be said that there is a regional concentration in this category.

In the provinces in the third category (light color), the participation rate is high, but Internet access rate is low. There are 20 provinces in this category and these provinces are concentrated in central Anatolia from north to south except Manisa Burdur, Isparta, Düzce, Kahramanmaraş, Siirt, and Erzincan.

In the provinces in the fourth category (lightest color), Internet access and election participation rates are low. There are 21 provinces in this category, and all provinces are located in the east of the country from north to south except Sinop and Osmaniye. Therefore, it can be said that there is a regional concentration in the western part of the country in terms of both Internet access and high political participation rates.

In this framework, the relationship between the categorization made in Fig. 5 and the rate of Internet access (Figs. 2 and 3) and participation rate in local elections (Fig. 4) is presented comparatively on 81 provinces. Although the political participation rates differ, the Internet access rate is high in 40 provinces in the first and second categories, and low in 41 provinces in the third and fourth categories. However, although Internet access rates differ, the political participation rate is high in 48 provinces in the first and third categories and low in 43 provinces in the second and fourth categories. As assessed within the scope of Fig. 4, although there are differences between provinces, the average participation rate of 81 provinces in local elections is quite high. Accordingly, the fact that political participation is already on a local scale is an indication that the public has the character of a political actor in the local sense. However, despite being a political actor, the restriction caused by Internet access opportunities causes the contact with the politicization opportunities provided through digital media to be limited.

Fourth Stage: Digitalization Opportunity of Political Participation by Province

At this stage, the inclusion of the city councils' web pages as a factor that facilitates the political participation of the local people is shown and evaluated on a map by dividing into 81 provinces.

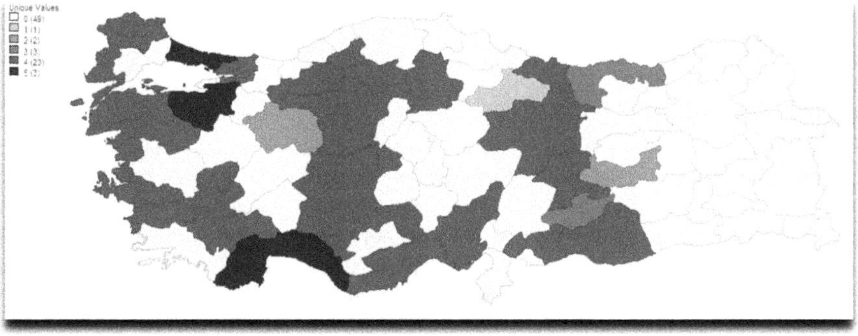

Fig. 6: Digitalization level of city councils

The level of digitalization of city councils increases as you move from light to dark. Accordingly, the provinces where broadband Internet subscription is high are located in the western and southern regions, while the provinces with the lowest are located in the eastern and northern regions. However, when the provinces are examined in comparison with the neighboring provinces, it can be said that the provinces are in different groups than at least one neighbor. Therefore, despite the concentration at the regional level, there is diversity on a provincial basis.

The provinces with the highest digitization level are Bursa, Antalya, and Istanbul when the province is evaluated within the scope of the distribution. Necessary level of information is provided on the web pages of city councils in these provinces. In addition, while there is an opportunity to access other information in the city council of 23 provinces other than working or publishing the annual report, it is possible to reach very limited information in Giresun, Trabzon, Adıyaman, Eskişehir, Elazığ, and Tokat. In 49 provinces, the city councils do not have a website, and limited announcements are made through the municipalities in these provinces.

5 Conclusion

Enabling digitalization within the administration–citizen relationship has six basic requirements: technological, organizational, legal, ethical, sociocultural and economic requirements. Of these, particularly legal and technological requirements constitute the basis of digitalization. While the legal requirement is the preliminary factor provided by the administration and allowing for contact between the citizen and the digital context, technological requirements have a context related to both the administration and the citizen. If the necessary technological infrastructure is presented to the citizens by the administration, whether the citizen makes use of this service is related to whether the citizen is or can be included in the social process, therefore, the political process.

As a result of the technicalization of political financial processes and the deepening of this technicalization with digitalization, a social result is encountered as the passivation of the citizens in the public financial management. When this situation is approached on the basis of 81 cities in Turkey, it is seen that the opportunity to access digital media via the broadband Internet subscription differs between the regions but exists considerably. Considering the citizen's opportunity to contact the digital media as well as the level of participation in the local elections, which directly concern the citizen in the political context, it is concluded that the level of political participation is substantially high, but there are differences between the regions. When the citizens' access to the digital and behaviors toward participation in elections are taken into consideration comparatively, the regional differences become clear, and there are small differences between the regions when citizens are vested with political actors; it is seen that the restriction caused by the Internet opportunities limits the contacts with politicization opportunities provided via digital media. It is observed that there is a significant difference between the east and the west of the country regarding this limitation, and this limitation has emerged clearly in the east of the country. On the other hand, when the city councils that allow the local people to be included in the political processes by helping with solving the technicalization problem related to financial processes are considered in the digital context, a significant problem is faced. Accordingly, the great majority of city councils, which are the main actors between the public financial management process and the citizens, do not have a digital platform. Similar to the other results, the absence of digital platforms, especially in city councils in the eastern cities of the country, reveals a significant difference between the east and the west of the country in terms of fiscal transparency.

In this sense, when digitalization is evaluated in the context of the social dimension of fiscal transparency on a regional basis in Turkey, it is concluded that the technological factor problem, which is basic in the formation of digitalization, is primarily apparent. The fact that the participation of the citizen in the public financial management process as a political actor is primarily a technical knowledge problem and the structures that allow for solving this problem do not have a digital infrastructure is the basic problem in the social context of fiscal transparency. Accordingly, the political actor quality of the citizen in the context of public financial management is substantially affected by the technicalization of the process. Since this technical process is digitalized and this digitalization does not have a level enabling the participation of the citizen, fiscal transparency is not provided for the citizen with the effect of digitalization although it seems like it is provided with its administrative dimension. Under the current circumstances, the citizens become passive actors within the scope of the public financial management due to the lack of the citizen context of digitalization and, thus, digitalization. In this sense, two basic policy recommendations can be made. First, since fiscal transparency is related to the overall country, priority should be given to specific regional policies, allowing for the elimination of the regional differences arising from the level of digitalization. Second, the digitalization of the city councils, which serve as tools to ensure fiscal transparency, should be another privileged policy.

Bibliography

Anselin, Luc, (1992), SpaceStat tutorial. *A Workbook for using SpaceStat in the Analysis of Spatial Data,* Regional Research Institute, West Virginia University, Morgantown.

Anselin, Luc, (1998), “Interactive Techniques and Explanatory Spatial Data Analysis”, (P. Longley, M. Goodchild, D. Maguire, and D. Wind, Eds.), *Geographical Information Systems: Principles, Techniques, Management and Applications,* New York: Wiley, pp. 253–265.

Bailey, Trevor C. and Gatrell, Anthony C., (1995), *Interactive Spatial Data Analysis.* Harlow, UK: Addison Wesley Longman Limited.

Barcevičius, Egidijus, Cibaitė, Guoda, Codagnone, Cristiano, Gineikytė, Vaida, Klimavičiūtė, Luka, Liva, Giovanni, Matuleic, Loretta, Misuraca, Gianluca, Vanini, Irene, (2019), “Exploring Digital Government”, Spain: European Commisson,https://publications.jrc.ec.europa.eu/repository/bitstream/

JRC118857/jrc118857_jrc_s4p_report_digigov_soa_04122019_def.pdf (15.15.2020).

Dikmen, Ahmet Alpay, (2003), "Kamu Yönetimi Temel Kanunu Taslağı Hakkında Sorular ve Yanıtlar", *A.Ü. SBF GETA Tartışma Metinleri, Özel Sayı*, Vo. 59, pp. 35–46.

Doğan, Esra, (2019), "Toplumsal Katılımı Belirleyen Faktörler: Türkiye'de Katılımcı Bütçenin Uygulanabilirliğine Dair Bir Sorgulama", *Ankara Hacı Bayram Veli Üniversitesi İktisadi ve İdari Bilimler Fakültesi Dergisi*, Vol. 21, N. 3, pp. 159–184

Eroğlu, Erdal and Demirbaş, Tolga, (2016), "Mali Saydamlığın Önemi ve Türkiye'de Kamu Mali Yönetiminin Mali Saydamlık Açısından Değerlendirilmesi", *Atatürk Üniversitesi Sosyal Bilimler Enstitüsü Dergisi*, Vol. 20, N. 1, pp. 287–312.

Fiscal Transparency (2020), "The Global Initiative for Fiscal Transparency: Public Participation Principles Guide", http://www.fiscaltransparency.org.mk/index.php/public-participation-principles-guide, (15.01.2020).

Griffith, Daniel, Layne, L.J, Ord, J.K. and Sone, A., (1999), *A Casebook for Spatial Statistical Data Analysis: A Compilation of Analyses of Different Thematic Data Seti*, Oxford University Press.

IMF, (2018), "Fiscal Transparency Handbook 2018", Washington D.C. https://www.elibrary.imf.org/doc/IMF069/24788-9781484331859/24788-9781484331859/Other_formats/Source_PDF/24788-9781484348598.pdf, (15.01.2020).

Information Technologies and Communications Authority, (2020), "Information Technologies and Communications Authority Annual Statistical Bulletin on the Provincial Basis of the Electronic Communication Sector 2018", https://www.btk.gov.tr/uploads/pages/yillik-il-istatistikleri/elektronik-haberlesme-sektorune-iliskin-il-bazinda-yillik-istatistik-bulteni-2018.pdf, (15.01.2020).

Information Technologies and Communications Authority, (2020), "Information Technologies and Communications Authority Annual Report 2018", https://www.btk.gov.tr/faaliyet-raporlari, (15.01.2020).

Narin, Özgür, (2018), "Sanayi 4.0'ın İma Ettikleri ve Sınırlarına Dair Bir Tartışma: Çevrimiçi Bir Kapitalizm Olabilir mi?", *Toplum ve Bilim*, N. 144, pp. 233–252.

OECD Reports, (2006), "Implementing E-Government in OECD Countries", Paris: OECD., http://www.oecd.org/mena/governance/36853121.pdf, (01.02.2020).

Open Budget Survey, (2019), "Open Budget Survey 2017: Turkey", https://www.internationalbudget.org/wp-content/uploads/turkey-open-budget-survey-2017-summary.pdf, (15.01.2019).

Open Budget Survey, (2019), "Open Budget Survey 2019: Turkey", https://www.internationalbudget.org/open-budget-survey/country-results/2019/turkey (15.01.2020).

Popper, Karl, (1994), *Alles Leben ist Problemlösen: Über Erkenntnis, Geschichte und Politik*, München.

Söyler, İlhami, (2006), "Kamu Mali Yönetim Sisteminin Bütçe İlkeleri Açısından Değerlendirilmesi (Eleştirel ve Normatif Bir Yaklaşım)", *Gazi Üniversitesi İktisadi ve İdari Bilimler Fakültesi Dergisi*, Vol. 8, N. 1, pp. 187–208.

Turkish Republic Presidential Legislative Information System, (2005), "Municipal Law No. 5393", https://www.mevzuat.gov.tr/MevzuatMetin/1.5.5393.pdf, (15.01.2020).

Turkish Republic Presidential Legislative Information System, (2003), "Public Financial Management and Control Law No. 5018", https://www.mevzuat.gov.tr/MevzuatMetin/1.5.5018.pdf, (15.01.2020).

Turkish Statistical Instute, (2020), "Regional Statistics", https://biruni.tuik.gov.tr/bolgeselistatistik/degiskenlerUzerindenSorgula.do, (15.02.2020).

Defne Nazlı BAŞ

Digitalization of Sports: The Case of Video Referee System in Football

1 Introduction

Today, rapid advances in technology bring changes and transformations in sociological, cultural, and economic fields as well. Digital technologies, which expresses the fourth Industrial Revolution that started in the 1960s and gained speed in the 1990s, play a major role in shaping the twenty-first century, especially in the field of communication, eliminating the problem of distance and time all over the world. The twenty-first century, called the "information age" or "digital age," has both brought technological innovations to the service of humanity and enabled the development of new scientific methods in different disciplines.

Sports used to be an activity of a certain social class, but now it is a phenomenon that is widely esteemed. The fact that the sport field has spread over a large area is an indication that it has become a massive phenomenon. It makes the possibility of examining a micro miniature of the social structure we live in through the sports institution more available. Sports includes in itself many elements such as power relations, contradictions among classes, the way the political structure works, gender discrimination, and how the existing economic logic works. Sports is a field that should be investigated at least as much as other institutions (Talimciler, 2018: 169–178).

To what extent sports uses technology, how it adapts, and the problems caused by the technological systems that are being implemented as the return of the digitalization are important in the sports industry gaining a new dimension day by day. The involvement of the sports in this process has made it compulsory to improve conditions and make it a functional area for providing convenience and better service to the fans, players, and people involved in the match.

With the rapidly developing technology, the digitalized society is gradually transforming the existing society structure into surveillance society. New surveillance techniques based on data collection and processing have created a new system in which everyone in the domain is constantly checked, monitored, examined, evaluated, and judged. Technologies used as a surveillance tool in sports have brought transformations like providing high security, minimizing errors and maximizing data security.

Technology, which has an effect in many areas, is gradually increasing its effect in sports, and with the beginning of the 2000s, technological tools and equipment have been developed in order to provide the best conditions for the audience, players, and competitions and some studies have been started to carry out in this area. When we look at the technologies used in sports, mobile ticketing, diversity in media organs, wearable technology, Hawk eye, virtual lines drawn on the football field, smart stadiums connected to the computer and electronic system are seen as reflections of the fourth Industrial Revolution.

In this study, it is aimed to point out how sports, which has been gradually affected by technology, has transformed with digitalization and how the video assistant referee (VAR) system implemented in Turkey is represented in media. Despite the use of technology, it is significant to consider what kind of problems the video referee system, which is interrupted by human interference, that is, the final decision of the referee is effective, causes and what it facilitates in the sports community.

2 Sports as a Concept and Historical Development of Sports

Sports is a phenomenon that is influenced and fed by the society in which it is performed. Considering that sports is a socially constructed area, it becomes stronger to establish a connection with social life through sports and transfer of values becomes possible. As sports is a part of sociological integrity, it gains value as one of the indispensable areas of sociology. The increasing importance attached to sports has made it a kind of activity. We have the opportunity to examine a micro miniature of the social structure we live in through the sports institution (Talimciler, 2016: 170). The sports concept, which is a different stage of the changing and transforming structure of play, and its historical process are discussed. The word sports has originated from Latin. In Latin, the words Desporta re and Isportus mean entertainment, enjoy oneself, and dally. Having loaned from French to English in the eleventh century, *desport* means entertain, amuse, and relax. In English, the word sport has started to be used as idle away, entertain, dally and hobby (Talimciler, 2018: 15). According to Diem, sports "is a physical exercise that depends on the individual's free will, contains certain values, has a unique ceremony, is an exercise in the form of competition, is performed in a festive and joyful manner, and is foreseen to become competent with developed and specified rules." According to Sports Pedagogue Neusel, sports "is an action that includes all motoric activities organized through leisure time, rest, being in good condition or competition." For sociologist Heinmann, sports 'is a physical activity. Sports is regulated according to social norms, it is not intended for production' (Orhun, 1991 quoted in Talimciler, 2018: 15). In the process that play has changed by altering its shape over time, it is revealed that the sports industry, which affects a wide area, as well as the sports institution have turned into a business. Play, which Huizinga defines as older than culture, has undergone a transformation. Besides being a voluntary activity, play is distinct from ordinary life both to locality and duration (Huizinga, 2018: 9–23).

The birth of sports can be traced to the Neolithic Age. It is believed that these first movements of the individual who learnt defense and attack to meet their basic needs are also the origin of sports. It is thought that the physical struggle with confronted animals, wrestling and running away from predators led to sports such as javelin and archery as tools developed over time (Bozyiğit, 2016: 3–6). In ancient times, thought, philosophy and art activities in Greek civilization, which shaped the basis of Western civilization, formed the basis of modern culture as well as sports. The physical movements, which began in the form of a struggle at first, then continued as entertainment, and the superiority of man turned into sports with the development of rules. The ancient Olympics, which took place between 776 BC and AD 393 in the ancient Ages, were the prominent sports events in the world. Therefore, the first sports competitions in history emerged in Greek city-states where every other person was a slave (Talimciler, 2018: 17). During this period, women were not able to watch the Olympics since competing players did sports in a completely naked way. Women watching the competition secretly could be killed. Women could not participate in the games to compete, either. In addition to many features that have remained unchanged since the day sports first appeared, this sexist dimension has not changed much: "Sports has been one of the main tools by which masculinity has been built against femininity to provide a distinctly hierarchy of cultural power" (Rowe, 1996: 220). Similar to ancient times, in modern Olympics, the players swear, and sports take place in a sequence. There is the depiction of the Greek goddess of victory Nike on one side of the coin won by players (Tunalı, 2017: 13–14).

In ancient Greece and Rome, especially officers had to be good players. The elite ruling class in the Ottoman Empire underwent a similar educational process as well. In the training received by the prince, sports such as arrow shooting, hunting, javelin, wrestling, and horseback riding were taught according to his wishes and abilities (Talimciler, 2018: 20). Unlike the Greek Empire, sports in the Roman Empire were carried out not by the free Roman citizen but by private professional people trained by prisoners and slaves. Body training took its place in the educational plan of the high class in Egypt. In Iran, however, sports had its place in a military physical education plan in the education of the children of the noble class. Arrow shooting, which was the test of nobility in China, was a musical ceremony with extremely measured movements (Gür, 1979 quoted in Talimciler, 2018: 23–24).

In the Middle Ages, sports lost its old order. In fairs and seasonal festivities, men competed to lift rocks or grain sacks and women ran. The urbanized citizens of the Middle Ages and Renaissance had fun with arrow shooting races. With the Renaissance, sports became a completely nonreligious event. In the fifteenth and sixteenth centuries, harmonious movements and aesthetics gained importance in sports. In its present form, competitive playful sports first appeared in England in the seventeenth century. Since the nineteenth century, sports have become a mass leisure activity with the effect of Modernism. There were some social and economic transformations in the nineteenth century. First of all, the fact that large publics had leisure time allowed them to participate in sports at both the player and audience level. Secondly, in the economic crisis of 1871, sports developed, participation in sports increased very much, and people turned to sports increasingly in those dark years to raise their morale (Arık, 2004:111–115).

Programmable machines, developed for the first time in 1968, were the preparators of the third Industrial Revolution. Redefined in the continuation of the "leisure time" approach brought by the Industrial Revolution to human life, sports have become an indispensable sector of the capitalist economy as part of the profit process. Along with the industrialization, the sports industry is organized to provide various goods and services to the participants and the audience. The marketization process of sports has accelerated, especially with the neoliberal policies that have been implemented since the late 1970s. The neoliberal approach is basically based on the view that uninterrupted flow of capital is the key to development and that social and individual values are based on economic indicators (Ertuğran Öğüt, 2014:32).

The commercialization of sports in the twentieth century has been accompanied by the development of Fordism, which is defined as the industrial and social system that develops mass-produced, advertised, and consumed commodities by mechanizing the tasks at work and distributing them to the production regions that are spatially concentrated. Examples of these are sports competitions becoming the subject of mass media and the mass advertisement and consumption of sports brands, scale and standardization. The status of sports in the face of the culture industry has found its place with the agreements made in the second and third Industrial Revolutions it supports. Conscious targeting of sports leisure practices and identities, organizing the marketing to the specific audience based on image consciousness, forming dispersed production methods globally with its regions and the industrial production in sports have made sports a post-Fordist structure (Rowe, 1996: 196–197).

Industry 4.0 (fourth Industrial Revolution), which is discussed today and was first voiced in 2006 in the United States but with a stronger voic, at the Hannover Fair in Germany in 2011, shows that a new paradigm has emerged in production (Alçın, 2016: 47). This new revolution is called the Industrial Internet Consortium in America. These applications, which consist of essentially similar content, express that cyber systems, objects, and services that contribute to the simultaneous exchange of information in production processes create value through the Internet (Stock and Seliger, 2016: 538 quoted in Tekin and Karakuş, 2018: 2106). Smart phones, tablets, computers, and white goods that save time; machines to ensure mass production; and finally robots that will replace human labor are developing day by day and are included in our lives (Demirkol and Tis, 2018: 536).

As it can be understood from Industry 4.0 applications in the sports sector, based on certain technologies, new types of sports products, types of clothing using conductive polymer structures, smart sensors, smart stadiums and full HD scoreboards, and 4.0 applications such as Internet of Things, cloud computing, augmented reality, autonomous robots–smart machines, and layered production-3D printers are also used in the sports industry (Tekin and Karakuş, 2018: 2115). These extraordinary changes experienced in technology have affected every field as well as sports and led to the emergence of new developments (Argan, 2006). Sports has always existed in human life. The interaction of sports, which is affected by every development in the world, with different disciplines is also inevitable. And this situation has extended the scope of the sports sector (Murathan and Murathan, 2019: 65).

With the beginning of the 2000s, the economic share of sports activities has started to grow and sports competitions have become entertainment products. As a result of this, studies have been carried out to provide the best conditions for the audience and players. Improvements have been made with innovations such as favorable conditions of the field where the players train, better quality of the equipment they wear, and reconstruction of stadiums with state-of-the-art infrastructure. With the improving economic share, there has been a serious increase in the players' wages, and the amount of testimonials spent on player transfers has peaked. In the light of these all, today's sports industry has affected all areas (Akgöl, 2019: 209).

2.1 Digitalization of Sports

The transition from industrial societies to information society as a result of technological developments is another phase of change in the history of humanity. Information technology, which is one of the important development areas in the twentieth and twenty-first centuries, has created an information explosion that multiplies the knowledge of the past centuries and has enabled the produced information to be managed by information control tools and transmitted via Internet technology to reach large masses. This progress has led to digital transformation (Yankın, 2019: 8–9). Digitalization is the structure that connects networks in all data and mass media. Digitalization has made it possible to transfer and connect the best results and all connection between the transmitter and the receiver quickly and without interference. Here all data are processed easily; sounds and images become better quality. Most importantly, with digitalization, data transmission and computer technology have become the dominant factors in all communication infrastructures (Dijk, 2016: 78–79).

Digitalization is basically the use of digitalized data to automate business functions and communications. It means transforming interactions, communications, business functions, and business models (more) into digital media. Digitalization is about implementing technology to the current job. Digital transformation is a broader term than digitalization. Digital transformation is fundamentally the effect caused by the digitalization process. The key goal of digitalization is not only to "go digital" but also add more value or opportunity to the business with modern tools. This may be a small- or large-scale change. Technology brings with it a process that digitizes the tools it creates. Digitization is the process of converting information from a physical and analog format to a digital format. Digitization means computerizing systems and jobs to make them easier and more accessible (KOBİ Bilgi Teknolojileri, 2017).

Digital technologies create and maintain a transformation in all areas of the social system. The fact that sociological digitalization is sustained through the Internet infrastructure spread all over the planet has changed the current production, consumption, and even power relations. Digital products, which can be served over the Internet infrastructure, have accelerated changes in individuals' daily life fiction as they can be easily distributed. Societies create new technologies of the period based on their needs, values, and norms. Digital technologies also reconstruct societies by creating new opportunities. Each new technology transforms the society as "a new input" (Önür and Kalaman, 2016: 271–273). Digitalization, which has intense effects on our lives, especially from the late 1980s to the present, not only produces new forms of perception but also provides new possibilities for the expression of both existing and deriving and evolving the perceptions. Digital media and tools, which take their data from an abstract area, also offer the opportunity to analyze, interpret, transform, and even reconstruct the nature of the physical reality that we perceive as limited to our reference frames (Kuruüzümcü, 2010: 93 quoted in Ekin, 2013: 9). For example, through "robot eye" or simulation applications, environments created by technological developments and drawing the real–virtual boundary and sometimes exceeding the boundary, transfer to the virtual environment in the form of indicators related to the players' performances and technical features every day in the context of "sports events," and after these indicators are made meanings, the answer to the question of what kind of a healing environment is created "in reality" started to be questioned (Ekin, 2013: 9).

When looking at the technologies used in sports, it has incorporated many technologies, especially in football, from wearable technology to the display systems used. From crampons to jerseys worn by football players, goalkeeper gloves to field turf, products that are carefully developed by companies with space technology are used. With the technologies that can determine the possible injuries of players, the types of training that they should do are determined and their biological deficiencies are eliminated.

Many countries develop systems producing sports technologies. Today, innovations such as satellite broadcasts that enable sports organizations to reach more audiences, fiber optic cable communication tools and developed Internet networks have been made. Technology-related applications in the sports industry are followed by all sports product consumers, especially fans. New developments in the sports industry come to the fore with sports that are popular with millions. For example, virtual lines drawn on the football field, smart stadiums connected to the computer and electronic system, smart sports shoes, artificial grass, wearable nanotechnology products, and computer programs compatible with mobile devices are the reflections of the fourth Industrial Revolution experienced today (Karakuş and Tekin, 2018: 2110–2115). Technology in many sectors simplifies and modernizes jobs. Thanks to scientific methods and technology, which is in continuous motion, things have become much more modern and easier in many sectors. The integration of sports, which has become a global phenomenon and has an important economy, with technology seems inevitable in this respect (Akgöl, 2019: 210).

3 The Representation of Sports in Media

Media are all the visual, printed and audio mass media tools that transfer information to individuals and communities and have basic responsibilities such as entertainment, information, and education. In addition to learning what is happening by following the media, people learn how important the issues or problems are with the media's way of presenting the news. Furthermore, it can be said that the need for the emergence of the media originates from the "need for a way of thinking that directs people to reach the right results" (Yılmaz, 2013: 234–235). Media are the tools that have a very important place in our lives in terms of reaching large masses and affecting them easily. The development of the media in the press, the spread of visual and audio tools such as radio and television, and the fact that the rapid developments in technology cannot be expressed with the concept of classical press require a concept to express all of them.

The media play a major role in the development of sports as a social phenomenon. Especially, television, which is one of the mass media, maintains its dominance as the most important tool in the industrialization process of sports with the display of the competitions on the screen. Television and digital opportunities also turn into tools that contribute to the economic field of sports, which will enable the sports industry to reproduce itself at the highest point. Sports contributes greatly to the income of the media as well (Şahan and Akpınar and Ulukan and Akpınar, 2008: 161). Sports and media are in interaction. The media, especially the print media, helps to form the thought. It not only informs but explains and comments on the news as well. It is quite difficult to distinguish between news and commentary precisely. The media, in the way it chooses and presents the news, has a strong influence on the public and affects the flow of events. In business, politics, and sports, there are numerous people who owe their success and failure to news or commentaries. Initially, it was necessary to create a massive interest that would encompass large masses rather than a handful of wealthy people, in order to increase the income and potential of watching sports and make this potential real. Considering that sports have become a marketable commodity and the media market it, this relationship has become inseparable (Demir and Demir, 1995: 285–288).

4 Video Referee System and Its Principles

The video referee application is a system in which another referee or referees watch the replay of a position through the video and assist the head referee in the position where the head referee has difficulty in making the decision while the match is in progress. When we look at the way the video referee application works, the VAR watches the game from the TV screens in the video operation room (VOR). The VAR automatically "Checks" every situation/decision to see if any potential obvious errors are made in a situation that changes the course of the match (goal, penalty, red card, card to the wrong player) or if a serious incident/violation is missed. If a review is not required, there is no need to contact the referee. This is called "Silent check." If a 'Check' indicates that an incident should be reviewed, the referee is informed immediately. If the referee suspects that an important error has occurred or something serious has been missed, they may request a review (Turkish Football Federation, 2018).

Only the referee starts the review in the VAR. The referee may decide that they and their assistants clearly see the incident, and therefore, no review is required. If the referee requests a review before stopping the match, the match is stopped as soon as the team reaches a "neutral" zone/situation where there is no possibility of a good attack. The referee clearly shows that the review process has been started by visually showing a TV screen frame. The restart of the game is delayed until the end of the review. The referees prevent the team/player who "violates" from attempting to restart the game to avoid the review. In case of violations of goals, penalty positions and obvious goal chances, the referee can examine how the ball is possessed from the beginning of the offensive movement leading to the event in question and, if relevant, at the beginning of that stage of the game. The play rules do not allow changes to restart (corner kick, throw-in, etc.) after the game is restarted; therefore, restart decisions cannot be reviewed (Turkish Football Federation, 2018).

At the end of the review process, the referee makes the final decision and clearly shows the result of the review; they give/change/cancel (if appropriate) any disciplinary action and ensure the proper restart of the game. Although the VAR system first appeared as an application that was tried in the friendly match between Italy and France on September 1, 2016, it is currently implemented in a total of 16 countries, which are as follows: Austria, Belgium, Brazil, China, the Czech Republic, England, France, Germany, Italy, South Korea, the Netherlands, Poland, Portugal, Qatar and the United States. In addition to these countries, Turkey uses the VAR technology (Turkish Football Federation, 2018).

There are some principles of the VAR system. When we look at these principles, first of all, video technology is used only to correct obvious errors in the decisions that change the course of the match and in serious missed incidents: goal, penalty/not penalty, directly discarding and giving the card to the wrong player (for example, the referee warns/discards the wrong player). Secondly, the referee always makes the final decision. Third, only the referee can start the review; the VAR can only recommend the review to the referee. Fourth, there is no time pressure to quickly review the decision, regardless of the review process since accuracy is more important than time. Players and team officials should not wrap around the referee or attempt to influence the review process or the final decision if a decision is to be reviewed. A player using the "review signal" receives a yellow card. No one is allowed to enter the referee review area at any stage. The player who does this receives a warning (yellow card), and team officials are taken out of the technical area. The referee penalizes the player who signs the TV screen with a warning. The VAR continuously "checks" every situation/decision to see if any potential obvious errors are made in a situation that change the course of the match, or if a serious incident/violation is missed. Therefore, it is strictly forbidden for coaches or players to request a review since anything that is missed can be seen by the VAR (Turkish Football Federation, 2018).

5 Methodology

The method is the set of techniques used to identify cases, measure, and observe different aspects of social life; collect and screen data and study data. Qualitative content analysis was used as a method in the study. Themes and concepts were created as a means of generalizing with qualitative data. With qualitative content analysis, it is also assumed that there are deeper social structures or relationships under the outer surface of reality. Thus, the findings reflect the deeper structures and forces that can remain underneath the surface (Neuman, 2012: 3–471). The raw data used in this study are 427 news between the dates of November 10, 2016 and February 29, 2020 in the sports arena section on the website of Hürriyet newspaper and 652 VAR news between December 9, 2016 and January 31, 2020 on Fotomaç newspaper's Internet source, all of which were converted into systematic data as a result of scanning. Three different categories have emerged with the findings. When we look at our categories, they are presented as "Correction of Mistakes in the Game," "Response to the VAR System," and "System Affecting the Results."

6 Findings

In this section, the findings are within the scope of the qualitative content analysis study coding categories; they are presented under the headings of "Correction of Mistakes in the Game," "Response to the VAR System," and "System Affecting the Results" and supported by quotes from sports news directly.

6.1 Correction of Mistakes in the Game

The referees decide hundreds of times in competitions, but with the use of VAR they only need to report whether there is a violation in order for the goal to be invalid before the goal took place. To report whether there is a clear mistake in a penalty decision made or not. It is to give information about the referee's state of warning or sendoff towards the wrong player in the violating team or which player should be punished and whether there is a clear error in the sendoff decision made or not.

The Turkey Football Federation (TFF), the Central Committee of Arbitration (CAC) President Yusuf Namoğlu, stated that although the referees begin the Spor

Toto Super League season well, later they make some mistakes. For the VAR system, which will be effective in reducing errors he used the following expressions:

> *VAR is a technological event. In other words, a system that helps the referee. Just like goal line technology. We are also inspired by this and we are working to bring our observing institution to a different point towards a system in terms of technological observation. I hope we will implement this as soon as possible...* (HürriyetNews_a, 2017)

In the presentation describing the VAR system, presentations were made to Super League referees and top-class assistant referees with the headlines of the illegal use of hands and arms, the cutting of the promising attack, the action to prevent the obvious goal position, hand-playing, offside and objections. He emphasized that the VAR system will bring football to a different point, and technological development will be useful in correcting the mistakes. In the article titled "*This system will also increase the quality,*"

> *Where people are, there is always a mistake, but technology should be used to minimize the error. Many leagues will use this beginning with the World Cup. From now on, this will become the norm... It is very important for being innovative and the use of technology. I attach great importance to technology... In many games last year, many clubs complained about the referee mistakes. Last year, mistakes were made more than the average... I think that this technology will minimize the errors, the debates will decrease in televisions and newspapers, and so the tensions of the people and the fans will decrease. I attach great importance to this. The more mistakes there are, the more tense there is, and this is reflected in the stands. So, it is a correct investment.* (HürriyetNews_e, 2018)

Although there are rules in the game, mistake seems to be inevitable where there is human intervention. Along with the concept of competition, which is always present in sports and life, the concepts touched upon are the teams' desire to win, the fact that the game is score oriented, and the referees being more devoted and diligent in decision-making mechanisms in matches. In sports competitions, referees are always expected to take a fair and equitable position within the framework of ethical, moral rules when making decisions about the game. Athletes, coaches, and referees also support VAR's decision when they make mistakes and fail to see the position fully.

With the switch to the VAR system, which is the technological profile of football, minimizing some invisible errors in the game comes to the fore. Critical referee decisions that affect the match, such as penalties and red cards, which are discussed for days or even weeks in sports programs, are discussed. With the VAR system, the important events, errors and mistakes that affect the game are corrected and the decision-making referees of the game are both seen as being easier and a more "fair" and "equal" game takes place. In the news article titled "*Call for Video Referees from Gümüşdağ,*"

Not only in Turkey, now we have to accept that change the values of football in the world. It is time to move away from old beliefs such as "Referee mistakes are the beauty of football." Unfortunately, we all know that there are referee errors behind the loss of points that can cost us a lot... That's why the Video referee application is a revolution that must be realized without losing time with some outdated criticism. Football is a game that deservers must win. It must be promoted to a system that will not allow good faith or malicious mistakes. The beauty of football should be revealed with regular goals and error-free whistles... I request that all our stakeholders take the video referee application very seriously, whether they postpone it or not, and remember that the future of football will be one of the most important elements of the brand value. (FotomaçNews$_a$, 2017)

When we look at the prominent situation in the news, it is directly affecting the match when making decisions in positions is considered important. Even the most successful referees in the game can misread a position. Making these overlooked situations visible with the VAR system has been deemed necessary for a more regular and systematic progress of the game.

When the comments of the people in the football community from the news about the VAR system are examined, they come to the fore as it is normal, it contributes to the justice in football, and the technology used contributes to the development of football; although the technology will be used, it is the referee who will make the final decision.

6.2 Response to the VAR System

While the video referee system prevented the goals scored or taken due to a wrong referee decision, this created a subject of discussion on social media. Some football fans argued that the VAR system is opposite to the spirit of football, while others argued that this system would determine the results fairly. Comments come to the forefront that the referees will not be at risk, that it is seen as against the nature of football, and that the fans are reacted in the losing matches where the VAR system is used. In an article entitled *"Shocking words! If a decision is made with VAR in the final minute of the championship game...,"*

VAR is a system that needs to be discussed. We accept in principle. We are in the direction of implementation on certain issues. But there is a fact that is ignored here. The referees will make a mid-way decision and will not take risks, relying on the VAR in most controversial positions. While recommending VAR, FIFA recommended that this system will help the referees. He said that this system could assist referees in a limited number. For example, goal decisions, red card decisions, penalty decisions... They said that it can be used for these situations. It will not be applied in decisions such as corners, throw-ins and offsides. In order not to affect the flow of football. But my concern, I think it may affect these referees negatively. Referees will not be at risk. The referees are directing the match under pressure, they may manage a mid-way match, (HürriyetNews$_b$, 2017) *he said.*

German referee Bernd Heynemann, the former FIFA cockade referee, who made evaluations about VAR, criticized that the decisions of the referee on the field were not taken seriously due to the VAR and that he was "*made poor.*" The managers of the Bundesliga clubs, in their recent statements, stated that VAR caused confusion of authority and the decisions started to be discussed more. Augsburg's manager, Stefan Reuter, called for the parties to meet and put the necessity of the VAR system back on the table (HürriyetNews$_c$, 2017). Stating that he is not satisfied in the VAR system, Sivaspor's player Robinho said the following:

> *Frankly, I am not personally satisfied. It covers the beauty of football. It is against the nature of football, so we have to get used to it. There is nothing we can do, we will adapt. We will survive the adaptation quickly.* (HürriyetNews$_f$, 2018)

The VAR System has created a controversy in the media since its implementation. When looking at the prominent themes for the category, those who oppose the use of technology are worried that it will slow down the pace of the match. Another concern is the risk of reducing the concentration of those in the stadium. It is stated that the viewers in the stadium will not be able to see the repetition of the position in which the VAR takes over and they will be excluded from this controversial process. It is also among the reactions that the audience can break away from the match (BBCNews, 2018). In the article titled "*The Video Referee Debate Continues*,"

> *The video referee app split the super league into two. Most of the technical directors said, "The referees' job is very difficult. This application will definitely contribute. It is a situation that should be in football." The opinion of the 4 teachers is very clear: "We are against the halt of the game. We have to trust each other. We should not pollute the football climate. We do not lean towards it...* (FotomaçNews$_b$, 2018).

In the news about the video referee system, it is emphasized that it disrupts the football's game flow and tempo. The game only stops in important positions, in situations such as penalties, red cards, etc. Considering the audience profile of football, which is prominent in the news, since the commitment to the team is an indication of fanaticism, the idea of not accepting the VAR even if they follow a situation against the team, and the decision on technological devices, even if they see the event directly. In the news article "*VAR Will Eliminate the Fudge*,"

> *"I think VAR will rule out the fudge. When you visually show something to people, they can be satisfied, but when you do not, there is always a polemic. I think it will bring a huge plus. Everyone should believe, respect. The clubs who think that they are mistreated, will no longer think as such. Another step towards winning of the one who deserves will be there* (FotomaçNews$_c$, 2018).

In terms of transparency of the game, the use and development of technological equipment in the digital platform can relax the sports spectators and teams. Because the referee in the room sees what the referee in the field cannot see. In every sense, there is a control and supervision mechanism inside and outside the field. It is not essential for the losing teams to say that there is an injustice in their objections in a game where control and supervision is so intense.

6.3 System Affecting the Results

In controversial situations that may affect the score of the match with the VAR system, the referees now have the chance to change their decisions by watching the position again. The VAR practice in Turkey is that the VAR team will not be involved in gray positions. The VAR will intervene in all positions that are open and clear. The VAR will not decide; it will allow the referee to follow the position again. The decision will be made by the referee of the match again. In the news article titled "*Support from Maradona to the Video Referee Application*," Maradona, a legendary name of football, said that the goal he scored against England in 1986 with the Argentina jersey and called the Hand of God would not count if there was the video assistant referee. Speaking to FIFA, Maradona said he supports the VAR:

> *Technology brings transparency and quality. Thanks to this new technology, teams get positive results. The goal I scored against England in 1986 would not be counted. I even used my hand to pull the ball off the line against the Soviet Union at the 1990 World Cup. We were lucky the referee did not see. At that time there was no technology, but today there is a different story* (HürriyetNews$_{d,}$ 2017).

He states that the VAR system will change the course of the match, and the goal he scored would not count if the Video Assistant Referee was present. Maradona's goal against England, which he called the "Hand of God" at the world cup in Mexico in 1986, has been the subject of many years of debate due to reasons such as insufficient number of referees and limited technological opportunities. Nowadays, it is impossible to decide in favor of situations like Maradona's goal. Because there are two referees and two assistant referees on the field, a total of four referees. In fact, in the last European Championship, there were six referees with the addition of line referees. With the introduction of the video referee system, the decision of the wrong referee decreases and intervention with the VAR comes to the fore. In the news article titled "The *VAR stamp for the First Half of the Super League*,"

> *The Video Assistant Referee (VAR) system, which started to be implemented in Turkish football this season, allowed 98 decisions to change in the first half of the Spor Toto Super League... In 24 positions when the ball went to the net the referees detected an offense, in opposition of the scoring team, such as an offside, foul or handball before the goal was scored. After consulting VAR, the referees changed their decisions and granted the goal... In 27 Penalty position in the first half of Super League, referees changed their decisions after consulting with VAR... Referees cancelled 16 red cards appeared in the first half of the season after the examination with VAR... In addition, referees showed yellow card 23 times after viewing the video in VAR and changed a corner decision to fault, resulting in penalty...* (FotomaçNews$_{d}$, 2018)

It is emphasized that the video referee system will make decisions that can significantly change the course of the game. Especially in positions such as penalties and red cards, decisions were taken directly for or against the teams because of consultation to VAR. The decisions that changed as a result of the VAR room's decisions in positions that the field referees could not see during the match were seen as very determinant in the axis of winning and losing the game. Positions such as penalties not given in some matches and red cards not shown without consulting the video referee system were discussed in the sports media. When the video referee system was introduced, comments were made regarding the transparency of all aspects of the game. In the article titled "*VAR does not kill football*,"

> *A position that is penalty for everyone, that is for you, for me, for everyone; and issues such as sendoff, VAR will take over immediately and make the right decision, given the referee made a mistake. If this is the case, we will not consult it every time, the match will not stop, it can be completed with very little pause in the match. Of course, we have seen examples in the world: wrong examples such as 2 minutes have passed, blah blah, cannot prevent a correct system. It is an implementation error if you ask me. If the implementation is done correctly, this system fits better with a good education and finalizes with the right decisions. To me, it doesn't kill football. By making the right decisions there, those who have the right in the field are ensured to get the result.* (FotomaçNews$_{e,}$ 2018)

In the news that stands out against this news article, there is no stop at every position in every match when consulted to the video referee system. As for the implementation and method, the federation presidents and representatives of the arbitration committee of each country should provide continuous training to the referees. Today's referees are now seen as more knowledgeable, equipped and referees from the sports culture. Referees are expanding the game rules and information network through national and international seminars. As the referees regularly update themselves about the VAR system in the digital arena and receive information, there are reports that there will be no problem in implementation.

7 Conclusion

In this study, the developments, problems, and discussions created by the VAR system used in football competitions were examined. The study is limited to raw data obtained from the Internet news sites of *Hürriyet* and *Fotomaç* newspapers. As a result of the qualitative content analysis, three coding categories were evaluated under the headlines: the correction of the mistakes in the game, response to the VAR system, and the system affecting the result in terms of how the sports adapted to the technology with the video referee system, supported with citations from the news.

In the news articles, although the technology will be used in the category of correcting the mistakes in the game, it is the opinion that it is normal that there may be a mistake in the place where the human is, since it is the referee who will make the final decision. Again, this system contributes to the justice in football. The themes of use of technology, bringing football forward, minimizing the mistakes made by the referees, and reducing the reactions of the fans in the decisions made come to the fore. In the category of response to the VAR system, it is seen that issues such as the opinion of referees is to make a mid-way decision and not take risks by relying on VAR in most controversial positions come to the fore. Discussions are reflected in the news as the idea of covering the beauty of football and contrary to the nature of football, temporally decreasing and slowing the tempo of the match, and reactions of the fans to the VAR in case of defeat. In the last category, the system that affects the result, it is seen that there is a discussion that all positions will directly affect the outcomes in critical points. *FIFA said later that the correct decision rate was 99.3 percent, thanks to the VAR in positions that could affect the fate of the match. Pierluigi Collina, Chairman of the FIFA Referee Committee, stated that the system "is very, very close to perfection"* (BBCNews, 2018).

According to a report released by the Football Federation of Turkey, in more than 800 matches and 20 tournaments, it was considered that the VAR system will interrupt the game very often and it is expressed that it was wrong. "*750 decisions were made by the referees in 3 matches. 15 of them were rechecked by VAR. 1 decision was revised.*" Secondly, it was said that the discussions concerning the affection the time played were wrong. It was stated, "*When VAR is not used, the duration of the match is 94:47 minutes and the added loss time is 287 seconds, when using VAR, this time is average 95:47 minutes and the added loss time is 347 seconds. The VAR system extends the match duration on average only 60 seconds. This period constitutes only 1.34 % of the total time*". Thirdly, it has come to the fore in the news that the VAR system will not contribute to the referees' decision. It seems that this is also wrong: "*It makes the decisions fairer and more accurate.*" Finally, according to the TFF result, 81 % of football fans believe that the VAR system will develop the game (Turkey Football Federation, 2017).

Digitalization has become the most current subject of sports. Digitalization also changes the relationship between the sports spectator, fan, and the club or athlete. Video technology allows coaches and athletes to gain a deeper understanding of their performance. Video recordings also allow live information on the side field by technical directors or coaches at the time of the competition.

As can be seen from the research, the VAR technology has brought the discussions with it from the moment it was implemented. Those who oppose the use of this technology are concerned that it will slow down the pace of the match and disrupt the nature of football. Supporters expressed the view that the mistakes of the referee will decrease. It is inevitable for sports, which has become a global phenomenon and has an important economy, to be integrated with technology.

Bibliography

Akgöl, Oğuzcan, (2019), "Spor Endüstrisi ve Dijitalleşme: Türkiye'de Espor Yapılanması Üzerine Bir İnceleme", *Trt Akademi Dergisi,* Vol. 4, N. 8, pp. 206–225.

Alçın, Sinan, (2016), "Endüstri 4.0 ve İnsan Kaynakları", *Popüler Yönetim Dergisi,* N. 63 pp. 46–47.

Arık, M. Bilal, (2004), *Top Ekranda-Medya Çağında Futbol ve Televizyon Arasındaki Kaçınılmaz İlişki,* İstanbul: Salyangoz Yayınları.

BBCNews, (2018), "Video Yardımcı Hakem (VAR) Sistemi Nedir: Uygulama Türkiye'de de Başlıyor", https://www.bbc.com/turkce/spor-45038263, (18.02.2020).

Budak, Emrah, (2019), "Türk Medyasında Dijital Spor Yayıncılığı Üzerine Bir Araştırma", *Trt Akademi Dergisi,* Vol. 4, N. 8, pp. 226–245.

Dandıl, Emre and Özkul, İnanç, (2019), "Futbol Maçları İçin Bilgisayarlı Görü Destekli Gol Karar Sistemi (GolKaSis): Bir Prototip Çalışma", *Gazi Üniversitesi Fen Bilimleri Dergisi,* Vol. 7, N. 1, pp. 213–224.

Demirkol, Nuray and Tis, Gizem, (2018), "Endüstri 4.0'ın İnsan Kaynakları Yönetimine Etkisi", *4. Uluslararası Kafkasya Orta-Asya Dış Ticaret ve Lojistik Kongresi,* pp. 535–547.

Dijk, Jan, Derk, (2016), *Ağ Toplumu,* İstanbul: Kafka Yayınları.

Ekin, Volkan, (2013), "Sanal Gerçeklik Ortamları ve Uygulamalar: Spor ve Sanal Ortam Göstergeleri", *Online Akademik Bilgi Teknolojisi Dergisi,* Vol. 4, N. 13, pp. 7–22.

Ertuğran, Öğüt, Esra, (2014), "Türkiye'de Spor Yönetiminin Neoliberal Dönüşümü", *Amme İdaresi Dergisi,* Vol. 47, N. 4, pp. 31–58.

FotomaçNews$_a$, (2017), "Gümüşdağ'dan Video Hakem çağrısı", https://www.fotomac.com.tr/medipol-basaksehir/2017/09/04/gumusdagdan-video hakem-cagrisi, (18.02.2020).

Fotomaç News$_b$, (2018), "Video Hakem Tartışması Sürüyorsa", https://www.fotomac.com.tr/superlig/2016/12/28/video-hakem-tartismasi-suruyora, (18.02.2020).

Fotomaç News$_c$, (2018), "VAR Boş Yorumu Ekarte Edecek", https://www.fotomac.com.tr/goztepe/2018/08/07/var-bos-yorumu-ekarte-edecek, (18.02.2020).

FotomaçNews$_d$, (2018), "Süper Lig'in Ilk Yarısına VAR Damgası", https://www.fotomac.com.tr/superlig/2018/12/25/super-ligin-ilk-yarisina-var damgasi, (18.02.2020).

Fotomaç News$_e$, (2018), "VAR Futbolu Öldürmez", https://www.fotomac.com.tr/superlig/2018/03/29/var-futbolu-oldurmez, (18.02.2020).

Güler, Lale and Demir, Vedat, (1995), "Spor ve Medya İlişkisi ve Türkiye'de Spor Medyası", *Marmara İletişim Dergisi*, N. 9, pp. 286–306.

Huizinga, Johan, (2018), *Homo Ludens*, Ankara: Dorlion Yayınları.

Hürriyet News$_a$, (2017), "Video Hakem Uygulaması Geliyor", https://www.hurriyet.com.tr/video hakem-uygulamasi-geliyor-40333056, (18.02.2020).

Hürriyet News$_b$, (2017), "Olay Sözler! Şampiyonluk Maçının Son Dakikasında VAR ile Bir Kara Verilirse..." https://www.hurriyet.com.tr/sporarena/olay-sozler-sampiyonluk-macinin son-dakikasinda-var-ile-bir-kara-verilirse-40631058, (18.02.2020).

Hürriyet News$_c$, (2017), " 'VAR' Bundesliga'da Kaos Çıkardı", https://www.hurriyet.com.tr/sporarena/var-bundesligada-kaos-cikardi-40637160, (18.02.2020).

HürriyetNews$_d$, (2017), "Maradona'dan Video Hakem Uygulamasına Destek", https://www.hurriyet.com.tr/sporarena/maradonadan-video-hakem-uygulamasina destek-40530767, (18.02.2020).

Hürriyet News$_e$, (2018), "Ali Koç: "Bu Sistem, Kalitenin de Artmasını Sağlayacak", https://www.hurriyet.com.tr/sporarena/ali-koc-bu-sistem-kalitenin-de-artmasini saglayacak-40915015, (18.02.2020).

Hürriyet News$_f$, (2018), "Robinho'dan VAR Sistemine Eleştiri" m.tr/sporarena/robinhodan-var-sistemine-elestiri-40940123, (18.02.2020)

Kara, Tolga and Aytekin, Çiğdem, (2014), "Kültür Endüstrisi Bağlamında Aktif Taraftar Çağı: Sosyal Medyanın Futbol Kulübü/Taraftar İlişkisine Etkisi", *INTCESS14 Uluslararası Eğitim veSosyal Bilimler Konferansı Bildirileri*, pp. 1656–1665.

KOBİ Bilgi Teknolojileri, (2017), https://www.kobibt.com.tr/dijitallestirme-dijitallesme-ve dijital-donusum/, (18.02.2020).

Murathan, Talha and Murathan, Fatih, (2019), "Spor Sektöründe Blok Zinciri Uygulamaları", *Gaziantep Üniversitesi Spor Bilimleri Dergisi*, Vol. 4, N. 1, pp. 64–74.

Önür, Nimet and Kalaman, Sefer, (2016), "Dijital Gündelik Yaşam: Yeni Toplumsallıklar ve Dijital Yabancılaşma", *Akademik Sosyal Araştırmalar Dergisi*, N. 35, pp. 269–289.

Şahan, Hasan, Akpınar, Selahatttin, Ulukan, Mehmet and Akpınar, Öznur, (2008), "Spor-Medya İlişkilerinde İletişim Teknolojilerinin Rolü", *Niğde Üniversitesi Beden Eğitimi ve Spor Bilimleri Dergisi*, Vol. 2, N.2 , pp. 155–164.

Tarkan, Oğuz, (2005), "Popüler Kültür ve Sporun Endüstrileşme Süreci", *Kurgu Dergisi*, N. 21, pp. 195–204.

Talimciler, Ahmet, (2007), "Futbol sahaları ve Spor Salonlarındaki Hak İhlalleri ve Taraftarlar Üzerindeki Etkisi", *Ege Üniversitesi Sosyoloji Dergisi,* Armağan Sayısı, pp. 279–291.

Talimciler, Ahmet, (2008), "Futbol Değil İş: Endüstriyel Futbol", *İletişim Kuram Ve Araştırma Dergisi,* N. 26, pp. 89–114.

Talimciler, Ahmet, (2016), "Endüstrileşen Spor/Futbolun Geleceği", *Ege Üniversitesi Sosyoloji Dergisi,* Armağan Sayısı, pp. 167–184.

Talimciler, Ahmet, (2018), *Sporun Sosyolojisi Sosyolojinin Sporu,* İstanbul: Bağlam Yayınları.

Tekin, Zeliha and Karakuş, Kübra (2018), "Gelenekselden Akıllı Üretime Spor Endüstrisi 4.0", *İnsan ve Toplum Bilimleri Araştırma Dergisi,* Vol. 7, N. 3, pp. 2103–2117.

Turkish Football Federation (2018), "Video Hakem Sistemi",https://www.tff.org/Resources/TFF/Images/MHK/VAR.pdf, (18.02.2020).

Yankın, Fahri, Bilal, (2019), "Dijital Dönüşüm Sürecinde Çalışma Yaşamı", *Trakya Üniversitesiİktisadi ve İdari Bilimler Fakültesi Dergisi,* Vol. 7, N. 2, pp. 1–38.

Yılmaz, Mesut, (2013), "Medya ve Siyaset İlişkilerinin Kamuoyu Üzerindeki Etkileri", *İnsan ve Toplum Bilimleri Araştırmaları Dergisi,* Vol. 2, N. 2, pp. 230–252.

İbrahim YÜKSEL

The Effects of Digital Media on the Ways of Thinking of Young People in Secondary Education

1 Introduction

Rapid development in communication technologies brings every culture in the world closer together. As a result, the interaction of every culture on earth with other cultures increases. In the past, it often used to take weeks or even months to hear an event occurring anywhere in the world in other parts of the world, and due to this delay, the effect of the event decreased over time. Nowadays, however, there is a completely opposite situation, and a statement made anywhere in the world can be heard all over the world within a few hours (Maalouf, 2009: 59). While this communication occurred on media tools in the past, today it takes place on information technology tools and mainly social media. The word media comes from the Latin word "medium" and it is the plural of "an intervening agency, means" (Oxford Dictionary and Thesaurus, 2009: 576). This concept, which is traditionally used for newspapers, radio television, and magazines, has recently been used on the Internet (Business Dictionary, 2020). The word social comes from the Latin word "socius" and means "companion" or "ally." In another definition, it is also described as living together or enjoying life in communities or organized groups (Collins English Dictionary, 2016: 762; Oxford Dictionary and Thesaurus, 2009: 882).

Social media has emerged as a new type of online media, with high levels of sharing unlike traditional media (Mayfield, 2008: 5). The most important feature of social media is probably that it allows users to create the content of social media themselves, while they are not allowed to interfere with traditional media content. On the social sharing page, people can prepare their own page and upload their photos or videos (Waters, 2010: 2, quoted by Balcı and Koçak, 2017). The Internet has started to become not a tool but an objective for people who organize all their daily life on social media. People have come to be in a situation where they follow news on the Internet not daily but instantly, leave a message on their personal page instead of calling each other, can video chat with their friends all over the world, can shop online on the virtual environments, can play games, or receive education (Hamamcı, 2015: 18). Nowadays, the use of social media applications has started to become necessary for many people in daily life (Masthi, Pruthvi an d Mallekavu, 2017: 207). With the increase in the frequency of using the Internet, the frequency of social media usage increases accordingly. In fact, it is thought that a significant part of Internet usage will be carried out via social media in the near future (Tektaş, 2014: 852).

Along with the change and development of communication opportunities, lifestyle and technology, the process of obtaining information has changed and taken a different dimension. In particular, the change of the media, the social media environment born with the new media has played a decisive factor in the diversification, dissemination, form, access, and internalization of information. Now individuals, especially young people, can easily follow political, social, cultural, economic, artistic, etc. information and agenda on the Internet and social media. The fact that the individual meets information, internalizes that information, produces new information with that information, and applies or uses this information in his life cycle is very effective in shaping their worldview and perspective on life (Kamiloğlu and Yurttaş, 2014: 132). Social media has an effect that can direct and activate people. Social media has increased its influence in the political field, like many other fields, in recent years. It has enabled individuals to be active in political participation and express their ideas more easily. Especially with the development of mobile technology, it has provided opportunities such as conveying an event or news to many people at the same time and mobilizing them and increased its power in terms of influencing political knowledge and shaping behavior day by day (Asığbulmuş, 2017: 93–94). The effective use of many different social media applications that will enable young people to communicate and share their ideas with each other has paved the way for studies to influence them (Aktas, 2004: 211).

It can be stated that media has a direct and controlling feature on people and their behavior (Katz and Lazarsfeld, 1995). The agenda-setting theory aims to explain the feature of media to determine what people will think. With this feature, the media has a dictating effect on people's thoughts. Similarly, it has positive or negative effects on young people's lives and development. Therefore, it has the same potential in influencing young people on social media (Lemish, 2013). Social networks enable individuals to have a profile page where they can share photos, videos, and information. In addition, they can create virtual communities related to their common interests in social networks or can be organized for a common purpose with individuals share common past, worldview, or political view (Filiz, Erol, Dönmez, and Kurt, 2014: 17). Today, as a requirement of the digital age, governments, local governments, public institutions, banks, and private companies have social media accounts, and it becomes important to show their presence on social media. Social media has a significant place in influencing community behavior (Kirschenbaum, 2004: 99).

As a concept, "social media" has begun to be in the middle of social life with computers, the Internet, smartphones, and other technological developments and thus has become the subject of scientific research. When social media is considered as a tool for individuals to express themselves, there is no sociocultural topic that it does not touch upon in terms of its effects. The effects of social media in every field, from the economic structure to cultural behavior, seem to be a subject that can be addressed (Konuk, 2019). According to Babacan (2014: 156), the relationship of social movements with the traditional ones has altered with the Internet and social media tools. "The relationship that social movements live with the Internet and social media tools shows that social movements are not only manipulated out of purpose, but they have undergone a radical transformation in the face of the atomized and fragmented reactions or the irregularity of the opposing stances from the social media tools that have many users." New forms of communication or social media tools, described as new media (Facebook, Instagram, Twitter), affect and transform society and social behavior (Konuk, 2019: 793).

Young people spend most of their time using social media and read many things that may affect their overall life and personality traits. Therefore, it can be asserted that social media is a very effective factor on young people's worldviews. The use of this medium has brought interpersonal interaction to digital media unlike the older generations. Government institutions, institutions of different structures, nongovernmental organizations and movements, private sector groups and their programs, and legal or illegal structures are on social media. All these groups can easily reach young people on social media and affect their worldviews (Gacumo, 2018). Young people born after 1990s have made technology and social media an integral part of their lives (Sheldon and Byrant, 2016). The use of social media arises from the feeling of belonging to something and the desire to display a character who is more different than it actually is (Nadkarni and Hofmann, 2012). With the use of the smartphone with the Internet being effective in human life, many researches have been started in this field. For example, studies by researchers such as Gross, Juvonen, and Gable (2002) and Hampton and Wellman (2001) show that young people are almost always online. The widespread use of smartphones and social media, especially among young people, has made it important to investigate the possible effects of these applications on young people.

2 Purpose of the Study

In today's modern societies, it can be claimed that the intensive use of information technology tools such as the Internet, mobile phones, and computers influences the way people think. Based on the fact that especially young people use new technologies and spend a lot of time with them, this study analyzes the effects of digital media on the way young people think. When the literature on youth and political participation is examined, it is seen that most of these studies are carried out on university youth. There are not many studies on secondary education youth. It is believed that the study will fill an important gap in the field. For this reason, this study aims to examine the effect of digital media in shaping secondary school students' interest in current issues and emerging their worldviews. To achieve this goal, answers to the following questions were sought:

1. What digital tools do secondary school students use to get information about current topics?
2. Which individuals or institutions do secondary school students approvingly and actively follow?

3. What are the opinions of secondary school students about the reliability of the posts on social media?
4. What do secondary school students think of digital media influencing their peers' thoughts?

3 Research Method

In this section, the method of the research, the working group, the type and source of the data, the data collection tools, and the statistics used in the research are given.

3.1 Research Model

Qualitative method was used in the research. It is a phenomenological study in terms of determining the influence of social media in shaping secondary school students' worldviews. Semistructured interview form was used as a qualitative data collection tool. In the qualitative study, maximum variation case sampling, which is one of the purposeful sampling methods, was used (Yıldırım and Şimşek, 2011). Face-to-face interviews were held with teachers in three different Anatolian high schools in Antalya. While these three Anatolian high schools were selected for the research sample, the student diploma achievement ranking in transition from primary education to secondary education was taken as a criterion. According to this criterion, based on the results of student success in transition to secondary education, a selection was made among the Anatolian high schools in the top, middle, and lower categories in Antalya city center. The study was conducted with thirty high school students from three different schools.

3.2 Study Group

The demographic information of the student group studied below is given in a table.

3.3 Data Collection Tools

The interview form developed by the researcher was used in the qualitative part of the study. In order to ensure the internal validity of the qualitative interview form, the interview form was delivered to two experts, and the form was given its final form after being examined. After a pilot interview with four teachers, interview voice recordings were converted into a written form on the interview transcript form created in computer environment in order to determine whether the questions were clear and understandable and whether the answers given reflected the answers to the questions asked. Another expert was requested to check whether the questions asked were clear and understandable, whether they covered the subject, and the possibility of providing the necessary information by examining the transcripts. The validity of the question items was determined at the end of this study. In qualitative research, internal validity is related to whether the

Tab. 1: Demographic information of the students

Code name	Grade	Age	Sex
S16	9	15	Male
S17	9	15	Male
S18	9	14	Male
S19	9	15	Female
S20	9	14	Female
S11	10	16	Female
S13	10	15	Female
S14	10	17	Female
S22	10	16	Male
S23	10	15	Female
S25	10	16	Male
S3	10	15	Male
S10	11	17	Female
S21	11	16	Male
S24	11	17	Male
S26	11	17	Female
S27	11	16	Female
S30	11	16	Male
S7	11	17	Female
S8	11	17	Female
S9	11	17	Male
S1	12	18	Male
S12	12	17	Male
S15	12	18	Male
S2	12	17	Male
S28	12	17	Female
S29	12	18	Female
S4	12	18	Female
S5	12	18	Female
S6	12	17	Male

researcher can actually measure the data they want to measure using the tool or method they use (Yıldırım and Şimşek, 2011).

3.4 Data Analysis

In the analysis of the qualitative data obtained, descriptive analysis and content analysis techniques were used together by utilizing the NVIVO 12 qualitative research program. In the descriptive analysis phase, a framework for data analysis was created primarily based on the conceptual framework of the research and the research questions. The narratives of the interviewees were also directly included (Yıldırım and Şimşek, 2011). In order to reveal the underlying concepts and relationships between the content analysis and the data, the interview records were listened and transferred to the electronic environment line by line. These texts were coded, and classification and themes were made based on the relationships among the codes (Creswell, 2017; Günbayi, 2019; Yıldırım and Şimşek, 2011). While coding students' opinions, the letter "S," symbolizing the student and the number code indicating the order, was given. Then analytical generalization was carried out based on the results of the research (Güler, Halıcıoğlu and Taşgın, 2013: 344). Two experts in the field of information technologies coded separately to determine the Kappa coefficient related to the study. Later, codec compatibility was reviewed and found at a very good level with .82.

4 Findings

In this section, the findings obtained as a result of the data analysis in the research are discussed in a comprehensive manner.

4.1 Digital Tools Used to Get Information About Current Issues

The findings about the opinions of secondary school students about digital tools used to get information about current issues are given below.

When Fig. 1.1 is examined, 93.3 % of the secondary education youth use the smartphone in the first place among digital devices. With a rate of 43.3 %, computer tablet is in the second place. With a rate of 26.6 %, TV is in the third place. The opinions of the participant students on this subject are given below:

> *"Phone and computer, but I use the phone the most"* (S7).
> *"Tablet, I sometimes use the computer as well"* (S12).

The social media applications used by students are given in the figure below.

When Fig. 1.2 is analyzed, 96.6 % of the secondary education youth use Instagram in the first place among social media applications. With a rate of 36.6 %, Twitter is in the second place. Then with a rate of 30 %, WhatsApp and YouTube are in the third place. They use Facebook in the last place. The students' views are given below:

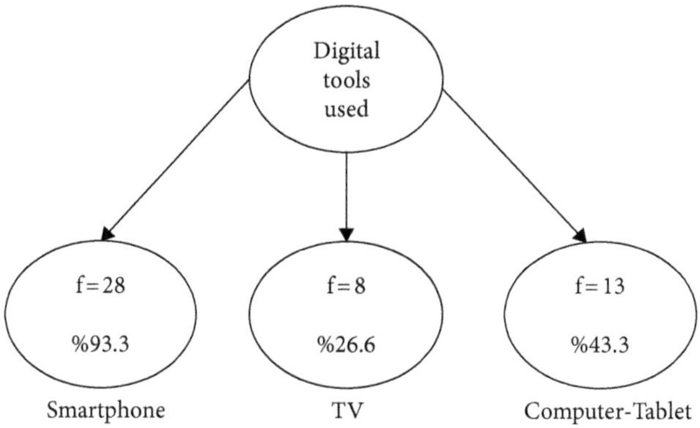

Fig. 1.1: Digital tools used by students to get information about current issues

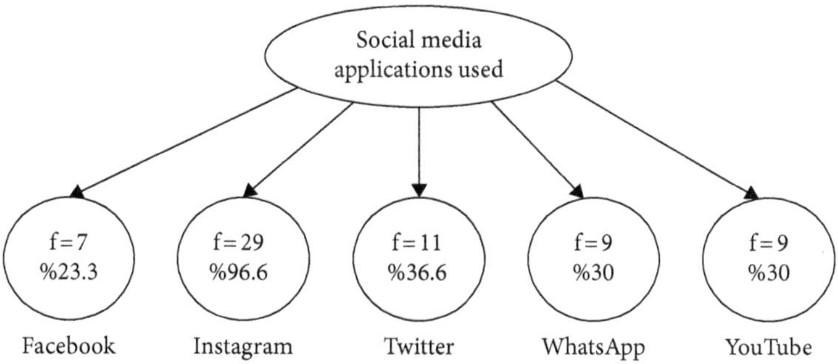

Fig. 1.2: Social media applications used by students

"I use Instagram more" (S4).
"I use Twitter" (S7).
"I use WhatsApp for communication" (S5).
"I generally follow the discussion programs on YouTube" (S2).
"I use Facebook to follow my relatives" (S25).

The time the students devote to social media in a day is given in the figure below.

When Fig. 1.3 examined, 40 % of the secondary education youth spend 3–5 hours a day on social media. Then, with a rate of 33.3 %, 0–2 hours is in the second place. Of them, 13.3 % stated that they spend 6 hours and more on social media in a day. The opinions of the participants on the subject are given below:

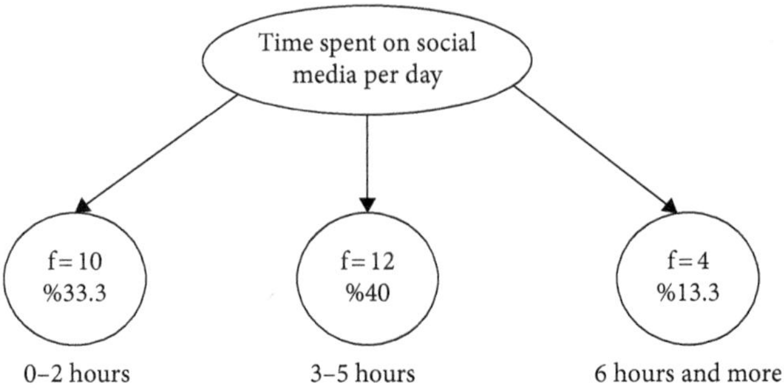

Fig. 1.3: The time students devote to social media in a day

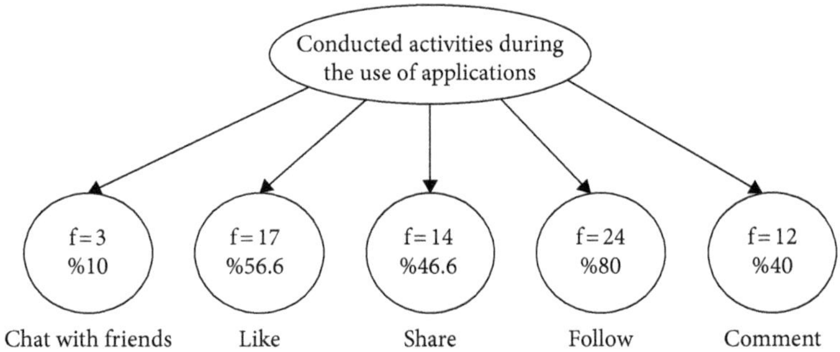

Fig. 1.4: What students do on social media applications

> *"I go to a private teaching institution, since I am in the senior year. That's why I usually spend a few hours on social media before bed. It takes an average of two hours in a day" (S1).*
> *"I can say that I am always on social media except for my personal needs. In other words, I spend four hours a day on average" (S7).*
> *"24/7, it is a bad thing for me, but I am always on social media, except for sleep and school, I never study. I spend time on the phone over an average of 10 hours" (S11).*

The findings about what students do on social media applications are given below.

When Fig. 1.4 is analyzed, with a rate of 80 %, the secondary education youth said that they "follow" on social media in the first place. With a rate of 56.6 %, they "like" in the second place. Then with a rate of 46.6 %, they "share." With a rate of 40 %, they "comment" in the fourth place. With a rate of 10 %, they said that they "chat" with their friends on social media in the last place. The opinions of the participant students on this subject are given below:

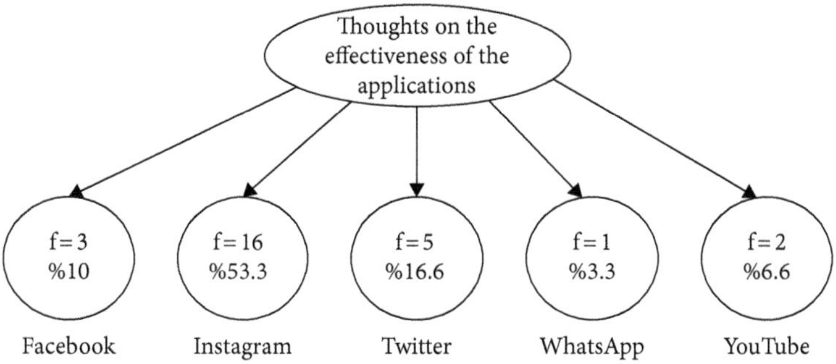

Fig. 1.5: The effectiveness of social media applications students use

> *"I mostly follow the agenda and my friends on social media" (S30).*
> *"I mostly like" (S13).*
> *"I share. I share my photos and videos. I share other things I like as well" (S19).*
> *"My friend shared a story about people who were under avalanche in Van, and I commented on it" (S16).*
> *"I chat with my friends" (S17).*

The findings about the students' opinions on the effectiveness of the social media applications they use are given below.

When Fig. 1.5 is analyzed, with a rate of 53.3 %, the secondary education youth claimed that they consider Instagram as the most effective application among social media applications. With a rate of 16.6 %, they see Twitter in second place. Then with a rate of 10 %, they believe Facebook is the most effective practice. With a rate of 6.62 %, YouTube is in the fourth place, and with a rate of 3.3 %, WhatsApp is in the last place. The students' views are given below;

> *"I only use Instagram. Young people influence each other in this regard. I guess my age group uses Instagram due to this influence" (S4).*
> *"I think Twitter is used effectively in following the news" (S5).*
> *"I think old people use Facebook. There are no proper Facebook users in my age group. I mostly follow my relatives on Facebook" (S17).*
> *"I don't think there is any application other than YouTube that can meet my requests. That's why I don't use other social media applications too much" (S2).*

4.2 Individuals or Institutions Approvingly and Actively Followed

The findings about people or institutions that students approvingly and actively follow on social media are presented below as a figure.

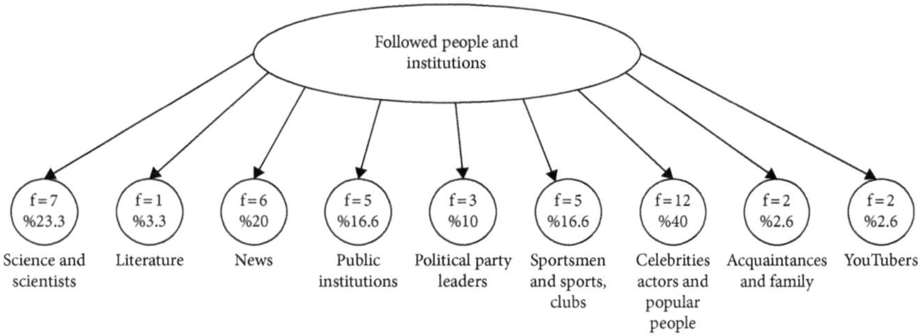

Fig. 2.1: People or institutions that students approvingly and actively follow on social media

When Fig. 2.1 is analyzed, with a rate of 40 %, the secondary education youth follow celebrities, actors, and popular people on social media applications in the first place. With a rate of 23.3 %, they follow science and scientists in the second place. Then with a rate of 20 %, it is seen that news accounts are followed. With a rate of 16.6 %, public institutions and sports and sports clubs are followed in fourth place. With a rate of 10 %, they follow political parties in the fifth place. With a rate of 2.6 %, YouTubers and relatives and family are followed in the sixth place. With a rate of 3.3 %, the secondary education youth follow literary accounts in the last place. Some of the participant students' views on this issue are given with direct references below:

"I follow movie and TV series actors and musicians" (S15).
"There are certain people I follow on YouTube. There is Oytun ERBAŞ in medicine. I also follow İlber ORTAYLI and Celal ŞENGÖR" (S2).
"I follow the accounts of news channels to be informed about the agenda. In general, I follow up for information" (S10).
"I follow the Turkish armed forces to get current news on Twitter. For example, I get information about the Idlib operation from here" (S14).
"I generally follow the NBA and NBA stars" (S21).
"I follow politics very closely. I follow the leaders of different parties" (S17).
"Barış Özcan shares exact information. For example, I learned from him about the claim that the design of our domestic car (TOGG) belongs to Italians. He also explained that Turkish designers work at TESLA" (S18).
"I follow my relatives and family" (S12).
"I frequently follow book accounts as they are beneficial to me" (S27).

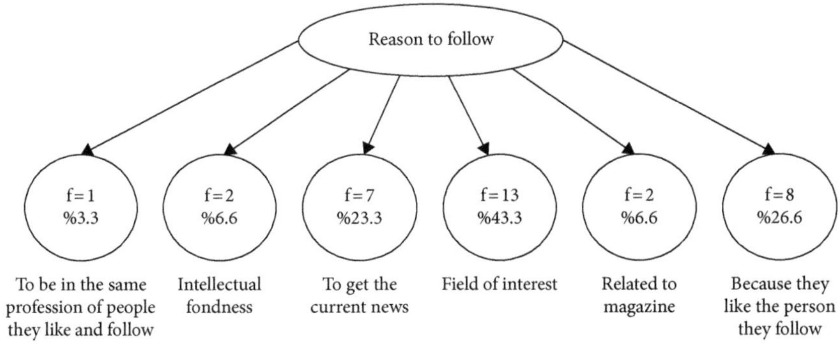

Fig. 2.2: The reasons for students to follow social media accounts

The findings about the reasons for students to follow social media accounts are given below.

When Fig. 2.2 is analyzed, with a rate of 43.3 %, the field of interest is in the first place among the reasons for students to follow social media accounts. With a rate of 26.6 %, they follow people they like in the second place. Then with a rate of 23.3 %, in the third place, the secondary education youth stated that they follow social media accounts in order to be informed about the developments. With a rate of 6.6 %, intellectual closeness and interest in magazines was expressed in the fourth place. With a rate of 3.3 %, the secondary education youth said that they use social media applications to like and follow people to learn and pursue their professions. The opinions of the participants on the subject are given below:

"I follow the technology accounts, since I am interested in technology. I also follow the actors because I like watching movies" (S15).
"I use social media to get information about the city I live in" (S29).
"I follow people who are close to my own political opinion" (S1).
"I follow F.Ü. because s/he attracts my attention due to having different personality like gossiping. F.Ü. fights with D.A. on social media and I like these" (S11).
"I want to be an actor and I want to learn something about this job. There are question and answer activities. I ask them questions, and they answer. I get information in that way" (S10).

The findings obtained from the interviews with the students about the influence of followed people or institutions on the young people's opinions are presented below.

When Fig. 2.3 is analyzed, while 73.3 % of the secondary education youth stated that the followed people or institutions on social media applications affect their opinions, a group of 20 % stated that they did not. The students' views are given below:

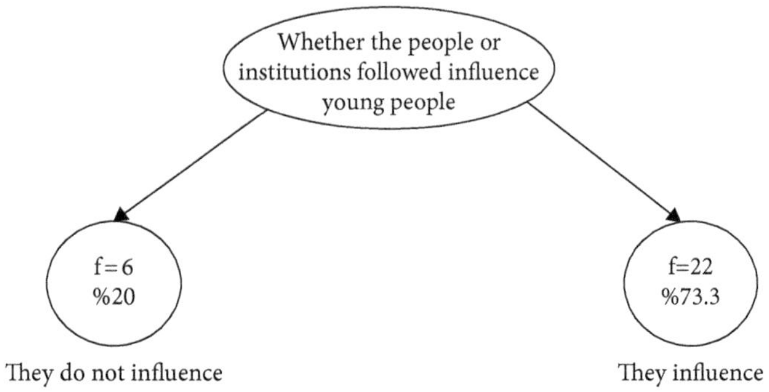

Fig. 2.3: The influence of followed people or institutions on young people's opinions

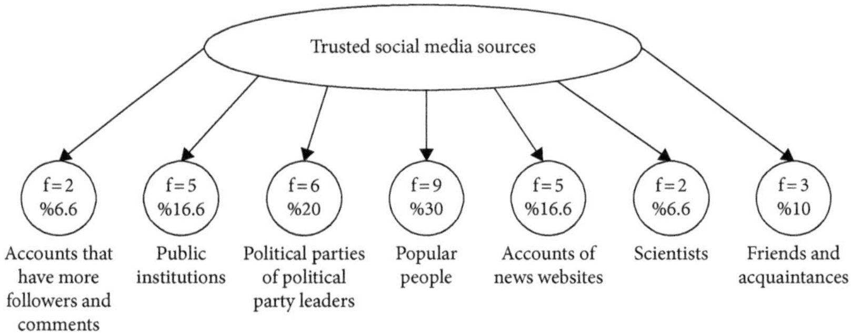

Fig. 3.1: The social media sources that students trust

"The shares of the people I follow affect me. I like their dressing style. For example, some celebrities have a shabby and marginal style. I like this. I want to be like them" (S19).
"It doesn't affect me much, after all, we don't live in the same way. They are older than me. I don't do what they do. They have night tours; they go to fashion shows. I'm not doing such things here. And, my age is young for these" (S5).

4.3 Opinions About the Reliability of Posts on Social Media

The findings about the social media sources that students trust are given below.

When Fig. 3.1 is analyzed, with a rate of 30 %, the accounts of public institutions are in the first place among the social media sources that students trust. With a rate of 20 %, the students stated that they trust the accounts of the news sites in the

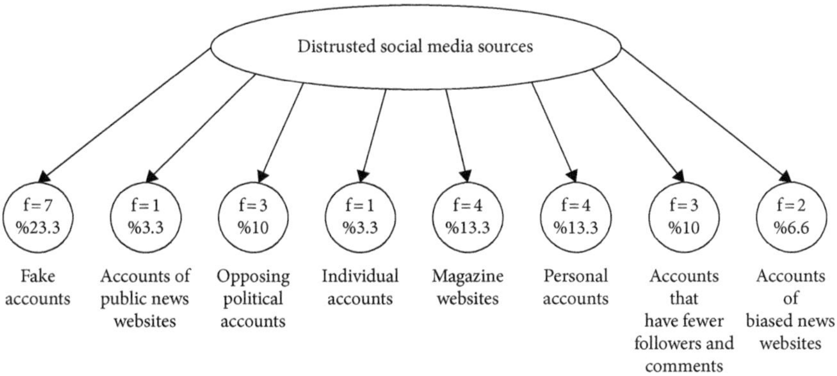

Fig. 3.2: The social media sources that students do not trust

second place. With a rate of 16.6 %, scientists and popular people are in the third place. With a rate of 10 %, there are accounts with a high number of followers and comments in the fourth place. With a rate of 6.6 %, the friends and relatives of the young people and the accounts of political parties and political leaders are in the last place among the social media sources that students trust. Some of the participant students' views on this issue are given with direct references below:

> *"I generally trust government institutions. I look at the health ministry's accounts of about health. I look at MEB's accounts about education. Because you may encounter false news in other channels" (S4).*
> *"I trust the accounts of news sites and approved accounts with blue ticks" (S21).*
> *"I trust it if it is information that has been confirmed by scientists" (S24).*
> *"I find magazine sources reliable. I believe in the account of Uçankuş site" (S11).*
> *"I try to understand its reliability by looking at the number of followers. If I will buy a shoe or dress, I first look at the number of followers. Then I look at the comments. I think I can trust it if the comments are positive" (S8).*
> *"Since I follow people close to my own opinion, I find their posts safe or they seem to me" (S1).*
> *"I trust my friends' posts" (S3).*

The findings about the social media sources that students do not trust are given below.

When Fig. 3.2 is analyzed, with a rate of 23.3 %, fake accounts are in the first place among the social media sources that students do not trust. With a rate of 13.3 %, the students stated that the accounts of the magazine sites and personal accounts that they do not trust are in the second place. With a rate of 10 %, opposing political accounts and accounts with a low number of followers and comments are

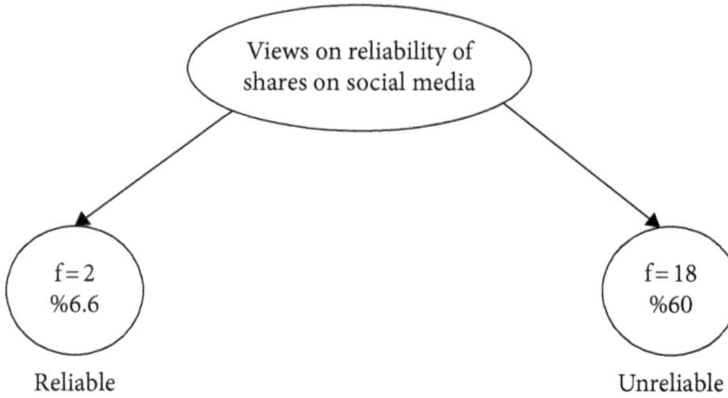

Fig. 3.3: Students' views on the reliability of the posts on social media

in the third place. With a rate of 6.67 %, there are accounts biased news sites in the fourth place. With a rate of 3.3 %, the accounts of public news sites and personal accounts are in the last place among the social media sources that students do not trust. The opinions of the participants on the subject are given below:

> *"Some accounts may be fake accounts. I understand from their followers. I understand from their photos. I don't trust them" (S11).*
> *"I do not trust personal accounts" (S12).*
> *"There are nonsense magazine accounts. There are accounts that constantly mess around, like on television. They share things that are not real. I just slide them" (S5).*
> *"For example, I don't trust the X party. I do not trust its posts on Instagram" (S17).*
> *"I don't trust X channels. I do not trust any channel broadcasting close to a party" (S12).*
> *"I don't trust the channels of the government very much" (S5).*
> *"I don't trust personal sites very much" (S29).*

The findings about the students' opinions on the reliability of posts on social media are given below.

When Fig. 3.3 is analyzed, 60 % of the secondary education youth thought that social media posts are not trustable, while a group of 6.6 % stated that they are. The students' views are given below:

> *"Not trustable. You can't fit everything into a small picture. People can be deceived on social media. You can only see what the picture shares. You can't know what's in the background. For example, they share a picture as if they were at a party, but they actually sit at home. We cannot know what they do" (S7).*
> *"If their accounts are not confidential, they are reliable" (S23).*

The findings about the sharing of the posts that the students follow on their social media applications are given below.

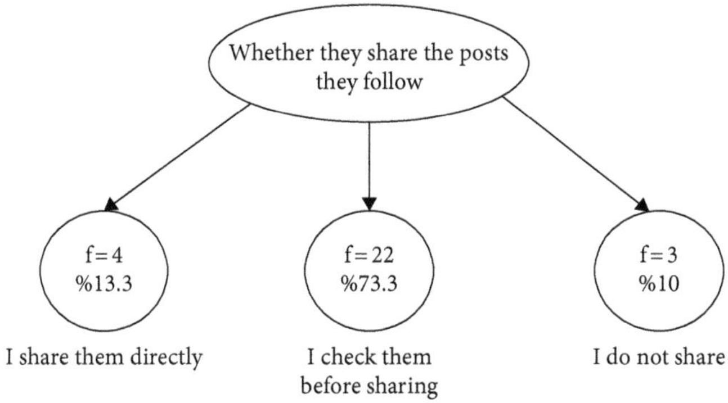

Fig. 3.4: The sharing of posts that the students follow on their social media applications

When Fig. 3.4 is examined, 73.3 % of the secondary education youth stated that they share the followed posts after checking. A group of 6.6 % said that they share them directly without checking. Of them, 10 % told that they did not share. The opinions of the participants on the subject are given below:

"I check at least four or five different sites to confirm the accuracy of a news. When I learner the news that the semester holiday was postponed in the Elazig earthquake, I confirmed that it was correct by looking at several sites" (S13).
"I don't care very much. I just share" (S7).
"I do not share easily" (S16).

4.4 General Comments on Digital Media Concerning Its Influence on Young People's Thoughts

The findings obtained about the general comments of the students on digital media influencing the thoughts of the youth are given below.

When Fig. 4.1 is analyzed, 93.3 % of the students in the secondary education stated that the digital media influences the thoughts of the young people in their age group, whereas no student expressed an opinion towards noninfluence. The students' views are given below.

"My peers think life on social media is real life. They believe in everything there. Everything on social media is fancy and colorful, but in reality none of their lives are such. My peers are influenced by the shares of the people they follow. For example, there is a

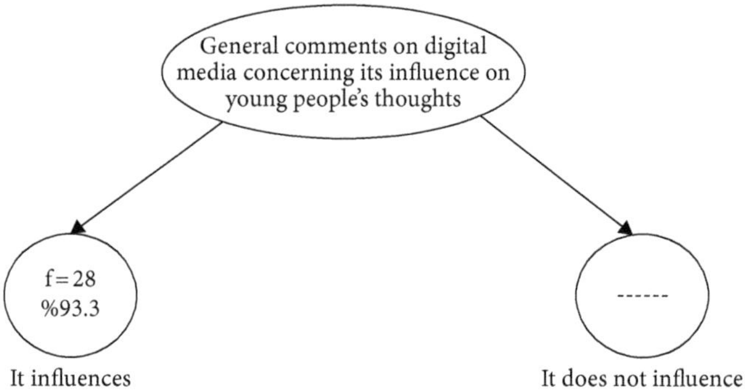

Fig. 4.1: The general comments of students on digital media influencing the thoughts of the youth

successful singer. They imitate her clothing, dressing, point of view, singing, behavior, everything" (S14).

"I think it influences. For example, the two YouTuber AB and MC, guys who have appeared recently, are putting words and speech styles into Turkish language that have never been used before. These affect the way of speaking and thinking of my peers" (S18).

"My peers spend a lot of time on digital media and they are influenced by what is written there. I try not to be influenced but my friends' ideas influence me when I'm with them. Some of my friends are exaggerating their social media posts so that they will even post their intimate moments in the toilet if they were not ashamed. My friends are influenced by the distortion and communication of the current political debates. In the Elâzığ earthquake, the help of a Syrian citizen to the earthquake survivors was on the news of social media, while the Azeri brothers'news was not brought up much" (S24).

"I think it influences a lot. There is such a pessimistic crew on social media. When they read and see them, they become depressed and share stuff like that. It also affects their daily lives. I do not want to enter politics very much, but my age group has a lot of discussion about it. For example, someone shares a funny post with political content and a fight occurs under the post. Although they do not know much about the topic under discussion, they believe in it because the post received much likes. This way they can blame Party A on that. I never get involved in these issues. I find it ridiculous" (S8).

5 Conclusion and Discussion

In this study, the digital tools used by secondary school students to get information about current issues, people, or institutions that they actively follow on social media, their views on the reliability of the posts on social media, and the general opinions of secondary school students on influencing the thoughts of young people in digital media are presented through descriptive analysis and content analysis methods.

According to the research findings, 93.3 % of the young people in the secondary education age use the smartphone in the first place among digital devices. The second place is the computer tablet with 43.3 % and the third place is the TV with 26.6 %. This result is similar to Onursoy's (2018: 1008) study results conducted on university youth. In the study, the use of smartphones is at the forefront in accessing information. There is a young audience who can actively access, sustain, and evaluate information when they need it. In other words, young people's adaptation to technological tools is high. Likewise, 81.1 % of the participants in the study of Sağır and Eraslan (2019) use social media applications on their smartphones in the first place and computers and tablets in the second place. The reason why young people prefer smartphones is the ease of use. The smartphone, which has easy and portable features, can perform many operations that can be done on the computer with its applications.

Of the youth in secondary education, 96.6 % use Instagram in the first place among social media applications. In the second place, they use Twitter with a value of 36.6 %, then WhatsApp and YouTube with a value of 30 %, and Facebook in the last place. When WeAreSocial 2020 World Social Media Usage Statistics are analyzed, 65 % of Instagram advertising audience is between the ages of 18 and 34. The percentage of users over the age of 35–44 is 16 %. This situation can be interpreted as Facebook has lost its old appeal especially among young people and Instagram has come to the fore. In addition, in the study by Sheldon and Bryant (2016), they concluded that Instagram users feel popular and cool. The degree of popularity is measured by the number of followers that the person has. In addition to this, it has been concluded that while presenting a person to other people, it is possible to present themselves in a way that leaves the desired image, not as they really are.

Of the secondary school age young people, 40 % spend 3–5 hours a day on social media. Then 0–2 hours is in the second place, with a value of 33.3 %. Of the youth, 13.3 % stated that they spend time on social media 6 hours a day or more. In the study conducted by Alakuş in 2019, 47.5 % of the participants stated that they spent an average of 3–4 hours a day on social media, while 25 % stated that this period was a minimum of 5 hours. While the number of participants who spend an average of 2 hours a day on social media is 20 %, the number of participants who spend only 45 minutes to 1 hour on social media is 7.5 %. In addition, Sağbaş, Ballı, and Şen (2016) state that one out of every three young people spend at least 3 hours a day on social media and that a large number of young people who go online are social media users. With the increase of time devoted to social media applications, young people can easily access the information they want in any part of the world, establish new friendships and consume them in a short period of time, and as a result, they become an impatient, quickly bored, and unsatisfied adult. This situation turns into an indication that in the next step, young people break with their social environment. Breaking away from social relations is an important context in the value shift of young people. The weakening of human relations causes social interactions to break and the communication between people does not occur even in physical encounters. It is seen that many values such as love, respect, and sharing, etc. have started to lose their meaning with the breaks experienced in these interactions (Sağır and Eraslan,2019).

Of the secondary school age youth, 80 % stated that they make follows on social media in the first place. They like in the second place on social media with a rate of 56.6 %. Then they share with a rate of 46.6 %. In the fourth place, they comment on social media with a rate of 40 %. They said that they chatted with their friends on social media with a rate of 10 %. Sağır and Eraslan (2019) stated that reaching high numbers in social media applications in terms of "follow, like, sharing, and comment" is an important situation for young people. As the young people share their content from their social media accounts, they prove their existence to the friends and society (Alakuş, 2019). This situation can be interpreted as the photos or videos with high likes, comments, shares, or follows affect the daily behavior of the youth.

Of the secondary school age youth, 53.3 % stated that they see Instagram as the most effective application among social media applications. Twitter is in the second place with 16.6 % value, then 10 % of them view Facebook as the most effective application. In the fourth place is YouTube with a value of 6.62 % and in the last place is WhatsApp with a value of 3.3 %. This result is in parallel with the finding that the youth in the work of Türkoğlu (2018) think that Instagram is an effective program as a social media application.

Of the secondary school age youth, 40 % follow celebrities, actors, and popular people in the first place through social media applications. In the second place, science and scientists are followed with a rate of 23.3 %. Then, it is seen that news accounts are followed with a rate of 20 %. In fourth place, public institutions and sports and sports clubs are followed with a value of 16.6 %. Political parties come in the fifth place with a value of 10 %. In the sixth place, YouTubers and acquaintances and family are followed with a rate of 2.6 %. In the last place, with a value of 3.3 %, literature accounts are followed by young people in the secondary education age. This result coincides with Hazar (2011)'s study at Gazi University, where 47.2 % of respondents use social media to get information the most and watch the news with 23 %. In Şener, Emre, and Akyıldız (2015: 93–94)'s study, one of the main goals of the participants to connect to social networks was determined to be following the agenda. According to the results obtained by Kamiloğlu and Yurttaş (2014: 147), 70 % of the students obtain the information they need with their own efforts. In other words, they cannot access this information from the education system. Young people reach the information they need or the information they get on any subject, 90 % from the Internet and 70 % from Facebook. These results coincide with the conclusion that young people follow current news, celebrities, and scientists with the use of Internet and social media.

Field of interest is at the top of the list, with a rate of 43.3 % in terms of following the accounts in social media applications. In the second place, there is the liking of the person they follow with a value of 26.6 %. Then, with a rate of 23.3 % in the third place, the youth in the secondary education age follow social media accounts in order to be informed about developments. In the fourth place, intellectual fondness and interest in magazines were expressed at the same rate of 6.6 %. In the last place, 3.3 % of the young people in the secondary school expressed that they liked and followed people to be in their profession. This result overlaps with the finding that social media in the Youth and Social Media Research Report by the Ministry of Youth and Sports in 2013 meant following the agenda/creating an agenda (51 %) for young people. Similarly, in Tarhan (2014)'s study on university youth's social media motivations before local elections, the concepts of seeking information and guidance, social escape, and entertainment have emerged. Again, another study on high school students shows that those who use Facebook functionally are more sensitive to social events while sharing funny and entertaining videos. On the other hand, issues addressing emotions and religious and political issues are less demanded among young people (Kamiloğlu and Yurttaş 2014: 145). It is seen that these results overlap substantially with the results of our study.

While 73.3 % of the students in the secondary education period stated that the people or institutions followed in social media applications influence their own thoughts, a group of 20 % stated that they do not influence. In the study carried out by Kamiloğlu and Yurttaş (2014: 145–148), it was found out that using Facebook has a significant relationship between the value that high school students add to their knowledge and there is a significant relationship between the value that high school students add to their personal development. Almost half of the youth went to a cinema and a concert, seeing and being influenced on Facebook. In addition, it was observed that 30–40 % of the youth attended a music course, sports activity, and computer, theater, and foreign language courses by seeing them on Facebook. Young people are informed about many social, cultural, and artistic activities through Facebook. These results seem to be in line with the result of our study that social media mostly influences the youth's views. In the study of Çağlar and Asığbulmuş (2017: 108), a positive, high, and significant relationship was found between following social media and the effect factor of social media. The highest relationship was found to be between these two factors. Here, a comment can be made that participants who follow social media are generally influenced by social media. It has been determined that there is a positive and significant relationship between social media's directive effect and follows and its effects. This finding can be interpreted as social media directing its followers.

Among the trusted social media sources, the accounts of public institutions ranked first, with a rate of 30 %. The students stated that they trusted the accounts of news sites with a value of 20 % in the second place. Scientists and popular people are in the third place with a rate of 16.6 %. In the fourth place, there are accounts with more followers and comments with a value of 10 %. In the last place, regarding trusted social media sources, with the value of 6.6 %, are the friends and acquaintances of the youth and the accounts of political parties and political leaders. İnce and Koçak reached the conclusion in their 2017 study that the participants trust social media at the medium level (x = 4.70).

Fake accounts ranked first with 23.3 % among unreliable social media sources. The students stated that they did not trust the accounts of magazine sites and personal accounts, at 13.3 %, placing it second. In the third place, opposing political accounts and accounts with fewer followers and comments at a rate of 10 %. In the fourth place, there are accounts of biased news sites with a value of 6.67 %. The accounts of public news sites and individual accounts occupy the last place as unreliable social media sources, at the rate of 3.3 %. In the study of Gedik and Çosar (2020), secondary school students have defined social media as a field with harmful and inappropriate content, a bad environment, and applications where hollow information is delivered to people.

While 60 % of the students in the secondary education period think that social media shares are unreliable, a group of 6.6 % believe they are reliable. In the study of Gyagri in Ghana in 2016, it was revealed that social media is the most used communication tool in the context of political information, but despite the fact that social media is frequently used, it is not reliable in political information and users do not trust the political information shared on these platforms. In the study by Jhonson and Kaye in 2019 (from Bal and Delal, 2019: 129), social networking sites were found to be the least reliable among the nine traditional and online sources examined during the political information process. This result seems to be largely compatible with the results of our study of social media applications.

Of the secondary school age youth, 73.3 % stated that they shared posts after checking. A 13.3 % student group said that they shared directly without checking. Of them, 10 % stated that they did not share. In the study conducted by Demircan (2015: 1529), it was determined that the young people interviewed were active social media users, but despite being active, they abstained for various reasons and did not share their real thoughts. The study carried out by Ertem and Korhan (2019: 230) reveals that a high number of young adults check the foreign policy news they see on social media for accuracy, but almost 60 % of them do this for control over websites and social media platforms. In the study carried out by Onursoy (2018: 1008) on university youth, it was found that the rate of use of digital tools by young people is high, but there is a minority who still cannot reach technological information, do not search in detail, nor do they check the information source or verify the information. Likewise, about a quarter of the participants state that they sometimes or almost never understand the fake news or information.

While 93.3 % of the students in the secondary education period stated that the digital media influenced the thoughts of the youth in their age group, no student who expressed their opinion on the subject commented that it did not. According to the results of Kamiloğlu and Yurttaş's (2014: 146) study on high school students, it has been revealed that Internet and social media environments determine the lifestyle and communication style of young people. In his study, Gacumo (2018) has reached the conclusion that most of the young people in secondary education age are influenced by their peers' decision-making processes and thought structures due to the fact that access to social media is easy and many information can be accessed from this environment. Another finding obtained in the study of Çağlar and Asığbulmuş is that generation Y is more inclined to social media than generation X. Considering the frequency of social media usage of X and Y generations, generation Y uses social media more than generation X. In these platforms, from the political content shared by individuals' friends or political actors, generation Y was found to be more influenced than generation X. Generation Y is more active in sharing political content than generation X. Therefore, it will be inevitable that generation Y, which uses social media more actively, will be more influenced by this media in their political behavior and preferences than generation X (Çağlar and Asığbulmuş 2017: 107–108). The study Ertem and Korhan (2019: 215) conducted with university students shows that 62 % of young adults in Turkey are influenced from the posts on social media about their thoughts on foreign policy. Majority of the participants (47 %) believe that social media posts are partly effective on the country's foreign policymaking, while 17 % believe it is not at all effective.

As a result of the study, it can be said that there are problems at the point of technology addiction when the devotion of the secondary school age students to social media applications and the actions they perform on these applications are considered. The findings obtained show similarities with TBM (Turkey Addiction Fighting Training Program), prepared by the Green Crescent (Yeşilay), and the subjects of "Technology Addiction and the Use of Social Media" in the context of "Neglect and Life Skills Against Abuse" project, prepared by Antalya Provincial Directorate for National Education. By expanding and enriching the content of these applications, it can be said that it will be beneficial to apply parts especially for parents more effectively. Since this study is a qualitative study limited to the central districts of Antalya, it will be possible to see the whole picture by carrying out such studies with more students in different provinces and districts. Considering the increasing impact of new media applications on young people, it can be said that determining the degree, direction, and weight of social media and Internet activity in order to understand young people, to examine their relations with the Internet and social media, and to communicate with them more effectively and healthily. It can be concluded that more comprehensive studies are needed in this regard.

Bibliography

Aktas, Hasret, (2004), *Bir Siyasal İletişim Aracı Olarak İnternet*, Turkey: Tablet Kitabevi.

Alakuş, Seda, (2019), *Sosyal Medyanın Gençlerin Marka Tercihlerine Etkisi: Özel Okul Öğrencileri Üzerine Bir Araştırma*, (Unpublished Master's Dissertation), İstanbul Üniversitesi, Istanbul.

Asığbulmuş, Hasan, (2017), *X ve Y Kuşaklarının Siyasal Davranışında Sosyal Medyanın Etkisi: Gaziosmanpaşa İlçesinde Bir Araştırma,* (Unpublished Master's Dissertation), Süleyman Demirel Üniversitesi, Isparta.

Babacan, M. Emin, (2014), "*Sosyal Medya Sonrası Yeni Toplumsal Hareketler", Birey ve Toplum Sosyal Bilimler Dergisi,* Vol. 4, N. 1, pp. 135–160.

Balcı, Şükrü and Koçak Mevlütcan, (2017), "Sosyal Medya Kullanımı ile Yaşam Doyumu Arasındaki Ilişki: Üniversite Öğrencileri Üzerinde Bir Araştırma", *1.Uluslararası İletişimde Yeni Yönelimler Konferansı,* pp. 34–45, http:// acikerisim.ticaret.edu.tr/xmlui/bitstream/ handle /11467/1598/M00800.pdf?sequ ence=1andisAllowed=y, (02.01.2020).

Bal, Enes and Delal, Özlem, (2019), "Siyasal Bilgilenmede Twıtter Kullanımı Üzerine Panoramik Bir Değerlendirme", *Social Sciences (NWSASOS),* Vol. 4, N. 3, pp. 118–132.

Business Directory Online, http://www.businessdictionary.com/definition/media. html (22.01.2020).

Collins English Dictionary, (2016), New York: Harper Collins.

Creswell, Jhon W. and Clark, Vicki, P.L., (2007), *Designing and Conducting Mixed Methods Research,* USA: Sage.

Çağlar, Nedret and Asığbulmuş, Hasan, (2017), "X ve Y Kuşaklarının Siyasal Davranışında Sosyal Medyanın Etkisi: Gaziosmanpaşa İlçesinde Bir Araştırma", *Süleyman Demirel Üniversitesi Sosyal Bilimler Enstitüsü Dergisi,* Vol. 4, N. 29, pp. 85–112.

Demircan, Birol, (2017), "Sosyal Medyanın Politik Bir Aradalığı Sağlama Potansiyeli", *International Journal of Social Sciences and Education Research,* Vol. 3, N. 5, pp. 1522–1532. https://dergipark.org.tr/tr/pub/ijsser/ issue/30374/321669r, (04.01.2020).

Ertem, H. Sarı and Korhan, Sevda, (2019), "Sosyal Medya ve Dış Politika İlişkisine Yakından Bakmak: Türkiye'de Üniversiteli Genç Yetişkinler Örneği", *Ekonomik ve Sosyal Araştırmalar Dergisi,* Vol. 15, N. 2, pp. 215–233.

Filiz, Ozan, Erol, Osman, Dönmez, Fevzi, and Kurt, Adile, (2014), "BÖTE Bölümü Öğrencilerinin Sosyal Ağ Siteleri Kullanım Amaçları ile İnternet Bağımlılıkları Arasındaki İlişkinin İncelenmesi", *Journal of Instructional Technologies and Teacher Education,* Vol. 3, N. 2, pp. 28–17.

Gacumo, R. James, (2018), "Rising Social Media Trends and Its Effects On the Opinions and Beliefs On Current Social and Political Issues of Specialized High School Students", https://www.researchgate.net/publication/335749790_ Rising_Social_Media_Trends_and_Its_Effects_On_the_Opinions_and_Beliefs_ On_Current_Social_and_Political_Issues_of_Specialized_High_School_Stude nts?channel=doiandlinkId=5d790c90a6fdcc9961c0b3f3andshowFulltext=true, (11.01.2020).

Gedik, Hatice and Cosar, H. Akbaş, (2020), "Perception of Social Media in Secondary Students", *International Education Studies*, Vol. 13, N. 3, pp. 6–17.

Gross, Elisheva. F, Juvonen, Jaana and Gable, Shelly, L, (2002), "Internet Use and Well-Being in Adolescence", *Journal of Social Issues*, Vol. 58, N. 1, pp. 75–90.

Güler, Ahmet, Halıcıoğlu, M.Bülent, and Taşğın, Serkan, (2013), *Sosyal Bilimlerde Nitel Araştırma Yöntemleri*, Ankara, Turkey:Şeçkin Yayıncılık.

Günbayı, İlhan, (2019), "*Nitel Araştırmada Veri Analizi: Tema Analizi, Betimsel Analiz, İçerik Analizi ve Analitik Genelleme*". Web: http://www.nirvanasosyal.com/h-392-nitel-arastirmada-veri-analizi-tema-analizi-betimsel-analiz-icerik-analizi-ve-analitik-genelleme.html, (03.02.2020).

Gyagri, Charles, Yaw, (2016), *Social Media As a Platform for Political Information: A Survey of University of Ghana Students*, University of Ghana http://ugspace.ug.edu.gh/ bitstream/handle/123456789/22824/Social%20Media%20as%20a%20Platform%20for%20Political%20Information%20A%20Survey%20of%20University%20of%20Ghana%20Students%20-%202016.pdf?sequence=1andisAllowed=y, (01.02.2020).

Hamamcı, Elif, (2015), *Televizyon ve Sosyal Medya Ilişkisi*, (Unpublished Master's Dissertation), Marmara Üniversitesi, İstanbul.

Hampton, Keith and Wellman, Barry, (2001), "Long Distance Community in the Network Society Contact and Support Beyond Netville", *American Behavioral Scientist*, Vol. 45, N. 3, pp. 476–495.

Hazar, Murat, (2011), "Sosyal Medya Bağımlılığı- Bir Alan Çalışması", *Gazi Üniversitesi İletişimFakültesi İletişim Kuram ve Araştırma Dergisi*, N. 32, pp. 151–176.

İnce, Mustafa and Koçak, Mevlütcan (2017), "Üniversite Öğrencilerinin Sosyal Medya Kullanım Alışkanlıkları: Necmettin Erbakan Üniversitesi Örneği", *Karabük Üniversitesi Sosyal Bilimler Enstitüsü Dergisi*, Vol. 7, N. 2, pp. 736–749.

Kamiloğlu Fatma and Yurttaş Ö. Uluğ, (2014), "Sosyal Medyanın Bilgi Edinme ve Kişisel Gelişim Sürecine Katkısı ve Lise Öğrencileri Üzerine Bir Alan Çalışması", *GSÜ İletişim Dergisi*, N. 21, pp. 129–150.

Katz, Elihu and Lazar, F, Paul, (1955), *Personal Influence*, New York: The Free Press.

Khan, Shadzah, (2012). "Impact of Social Networking Websites on Students", *Abasyn Journal of Social Sciences*, Vol. 5, N. 2.

Kirschenbaum, Alan, (2004), "Generic Sources of Disaster Communities: A Social Network Approach, International", *Journal of Sociology and Social Policy*, Vol. 24, N. 10/11, pp. 94–129.

Konuk, Nebiye, (2019), "Üniversite Öğrencilerinin Sosyal Medyada Şiddet Kavramına Bakışları", *The Turkish Online Journal of Design Art and Communication*, Vol. 9, N. 3, pp. 414–431. https://dergipark.org.tr/en/pub/tojdac/issue/46322/582139, (9.02.2020).

Lemish, Dafna, (2013), '*Children, adolescents, and media: creating a shared scholarly arena', Introduction, The Routledge International Handbook of Children, Adolescents and Media,* London and New York: Routledge.

Maalouf, Amin, (2009), *Çivisi Çıkmıs Dünya,* (Orçun Türkay, Trans.), İstanbul, Turkey: YKY.

Masthi, N. Ramesh, Pruthvi, S. and Mallekavu, Phaneendra, (2017), "A Comparative Study On Social Media Addiction Between Public and Private High School Students of Urban Bengaluru India", *Journal of Psychiatry,* Vol. 18, N. 2, pp. 206–215.

Mayfield, Antony, (2008), "What *is Social Media? Crossing eBook.* Retrieved from", http://www.icrossing.co.uk/fileadmin/uploads/eBooks/What_is_Social_Media_iCrossing_ebook.pdf, (01.05.2011).

Nadkarni Ashwini and Hofmann, Stephan, G. (2012), "Why Do People Use Facebook? Personality and Individual Differences", N. 52, pp. 243–249, http://dx.doi.org/10.1016/j.paid.2011.11.007.

Onursoy Sibel, (2018), "Üniversite Gençliğinin Dijital Okuryazarlık Düzeyleri: Anadolu Üniversitesi Öğrencileri Üzerine Bir Araştırma", *Gümüşhane Üniversitesi İletişim Fakültesi Dergisi,* Vol. 6, N. 2, pp. 989–1013.

Oxford Dictionary and Thesaurus, (2009), OUP: Oxford.

Sağbaş, Ensar, A, Ballı, Serkan, and Şen, Faruk, (2016), "Sosyal Medya ve Gençler Üzerindeki Etkileri", *Muğla Sıtkı Koçman Üniversitesi Uluslararası Gençlik Araştırmaları Kongresi Bildiri Kitabı,* pp. 153–163.

Sağır, Adem and Eraslan, Hasan, (2019), "Akıllı Telefonların Gençlerin Gündelik Hayatlarına Etkisi: Türkiye'de Üniversite Gençliği Örneği", *OPUS Uluslararası Toplum Araştırmaları Dergisi,* Vol. 10, N. 17, pp. 48–78.

Sheldon Pavice and Bryant Katherine, (2016), "Instagram: Motives for Its Use and Relationship to Narcissism and Contextual Age", *Computers in Human Behavior,* N. 58, pp. 89–97.

Şener Gülüm, Emre P. Öğün and Akyıldız Fatih, (2015), "Türkiye'de Sosyal Medyanın Siyasi Katılıma Etkileri", *Folklor/Edebiyat Dergisi,* Vol. 21, N. 83, pp. 75–98.

T.C. Gençlik ve Spor Bakanlığı, (2013), "Gençlik ve Sosyal Medya Araştırması", http://www.gsb.gov.tr/HaberDetaylari/1/3816/ genclik-ve-spor-bakanligi-turkiyenin-en-kapsamli-sosyal-medya-arastirmasini-yapti.aspx, (25.07.2015).

Tarhan, Ahmet, (2014), "The Role of Social Media to Political Information", 10th International Academic Conference, Vienna.

Tektaş, Necla, (2014), "Üniversite Öğrencilerinin Sosyal Ağları Kullanımlarına Yönelik Bir Araştırma", *Journal of History School (JOHS)*, Vol. 7, N. 17, pp. 851–870.

Türkoğlu, H. Semiz, (2018), "Sosyal Medya Üzerinden Mahremiyet Farkındalığı ve Değişimin Ölçümlenmesine Yönelik Bir Araştırma", *Istanbul University Journal of Communication Sciences*, N. 54, pp. 163–189.

Ünal, Serdal, (2017), "Ağ toplumunun Genç Yurttaşları: Dijital Sosyal Ortamların Siyasal Temelde Kullanımına Yönelik Karşıt Yaklaşımlar", *International Journal of Human Sciences*, Vol. 14, N. 3, pp. 2510–2540.

WeAreSocial 2020, 2020 *Dünya İnternet Sosyal Medya ve Mobil Kullanım İstatistikleri*. https://dijilopedi.com/2020-dunya-internet-sosyal-medya-ve-mobil-kullanim-istatistikleri, (15.01.2020).

Yıldırım Ali and Simsek, Hasan, (2011), *Sosyal Bilimlerde Nitel Araştırma Yöntemleri* (8. Edition), Turkey: Seçkin Yayınevi.

Murat ÇOBAN

Virtual Reality Technologies in Our Digitalized Education World: Its History, Usage Areas, Limitations and Future

1 Introduction

In our digitalized world, developments in information and communication technologies continue increasingly. The advantage of cheaper and faster programming of computer technologies, apart from the theories that accept traditional media and technology, also provides the opportunity to develop different technologies that can meet users' needs. Virtual reality (VR) technology, which is becoming more and more effective in many disciplines in our digital world, can undoubtedly be shown as one of the most important products of these technological developments. Thus, it is predicted that by 2022, VR technologies will grow to approximately 209.2 million dollars in the world market (Statista VR, 2020). Leading companies in technology such as *Alibaba, Samsung, Google, Facebook* and *HP* invest, develop and market for technologies such as VR. It is also emphasized that *Coca-Cola, McDonald's, IKEA* and many other leading companies have started to adopt VR technologies more in order to strengthen their relations with consumers (Hollebeek et al., 2020). In this context, it is predicted that VR technologies will be an important part of our digitalized world in the future.

It is emphasized that VR technologies will be used intensively in the education process, which is among the most important activities of our life. Because, VR technologies are seen as the learning tools that the twenty-first century aims (Rogers, 2019). In addition, the fact that VR technologies allow the visualization and interaction components to be integrated in the best way enable many educators and researchers to focus on these technologies (Radianti et al., 2020). In this context, researches on the effects of VR technologies on learning are conducted in many disciplines, such as literature, engineering, medicine, music, mathematics, aviation, architecture and space (Van Ginkel et al., 2019; Sopher et al., 2019; Frank et al., 2019; Innocenti et al., 2019; Akman and Çakır, 2019; Birt and Cownling, 2018). Thanks to the head-mounted displays (HMD) in the most advanced VR technologies like *HTC Vive* and *Oculus Rift*, students' experience of a high degree of reality has increased the interest in these researches. For example, a study shows that students who experience VR technologies are more successful in the process of remembering and applying information (Krokos et al., 2019).

In another study, it is observed that students devote more time to learning tasks and gain cognitive, psychomotor and emotional skills better in a VR environment (Jensen and Konradsen, 2018). Also, unlike the more traditional learning tools such as computers and tablets, it is determined that the environment of the VR, which offers the experience of reality, has positive effects on students' academic success (Krokos et al., 2019). In this context, the potential of VR technologies to improve learning can be shown among the reasons for researchers and educators to examine these technologies intensively (Radianti et al., 2020). Within the scope of the chapter, a general perspective is presented about the VR technologies' historical development process, usages in education, limitations and future.

2 What is Virtual Reality?

VR is defined as a computer-based simulation platform that enables users to see or experience a three-dimensional (3D) environment very realistically by using some special equipment (Linowes, 2015: 2). According to another definition, VR is a method for visualizing, manipulating and interacting complex data (Aukstakalnis and Blatner, 1992). Craig et al. (2009: 11) describe VR as environments that perceive users' positions and actions in an interactive environment involving 3D computer simulations, provide synthetic feedback to their one or more senses and enable them to experience the surrounding feeling. As a result, VR can be described as a computer-aided simulation technology that aims to give users a sense of reality in a dynamic and interactive environment.

In the VR environment, the user can move to any position in which they look at by using some tools, interact with objects in the location, define their physical properties or feel themselves as if they were in the real world by hearing sounds in 3D environment. In this context, it is stated that VR technology basically has four elements. These are virtual environment, virtual presence, feedback and interaction (Sherman and Craig, 2002: 6).

Virtual environment represents the virtual environment produced by the computer. It also contains the objects and rules in the environment. The objects and rules in the virtual environment are integrated by being associated with each other. The objects and characters included in the content of the virtual environment can be presented in visual, auditory and tactile dimensions. Like objects in our real world, objects in the VR environment have features such as shape, weight, colour, texture, density and temperature. These features can be observed by using different senses. For example, while the colour of an object can be detected only in the visual field, the texture of the object can be detected in both visual and tactile areas. Actions in the virtual environment usually limit the user to move within a small area. The user can observe and manipulate 3D objects in this area by using various interface elements (virtual buttons, keys, text, graphics) presented in the VR environment (Mihelj et al., 2014). Fig. 1 shows a user who experiences the VR environment and VR equipment (such as headset, hand control devices, sensor).

Virtual presence represents virtual sensitivity in which users can feel as if they are in the real world. The users can feel as if they are a part of the VR environment, and their participation in the activities within the environment can be provided with the feeling of immersion and presence (Passig et al., 2016). Because,

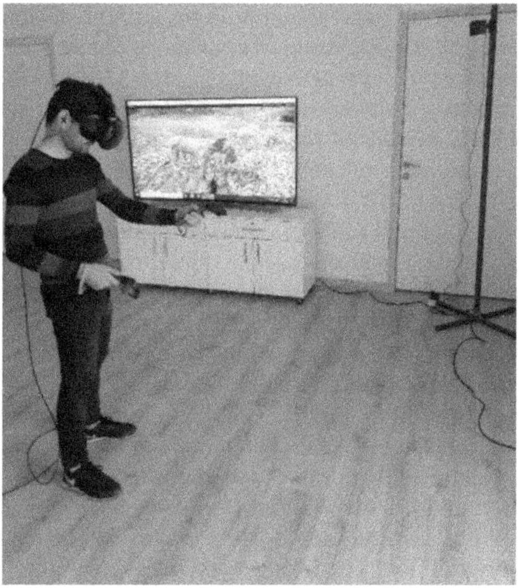

Fig. 1: A user who experiences VR technologies

it is emphasized that the basis of the VR theory is the immersion and presence components (Jensen and Konradsen, 2018). Immersion is defined as the sensitivity level of the sensory features offered by VR systems to users, and presence is described as the psychologically subjective responses of users to the virtual environment (Slater, 1999). In other words, presence is an individual response of the user to the virtual environment; immersion is depicted as total technological features that provide the users with a sense of reality by abstracting the richness, resolution and panoramic image of the users about the content from the other physical realities in the environment (Slater, 2018). As a result, immersion is considered as the technological characteristics of VR systems to provide users with a sense of reality (Radianti et al., 2020).

Feedback component represents the responses of the user as a result of interaction with the virtual world by using the input devices. Feedback can vary depending on the user's location, action or movement in the virtual environment. Feedback is an important component of VR. The VR system gives direct feedback based on users' physical location. In general, most feedback is provided with visual information. However, some environments can only use tactile information in the feedback process. In this context, the VR system must monitor the user's location in order to provide appropriate feedback to them. It should be able to automatically measure the position and direction of objects in the real environment (Mihelj et al., 2014).

Interaction represents the interaction of users with objects in the virtual environment. Considering that we use our five senses in our real-world experience, it is stated that the number of sensors provided by the VR environment is related to the number of our sensory channels that can interact (Klein, 2003). Therefore, it is argued that the high number of sensors in the VR environment is directly proportional to the provision of the probability of immersion and presence (Lombard and Ditton, 1997). Hence, the VR system must respond to the user's actions and appeal to more senses in order to be realistic. In other words, it should be interactive. Consequently, the ability of the user to influence computer-generated virtual environments represents a form of interaction (Mihelj et al., 2014).

Today's VR technologies can enable the users to navigate in 3D environment by moving their head, thanks to a head-mounted component (such as glasses) to view the 3D virtual environment. In addition, data can be entered by using special input components designed according to VR technologies, or 3D environment can be discovered using motion sensors. It can be beneficial to know the historical development process of these technologies in order to understand how the interfaces and devices of these technologies we use today are developed and what their sources of inspiration are. Also, information on the origin of VR technologies can help us understand the types of these technologies and predict their future evolution.

3 History of VR Technologies

In fact, VR technology is not a new technology. Based on their imagination, people have built their VR systems for decades before experiencing the first real applications. Although this technology was attempted to be hidden in military- and industry-related academic research laboratories, it already existed decades ago. The development process of VR technologies from the early 1900s to the early 2000s can be summarized as follows (Sherman and Craig, 2002):

One of the first and most important steps towards the foundation of VR technologies is shown as the head-based periscope-monitoring device, presented in Fig. 2. The patent for head-based periscope design was issued by the *American Patent Institute* to *Albert B. Pratt* in 1916.

In 1929, a simulator-based educational aircraft (*penguin*) was developed by *Edward Link*, with short wings and not much ability to rise above ground. Thanks to the simulator aircraft shown in Fig. 3, trainee pilots learned to fly and navigate after a few years of flight training by using the flight instructions in the cockpit.

In 1956, *Morton Heilig*, who was inspired by the widescreen image, developed the *Sensorama* device. *The Sensorama* device can be described as a multi-mode imaging system. In Fig. 3, *Sensorama* was designed for the user to experience the sense of reality by integrating the pre-recorded images (like motorcycle ride) with features such as sound, smell, vibration and flying.

In 1960, *Morton Heilig* obtained patent for the *Stereoscopic Television Apparatus* for individual use, which was very similar to the 1990s HMD, including mechanisms that appeal to auditory and olfactory senses. Thanks to the *Stereoscopic Television*

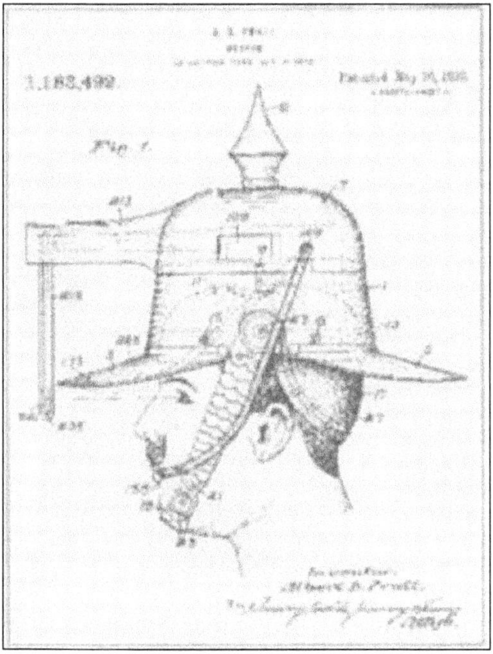

Fig. 2: Head-based periscope. **Source:** (Sherman and Craig, 2002: 24)

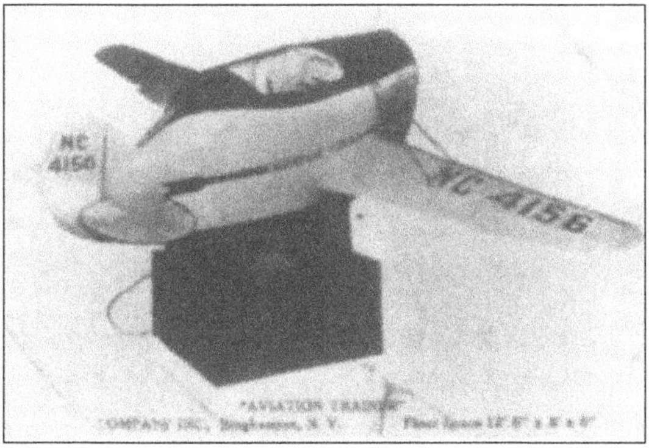

Fig. 3: *Penguin* (Roberson Museum and Science Centre). **Source:** (Sherman and Craig, 2002: 25)

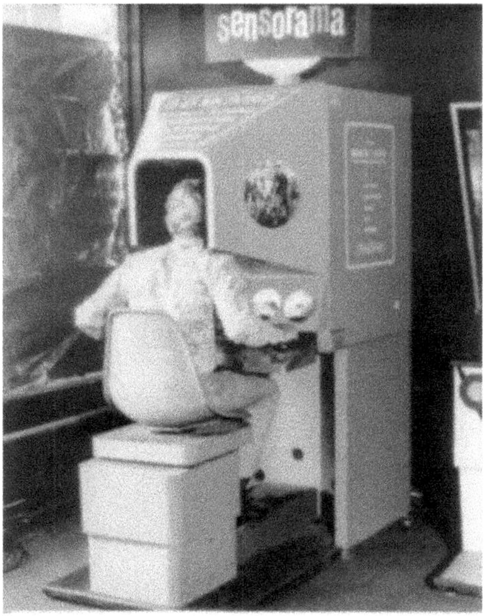

Fig. 4: A figure of the *Sensorama* device. **Source:** (Sensorama, 2020)

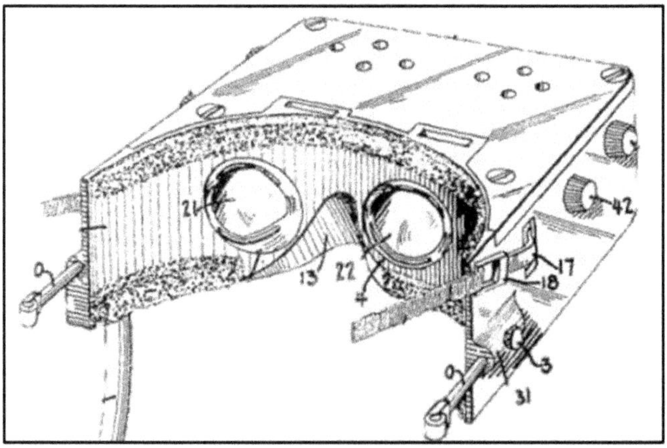

Fig. 5: Heilig's *Stereoscopic Television Apparatus.* **Source:** (Webster and Clark, 2015)

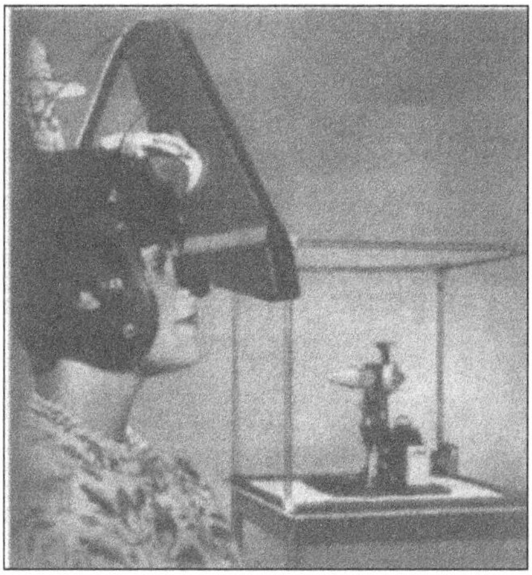

Fig. 6: HMD-integrated *Telepresence* System. **Source:** (Sherman and Craig, 2002: 25)

Apparatus in Fig. 5, *Heilig* enabled to actively stimulate the users' senses in order to improve their experience.

In 1961, *Comeau* and *Bryan* built an HMD device integrated into the remote video camera displaying system that tracks the user's head movements. Later, they established the company called *Telefactor Corparation* within the context of this system, as shown in Fig. 6.

The *Sketchpad* application was developed in 1963 by *Ivan Sutherland*; it gave the opportunity to draw graphics interactively in the computer environment. This application, which is used in the automotive industry, design field and many disciplines today, used a keyboard for data entry and a pen to perform operations such as selection and drawing. Then, in 1964, an interactive and computer-aided design project called *DAC* (design augmented by computer) was launched by *General Motors* to design in the automotive industry. In 1965, *Ivan Sutherland* put forward the concept of *"ultimate display."* This concept was based on the notion that "users can be present in an image environment where they can interact with objects without adhering to the physical rules in our real world." The definition of *Ivan Sutherland* included *kinaesthetic* or *haptic (tactual, touch, tactile)* devices as well as visual stimuli (Sherman and Craig, 2002). A *haptic* device can be defined as the human–machine interface device designed for enabling the user to actively interact in a computer-assisted simulation environment. The *haptic* device can be in the form of a mouse, steering wheel, fingertip, robotic arm, gripper, joystick, glove, magnetically moved wrist or even motion platforms that move the user's entire body (Haptic Device, 2020).

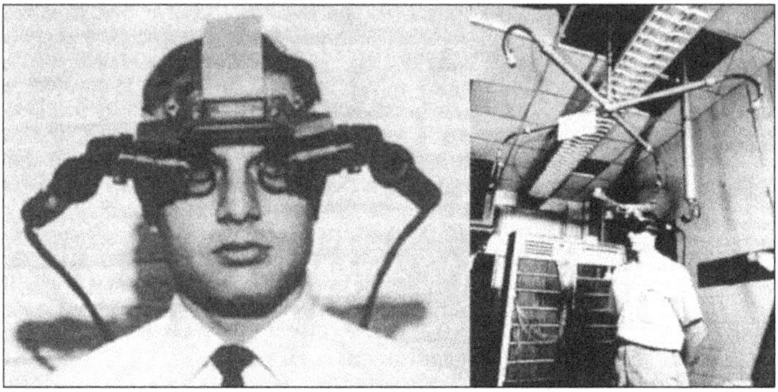

Fig. 7: The HMD device, invented by *Ivan Sutherland*. **Source:** (Linowes, 2015: 3)

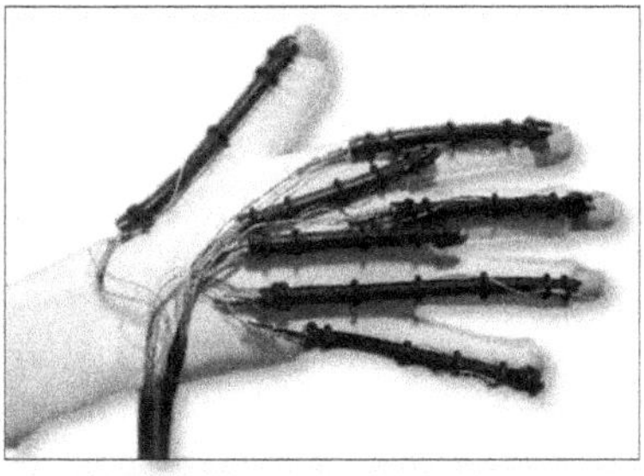

Fig. 8: *Sayre Glove* glove. Source: (Sayre Glove, 2020)

In 1967, *Fred Brooks*, who was inspired by *Ivan Sutherland*'s *"ultimate display"* concept, discovered that kinaesthetic interaction can be used as a tool to help biochemists feel interactions between protein molecules, under the GROPE project at the University of North Carolina (Brooks et al., 1990). Later, in 1968, *Ivan Sutherland* made an HMD device. As shown in Fig. 7, the device was able to present images to its user by using mechanical or ultrasonic motion options. Cathode rays were mainly used in the displaying process (Sherman and Craig, 2002). This device, which was quite large, cumbersome and expensive, could only be used by jointing to the ceiling, as seen in the figure (Linowes, 2015).

Thanks to the *Pong* game developed by *Atari* in 1972, users met real-time, multiplayer and interactive graphics. The company (E&S), created by *Evans* and *Sutherland* in 1973, produced the first digital flight simulation, *Novoview*. In 1976, *Myron Krueger* completed the prototype work called *Videoplace*. In this study, a virtual world platform was created where user movements can be controlled by cameras and other input devices. In 1977, a glove device named *Sayre Glove* was developed. The working mechanism of the glove seen in Fig. 8 was based on the logic of transmitting the amount of light according to the finger movements to the computer system with light conductive tubes.

In 1979, a system called *LEEP* (Large *Expanse Enhanced Perspective*) was developed by *Eric Howlett* to enable the image to be transferred to a larger area from a small screen. In 1981, the foundations of the graphical workstations in the VR technologies we use today have been laid and products with lower costs and faster products have been developed. In the same year, *Super Cockpit* was launched by *Tom Patterson. Super Cockpit* included a transparent HMD on the pilot's helmet. Thanks to this screen, pilots gained different information as they looked in various directions. In addition, studies on VR technologies were carried out in *MIT* laboratories. In these studies, researches were done on 3D drawing, architectural visualization and 3D computer chips.

In 1982, in her doctoral thesis, Sara Bly discovered that sounds can be used to represent big data (*Sonification*). This study has been the main basis for the use of computer-generated and computer-controlled voice in VR technologies. In 1984, *Dave Nagel*, chief of *NASA Aerospace Human Factors Research Division*, started working with *Scott Fisher* to build the virtual interface workstation. In 1987, Engineer *Jim Humphries*, who worked at *NASA*, developed a prototype from HMD-based displays called *BOOM*, which would be commercialized by *Fakespace* in 1990. After the VR word was termed by *VPL* Company in 1989, hardware and software products belonging to VR technologies started to enter the electronics market.

In 1991, *Virtual Research Systems* released the first possibly reliable VR flight helmet, *the VR-2 Flight Helmet*, under $10,000 (Sherman and Craig, 2002). In 1992, projection-based VR technology *CAVE* was developed as an alternative to HMD-based VR technologies. In the CAVE environment, the users were disconnected from the real environment outside and were provided to experience the virtual environment by fully integrating their audio-visual perception with the virtual environment. In 1997, the company *Virtual Technologies* introduced an application called *CyberGrasp*. Thanks to this application, the tactile and grasp sensation of the users in the virtual environment was increased. In 1998, *DisneyQuest*, which includ es projection- and HMD-based virtual technologies, was put into service by *Disney*.

In 2000, *a CAVE* application designed six-sided as a user-oriented was developed at *North America Iowa State University*. In 2012, *Palmer Luckey*, founder of *Oculus VR LLC*, presented a head-mounted VR display to *John Carmack*, the famous developer of classic video games such as *Doom, Wolfenstein 3D* and *Quake*. The community, running a successful *Kickstarter* campaign and waiting excitedly, released a developer kit called O*culus Rift Development Kit 1 (DK1)*. This developer kit attracted the attention of many investors, as well as *Mark Zuckerberg*, and in March 2014, the *Facebook* company bought the *Oculus VR LLC* company for $2 billion. As a result of this situation that helps to create a new category of consumer products, many companies including *Google, Sony, Samsung* and *Steam* have incorporated VR

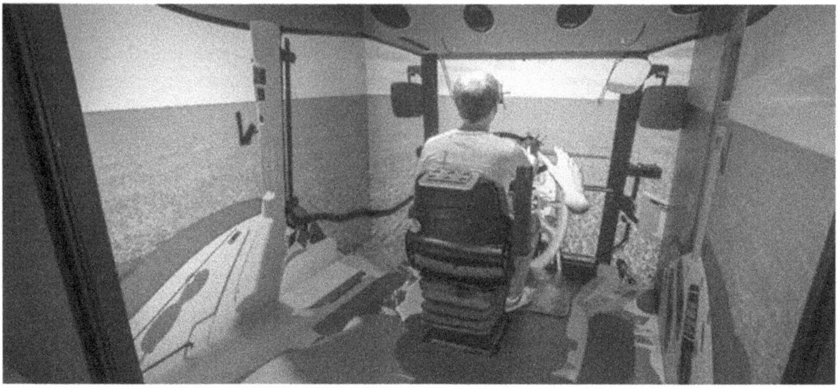

Fig. 9: CAVE simulator, a full-immersive VR application. **Source:** (CAVE, 2020)

technologies into their inventory. In addition, the latest innovations and devices that enrich our VR experience continue to be introduced. Nowadays, VR technologies are used in many platforms such as mobile, desktop, social networking platforms and 3D platforms. It is predicted that 3D websites based on VR technologies will be published more in the near future (Linowes, 2015).

4 Types of VR Technology

VR technologies are classified according to the different types of virtual sensitivity provided by the technological devices included in its content. In other words, VR technologies differ in their immersion characteristics. In this context, there are three different VR systems (Buttussi and Chittaro, 2018): CAVE (cave automatic virtual environment)–based VR, desktop computer–based VR and HMD-based VR.

4.1 CAVE-Based VR

The first *CAVE* system was invented in 1992 by a group of researchers in the Electronic Visualization Laboratory at the University of Illinois (Cruz-Neira et al., 1992). This type of VR technology requires a glove or hand control device, as well as an HMD that can change the image of the environment by tracking the user's head movements to send data to the virtual system to which the user is connected. CAVE VR system is also called full-immersive VR (Sherman and Craig, 2002). *CAVE* is a projection-based VR system. In this expensive VR system, the users are disconnected from the real environment outside and are provided to experience the virtual environment by fully integrating their audio-visual perception with the virtual environment. In this system, the user experiences the feeling of being fully

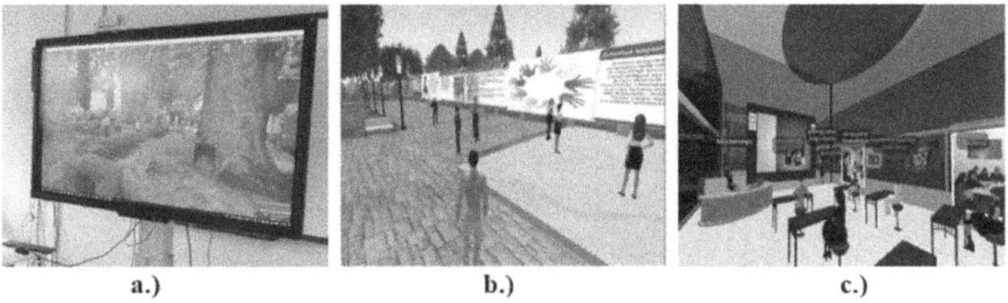

Fig. 10: (a) Desktop platform VR example, (b) virtual world platform, (c) an educational application in a virtual world platform. **Source:** a (Non-Immersive VR, 2020), b (Yılmaz et al., 2013), c (Yılmaz et al., 2013)

a part of the virtual environment (Cruz-Neira et al., 1993). An example of a full-immersive VR application is shown in Fig. 9.

The ideas of many researchers from different disciplines have been inspired to develop the *CAVE* platform. Modern *CAVE* systems can consist of six-wall or six-sided *CAVE* systems by projecting scenes onto the ceiling. The scenes displayed on the screens are reflected by high-resolution projectors and mirrors with rotating positions placed between the screens. In this way, a more realistic and immersive VR environment is designed by integrating the shadows of the users with the *CAVE* system (Muhanna, 2015). Today, *CAVE*-based VR systems are used in architecture, space research and education.

4.2 Desktop-Computer-Based VR

This VR technology, also called non-immersive VR, involves VR platforms in which the users have the opportunity to experience the VR environment on computer screen or other displaying devices without any input device (such as HMD, gloves, hand controller). (Sherman and Craig, 2002; Buttussi and Chittaro, 2018). In Fig. 10, examples of desktop-computer-based VR applications are presented.

These VR systems, which are lower in cost compared to CAVE- and HMD-based VR systems, can give the users a pleasant VR experience in a system with good graphic quality. However, in this VR system, the interaction of users with the virtual environment is in lower level. This VR type, which is mainly used in educational applications, can also be used in video games. Another use of this type of VR application is virtual world platforms. Virtual worlds can be called virtual platforms where users develop applications by using their own imagination (Alarifi, 2008). Examples of virtual world platforms are *Second Life, Active Worlds, Open Simulator*, etc.

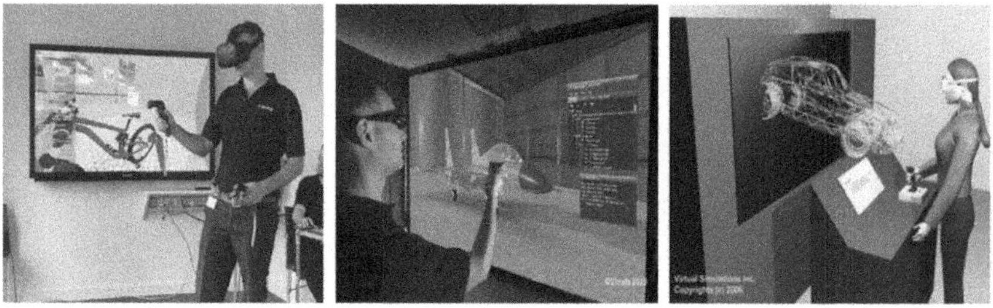

a.) b.) c.)

Fig. 11: (a) Example 1, (b) example 2, (c) example 3. Source: a (HMD-Based VR1, 2020), b (HMD-Based VR2, 2020), c (HMD-Based VR3, 2020)

4.3 HMD-Based VR

HMD-based VR involves VR platforms where desktop VR is combined with additional devices. For example, devices such as *Data Gloves* (data transfer gloves), mouse, keyboard, joystick and hand control devices can be considered as the additional devices. HMD-based VR platforms, which mainly contain the features of desktop-computer-based VR technologies, are platforms where high sensitivity and physical models are used. This VR system, also called the semi-immersive VR platform, enables textual expressions, graphics or other visual elements to be transmitted on a transparent screen so that users can interact with the real environment. (Alqahtani et al., 2017). HMD-based VR is the VR system that is mounted on a glass-like device with two LCD screens fixed according to the user's eye position and monitors the user's head direction and position, in some cases, through these glasses (Santos et al., 2009). Some examples of applications using HMD-based VR systems are shown in Fig. 11.

Thanks to many popular companies that produce HMD-based VR hardware (such as Oculus Rift, HTC Vive and PlayStation VR), a different window has been opened to our digital world (Gugenheimer et al., 2017). The opportunities that the users can navigate in a physical environment using their devices, interact with objects and experience virtual environments with immersion feature can be shown as proofs of this situation (HTC Vive, 2020). HMD- and CAVE-based VR systems have higher capacity in terms of providing the immersion feature to the users. In contrast, it is emphasized that desktop-computer-based VR systems have a lower capacity in terms of immersion (Meyer et al., 2019). In addition, it is argued that VR systems with immersion feature are more advantageous in terms of providing users' presence compared to desktop-computer-based VR systems (Mikropoulos and Natsis, 2011).

5 The Usage of VR Technologies in Education

VR technologies are used in areas such as medical education, video game industry, defence area, health sector, psychotherapy, aviation and space (Mihelj et al., 2014; Singh et al., 2020). In addition, it is stated that VR technologies support students to have realistic experiences by visiting a virtual museum (Lee et. al., 2011), to learn foreign languages (Chen, 2016; Alfadil, 2020) and to comprehend science subjects (Wu, 2018). VR technologies used in some current studies are presented in Tab. 1.

As presented in Tab. 1, it is seen that VR technologies are used in many disciplines. It is emphasized that with the sharp increase in the rate of applications and software developed for VR technologies, the costs of these technologies tend to decrease continuously (Jang and Park, 2019). In this context, VR technology, which is predicted to be a sub-field of computer technologies in the near future, is expected to offer great opportunities for educators and students (Dede, 2014; Akman and Çakır, 2019; Chang et al., 2020). Some meta-analysis studies in the literature also support this situation.

For example, Hirt and Beer (2020) conduct a meta-analysis study on how dementia should be treated, based on the use of VR technologies in the educational process. As a result of the study, they believe that VR technologies can be beneficial in the training process. Howard and Gutworth (2020) investigate the impact of VR-supported training programs in the development of social skills. According to the results, it is argued that VR-supported training programs are effective.

In the study of the effect of SG technologies on the knowledge, cognitive skills, attitudes and satisfaction of healthcare professionals, Kyaw et al. (2019) conduct a meta-analysis and review based on 31 studies. The results show that VR technologies improve employee knowledge compared to traditional teaching methods. Guedes et al. (2019) investigate the effect of VR technologies to teach minimally invasive procedures, which is a medical treatment method. According to the results obtained from 14 studies included in the meta-analysis, they emphasize that the training method based on VR technologies is more effective than the other training method (box-trainer).

Merchant et al. (2014) examine the effect of desktop VR-based instruction on students' learning outcomes in K-12 and higher education. During the study process, games, virtual worlds and simulations are evaluated as VR platforms. According to the results, it is emphasized that these platforms improve learning out comes. Haque and Srinivasan (2006) conduct a meta-analysis study including 16 studies on the educational effectiveness of VR-supported surgical simulators. As a result of the study, they point out that VR simulators reduce the errors in the process of performing a surgical task and enable surgical interventions to be performed in a shorter time interval. In the light of these studies, it can be said that VR technologies generally have positive effects. However, it should be taken into consideration that these technologies have some limitations.

Tab. 1: Some current implementations towards VR technologies

Study	**Discipline**	**Purpose of the research**	**Sample**	**Study result**
Chang, Hsu and Jong, 2020	Earth science	To determine the students' experiences about learning about an VR supported geological park developed for earth science lessons	51 fifth graders	Students participating in the VR design activity, along with the peer assessment learning approach, have higher learning experiences. It is also determined that students using the VR design system have higher self-efficacy and critical thinking tendencies than students who receive traditional teacher feedback.
Alfadil, 2020	Foreign language	To investigate the effect of developed VR game on foreign language (English) word acquisition of middle school students	64 middle school students between the ages of 12–15	Integrating VR technology into foreign language teaching is a promising strategy and can be more successful than traditional learning for word acquisition.
Innocenti et al., 2019	Music	To improve learning skills about musical instruments with VR-supported *VR4EDU* software and to inform primary school students about the history and theory of music	36 primary school students with an average age of 10	The use of mobile VR technologies in conjunction with traditional teaching methodologies can improve the music learning experience in terms of active learning, attention, and time.
Sopher et al., 2019	Engineering	To investigate the effect of different learning environments (Traditional studio environment and virtual reality studio environment) on the design process	10 undergraduate students with an average age of 22	Each environment supports different learning activities. It is also shown that immersive VR supports design convergence, which is considered a key skill.

Frank et al., 2019	Medicine	To determine the effectiveness of a VR-supported simulator for teaching arthroscopic skills	36 undergraduate students	The VR technology used reduces the time students perform tasks and improves their skills.
Tab. 1: (*Continuou*) Van Ginkel et al., 2019	Literature	To investigate the effect of VR supported software on students' oral presentation skills and oral presentation information compared to face-to-face presentation	36 undergraduate students	There is no significant difference between applications. In all three presentations, significant developments are demonstrated in terms of qualification component from pre-test to post-test. In addition, according to self-assessment tests, it is determined that students with VR support are satisfied with the detailed and analytical feedback they receive.
Moon and Ke, 2019	Psychology	To evaluate the effectiveness of virtual reality technology for the development of social skills of autistic children	15 students with an average age of 12	VR-based social skills training supports the treatment process.
Akman and Çakır, 2019	Math	To determine the opinions and experiences of students for a game based on virtual reality technologies developed to teach fractional numbers to primary school students	5 primary school students	The VR platform *Keşfet Kurtul (Explore to Survive)* is a digital learning environment where students have a lot of fun. In addition, it is determined that all tasks except the "Radio" component one, and the game in general, provide the flow experience.

(*continued*)

Tab. 1: Continued

Study	Discipline	Purpose of the research	Sample	Study result
Yang et al., 2019	Physical education	To compare the teaching of traditional physical movements with the teaching of physical movements based on virtual reality technology	High school students	In the field of physical education, a scientific framework is presented for the evaluation of teaching methods based on VR technologies.
Zhang 2019	Art	To examine and evaluate the effectiveness of virtual reality technology as a new technology in the field of art and painting	30 students at the undergraduate level	Students adopt the new teaching approach more. VR technology provides their attention significantly and increases their teaching efficiency.
Dede et al., 2019	Ecology	To evaluate a VR application developed in ecology in the deep learning process	280 primary school students	VR technologies can be beneficial to students on ecosystems including interdisciplinary information.
Chiaoa et al., 2018	Tourism	To investigate the effect of a virtual tour guide based on VR technology on learning and to determine the users' experiences	391 undergraduate students with an average age of 20	Students adopt learning activities and VR technologies in the education system. It is also determined that the interaction variable has an indirect effect on the use of technology. However, it is emphasized that the interaction between people and devices is important in a digital virtual tour environment.
Tarng et al., 2018	Nanotechnology	To compare the teaching of the fluorescence production in a laboratory based on VR technology with the traditional teaching method in terms of learning success	67 high school students	The virtual laboratory can increase learning motivation and effectiveness.

Tarng et al., 2018	Physics	To investigate the effect of an VR application developed to learn the photoelectric effect of LCD	109 undergraduate students	The virtual laboratory can increase students' learning interest and success. It is also shown that it can help students understand the photoelectric effect of LCD, thanks to high interaction and 3D visualization.
Sood and Singh, 2018	E-learning	To develop and evaluate an e-learning platform based on cloud and VR technologies	-	The results show that the optical network has important advantages for playing EEG-based games.
Rojas-Sola and Aguilera-García, 2018	History	To investigate the effect of VR and AR applications on students' satisfaction for teaching historical inventions	10 people	The students are very satisfied with the VR environment.
Johnson, 2018	Religious Education	To reveal and discuss the results of an experience that teaches rituals about religion and empathy by using 360-degree video and VR technologies	38 undergraduate students	VR-supported vehicles can help attract students and meet basic learning outcomes such as empathic understanding, ethnographic analysis of religious space, ritual and behaviour in the light of religious theories.
Birt and Cownling, 2018	Architecture	To develop and implement a VR application used to develop spatial visualization and spatial skills within the framework of design-based research	26 undergraduate students	Not only the positive aspects of the approach but also the necessary improvements in the presentation of visualizations to improve the visibility and visual errors caused by the lack of mobile processing will be important.

(continued)

Tab. 1: Continued

Study	Discipline	Purpose of the research	Sample	Study result
Watson and Livingstone, 2018	First Aid	To evaluate a first aid simulation based on VR technology according to three different teaching methods in the context of teaching effectiveness, usability and motor skills	61 students	VR technologies provide ease of use to students as both learning resources and application activities and are adopted by students.
Pham et al., 2018	Security	To test the effectiveness of 360-degree videos based on VR technology developed to provide students with information about building safety	60 undergraduate students	The developed VR is a strong pedagogical approach for students to effectively practice and acquire security knowledge.
Coyne et al., 2018	Distance Education	To evaluate the teamwork of people using a distance learning platform based on the VR system	18 undergraduate students with an average age of 18	The participants' responses about the application are positive. It is emphasized that the potential of VR to be a useful tool for online, team-based studies is considerably high.
Makransky and illeholt, 2018	Biology	To discover the differences between desktop-computer based VR and VR application which has the immersive feature in the process of using a virtual learning simulation in terms of motivation, learning and other variables	104 undergraduate students with an average age of 23	The VR simulation with immersive feature has a significant effect on motivation. It is also observed that the immersion and presence features of the VR environment affect perceived learning outcomes. In particular, it is emphasized that the immersion feature allows positive emotional experiences and that this is valuable in terms of cognition.

Al-Saud et al., 2016	Dentistry	To examine the effect of recalling, transfer and learning basic dental skills related to teeth by depending upon the feedback provided by a dental simulator based on VR technology	-	VR technologies are effective for teaching basic dental skills to novice trainees. It is also demonstrated that the use of hand-controlled device-assisted VR technology can be useful in dentistry education.
Su, 2016	Rehabilitation	To investigate the success factors in the rehabilitation process of patients with knee prosthesis by using emotion model-based game rehabilitation system that combines VR and motion capture technology	34 adults with an average age of 65	VR technologies are an effective tool for rehabilitation. It is also suggested that patients should be operated on in the early stages and have regular exercise habits for better rehabilitation.
Cao, 2016	Navigation	To test a VR application designed to provide training on the driving of ships	–	The system can reflect the visual system, thanks to components such as wide field of view and instructor control station. It is also observed that the presented VR training environment provides a strong visual enhancement and a sense of immersion.
Bauerle et al., 2016	Mining	To examine the experience of mining workers on a VR application developed to provide rescue training under the mine	31 people	There is a positive reaction towards BG-4 test software in general, and the test is effective. It is also determined that training can be useful in the process of preparing trainees for real emergencies.

(*continued*)

Tab. 1: Continued

Study	Discipline	Purpose of the research	Sample	Study result
Gavish et al., 2015	Industry	To compare VR and AR applications during the teaching of basic information about industry	40 people with an average age of 33	The use of the AR platform should be encouraged in the learning process, and for this purpose, the use of the VR platform should be evaluated more.
Jagodzinski and Wolski, 2015	Chemistry	To compare VR and AR applications during the teaching of basic information about industry	100 high school and college students with an average age of 15	Students get higher grades in the academic success and become more interested in chemistry.
Webster, 2014	Military Defense	To investigate the experiences of students who are trained in a VR laboratory designed for chemistry education and to evaluate the VR system	25 people with an average age of 30	The results show that both forms of teaching will increase learning. However, it is determined that students get higher scores from the VR-based education and that there is a statistically significant interaction between the task type and time.

6 Some Limitations of VR Technologies

For many years, the cost was seen as the main obstacle to using technological environments such as VR (Merchant et al., 2014). Also, some VR technologies were difficult to use. Because, as seen in the historical development process, the equipment that people who want to use these systems should wear in the past was rather cumbersome and mostly lacked the immersive feature. However, thanks to advances in technology, VR systems have been reduced in size to be used in mobile devices (Wu et al., 2013) and made suitable for use in viewers such as *Google Cardboard* (Boyles, 2017). Nevertheless, VR systems have some limitations, which are presented below:

- There may be risks such as breakage and deterioration depending on the use in VR devices, as in other hardware devices (Choi, 2016).
- When using VR systems, most people may experience problems like movement problems, nausea and headache (Murray, 2017).
- VR systems can be costly in terms of time, money and workload (Duarte et al., 2010; Eichenberg, 2011).
- Some technical and hardware knowledge may be required to use VR systems. In this context, users may need additional time. For instance, an incorrectly attached VR head may cause images and text to appear blurry. As a result, it can cause cognitive load in users (Boyles, 2017).
- Designing and developing programs that comply with VR systems are challenging processes. In this context, users who will develop VR environments may need some technical information (such as scenario creation, programming, design, drawing) (Murray, 2017).
- Integrating VR technologies with a training program may bring about a difficult process. For example, some teachers may resist using new technologies such as VR (Huang, 2016).

7 Conclusion

In our digitalized world, it is seen that the integrated use of VR technologies in many devices such as mobile phones, computers and tablets will become widespread and will create new markets by affecting the current technology marketing sectors. Accordingly, it can be said that many fields such as medicine, military defence, architecture, engineering and education will be affected by these developments. Due to the decrease in the cost of VR technologies and technological developments, it can be stated that these new technologies will be used more intensively, especially in the field of education, in the learning-teaching process. It is emphasized that these technologies have the potential to motivate students in the process of distinguishing real and virtual objects. In addition, VR technologies can provide powerful and unique educational experiences for educators and students, enable students to interact both in the virtual and real world and support their learning process.

It should be noted that VR systems have some disadvantages in terms of technology, economy and health. However, it is claimed that these limitations will gradually decrease with the developing technology. For example, *2019 oculus Quest* has been developed as a solution to the cable problem in HMD-based SG technologies (Radianti et al., 2019; Oculus Quest, 2020). VR technologies can be used by educators, and these systems have a high chance of contribution. Therefore, educators may be more courageous for applying VR technologies in learning activities. Considering the technological costs and limitations, the use of these technologies in small learning activities can be considered as the first concrete step. As a result, VR technologies promise a lot for the field of education as well as many disciplines. However, educators may need to blend various technical and pedagogical knowledge together with their field knowledge to be able use these technologies effectively in the future. Because, *technology will never replace great teachers, but technology in the hands of a great teacher can be transformational* (George Couros, 2020).

Bibliography

Akman, Emrah and Çakır, Recep, (2019), "Pupils' Opinions on an Educational Virtual Reality Game in Terms of Flow Experience", *International Journal of Emerging Technologies in Learning*, Vol. 14, N. 15, pp. 121–137.

Alarifi, S. Abdulaziz, (2008), *An Exploratory Study of Higher Education Virtual Campuses in Second Life,* (Unpublished Master's Dissertation), University of Nottingham, Nottingham.

Alfadil, Mohammed, (2020), "Effectiveness of Virtual Reality Game in Foreign Language Vocabulary Acquisition", *Computers & Education*, Vol. 153, N. 103893, pp. 1–13.

Alqahtani, A. Saeed, Daghestani, L. Foaud, and Ibrahim, L. Fattouh, (2017), "Environments and System Types of Virtual Reality Technology in STEM: A Survey", *International Journal of Advanced Computer Science and Applications*, Vol. 8, N. 6, pp. 77–89.

Al-Saud, M. Loulwa, et al, (2017), "Feedback and Motor Skill Acquisition Using a Haptic Dental Simulator", *European Journal of Dental Education*, Vol. 21, N. 4, pp. 240–247.

Aukstakalnis, Steve and Blatner, David, (1992), *Silicon Mirage; the Art and Science of Virtual Reality*, USA: Peachpit Press.

Bauerle, Tim, Brnich, J. Michael, and Navoyski, Jason, (2016), "Exploring Virtual Mental Practice in Maintenance Task Training", *Journal of Workplace Learning*, Vol. 28, N. 5, pp. 294–306.

Birt, R. James and Cowling, A. Michael, (2018), "Assessing Mobile Mixed Reality Affordances as A Comparative Visualization Pedagogy for Design Communication", *Research in Learning Technology*, Vol. 26, pp. 1–25.

Boyles, Brian, (2017), "Virtual Reality and Augmented Reality in Education", https://westpoint.edu/sites/default/files/inline-images/centers_research/center_for_teching_excellence/PDFs/mtp_project_papers/Boyles_17.pdf, (15.05.2020).

Brooks, P. Frederick, Ouh-Young, Ming, Battert, J. James, and Kilpatrick, P. Jerome, (1990), "Project GROPEHaptic Displays for Scientific Visualization", *Computer Graphics*, Vol. 24, N. 4, pp. 177–185.

Buttussi, Fabio and Chittaro, Luca, (2018), "Effects of Different Types of Virtual Reality Display on Presence and Learning in a Safety Training Scenario", *IEEE Transactions on Visualization and Computer Graphics*, Vol. 24, N. 2, pp. 1063–1076.

Cao, Fenghua, (2016), "A Ship Driving Teaching System Based on Multi-Level Virtual Reality Technology", *International Journal of Emerging Technologies in Learning*, Vol. 11, N. 11, pp. 26–31.

CAVE, (2020), "Virtual reality CAVE", https://www.avinteractive.com/news/virtual-augmented-mixed/virtual-reality-tractors-a-cave-and-pro-displays-ultra-black-screens-14-09-2015/, (03.06.2020).

Chang, S. Chen, Hsu, Ting-Chia, and Jong, M. S. Yung, (2020), "Integration of the Peer Assessment Approach with a Virtual Reality Design System for Learning Earth Science", *Computers & Education*, Vol. 146, N. 103758, pp. 1–15.

Chen, Yu-Li, (2016), "The Effects of Virtual Reality Learning Environment on Student Cognitive and Linguistic Development", *The Asia-Pacific Education Researcher*, Vol. 25, N. 4, pp. 637–646.

Chiao, Huei-Ming, Chen, Yu-Li, and Huang, Wei-Hsin, (2018), "Examining the Usability of an Online Virtual Tour-Guiding Platform for Cultural Tourism Education", *Journal of Hospitality, Leisure, Sport & Tourism Education*, Vol.23, pp. 29–38.

Choi, D. Hwa, Dailey-Hebert, Amber, and Estes, J. Simmons (Eds.), (2016), *Emerging Tools and Applications of Virtual Reality in Education*, Hershey: IGI Global.

Coyne, Leanne, Takemoto, K. Jody, Parmentier, L. Brittany, Merritt, Thayer, and Sharpton, A. Rachel, (2018), "Exploring Virtual Reality as a Platform for Distance Team-Based Learning", *Currents in Pharmacy Teaching and Learning*, Vol. 10, N. 10, pp. 1384–1390.

Craig, Alan, Sherman, R. William, Will, D. Jeffrey, (2009), *Developing Virtual Reality Applications: Foundations of Effective Design*, Burlington: Morgan-Kaufman.

Cruz-Neira, Carolina, Sandin, J. Daniel, De Fanti, A. Thomas, Kenyon, V. Robert, and Hart, C. John, (1992), "The CAVE: Audio Visual Experience Automatic Virtual Environment", *Communications of the ACM*, Vol. 35, No. 6, pp. 64–73.

Dede, Chris, (2014), *The Role of Digital Technologies in Deeper Learning, Retrieved from* https://files.eric.ed.gov/fulltext/ED561254.pdf, (10.05.2020).

Dede, Chris, Grotzer, A. Tina, Kamarainen, Amy, and Metcalf, Shari, (2017), "EcoXPT: Designing for Deeper Learning Through Experimentation in an Immersive Virtual Ecosystem", *Journal of Educational Technology & Society*, Vol. 20, N. 4, pp. 166–178.

Duarte, Emilia, Rebelo, Francisco, and Wogalter, S. Michael, (2010), "Virtual Reality and Its Potential for Evaluating Warning Compliance", *Human Factors and Ergonomics in Manufacturing & Service Industries*, Vol. 20, N. 6, pp. 526–537.

Eichenberg, Christiane, (2010), "Application of 'Virtual Realities' in Psychotherapy: Possibilities, Limitations and Effectiveness", (Jae-Jin Kim, Eds.), *Virtual Reality*, Croatio: InTech, pp. 469–484.

Frank, M. Rachel, et al., (2019), "Does Arthroscopic Simulation Training Improve Triangulation and Probing Skills? A Randomized Controlled Trial", *Journal of Surgical Education*, Vol. 76, N. 4, pp. 1131–1138.

Gavish, Nirit, et al., (2015), "Evaluating Virtual Reality and Augmented Reality Training for Industrial Maintenance and Assembly Tasks", *Interactive Learning Environments*, Vol. 23, N. 6, pp. 778–798.

Guedes, H. Goncalo, et al., (2019), "Virtual Reality Simulator Versus Box-Trainer to Teach Minimally Invasive Procedures: A Meta-analysis", *International Journal of Surgery*, Vol.61, pp. 60–68.

Gugenheimer, Jan, Stemasov, Evgeny, Frommel, Julian, and Rukzio, Enrico, (2017), "Sharevr: Enabling Co-located Experiences for Virtual Reality Between HMD and Non-HMD Users", in *Proceedings of the 2017 CHI Conference on Human Factors in Computing Systems*, pp. 4021–4033.

Haptic Device, (2020), "Haptic Devices", http://www.chai3d.org/download/doc/html/chapter6-haptic-devices.html, (02.05.2020).

Haque, Syed and Srinivasan, Shankar, (2006), "A Meta-analysis of the Training Effectiveness of Virtual Reality Surgical Simulators", *IEEE Transactions on Information Technology in Biomedicine*, Vol. 10, N. 1, pp. 51–58.

Hirt, Julian and Beer, Thomas, (2020), "Use and Impact of Virtual Reality Simulation in Dementia Care Education: A Scoping Review", *Nurse Education Today*, Vol. 84, N. 104207, pp. 1–8.

HMD-Based VR1, (2020), "Example 1", http://www.simulation.co.kr/entry/VR4CAD-article, (03.05.2020).

HMD-Based VR2, (2020), "Example 2", https://www.virtalis.com/blogs/2014/03/13/virtalis-launches-visionary-render-at-hannover-messe-2/, (02.05.2020).

HMD-Based VR3, (2020), "Example 3", https://virtasim.com/vsiproducts/, (12.05.2020).

Hollebeek, D. Linda, Clark, K. Moira, Andreassen, W. Tor, Sigurdsson, Valdimar, and Smith, Dale, (2020), "Virtual Reality through the Customer Journey: Framework and Propositions", *Journal of Retailing and Consumer Services*, Vol. 55, N. 102056, pp. 1–12.

Howard, C. Matt and Gutworth, B. Melissa, (2020), "A Meta-analysis of Virtual Reality Training Programs for Social Skill Development", *Computers & Education*, Vol.144, N. 103707, pp. 1–19.

HTC Vive, (2020), "HTC Vive Virtual Reality", https://www.vive.com/us/setup/, (05.06.2020).

Huang, H. Mei, Liaw, S. Sheng, and Lai, C. Min, (2016), "Exploring Learner Acceptance of the Use of Virtual Reality in Medical Education: A Case Study of Desktop and Projection-Based Display Systems", *Interactive Learning Environments*, Vol. 24, N. 1, pp. 3–19.

Innocenti, E. Degli, et al., (2019), "Mobile Virtual Reality for Musical Genre Learning in Primary Education", *Computers & Education*, Vol. 139, pp. 102–117.

Jagodziński, Piotr and Wolski, Robert, (2015), "Assessment of Application Technology of Natural User Interfaces in The Creation of a Virtual Chemical Laboratory", *Journal of Science Education and Technology*, Vol. 24, N. 1, pp. 16–28.

Jang, Yeonju and Park, Eunil, (2019), "An Adoption Model for Virtual Reality Games: The Roles of Presence and Enjoyment", *Telematics and Informatics*, Vol. 42, N. 101239, pp. 1–9.

Jensen, Lasse and Konradsen, Flemming, (2018), "A Review of the use of Virtual Reality Head-Mounted Displays in Education and Training", *Education and Information Technologies*, Vol. 23, N. 4, pp. 1515–1529.

Johnson, D. Christopher, (2018), "Using Virtual Reality and 360-Degree Video in the Religious Studies Classroom: An Experiment", *Teaching Theology & Religion*, Vol. 21, N. 3, pp. 228–241.

Klein, R. Lisa, (2003), "Creating Virtual Product Experiences: The Role of Telepresence", *Journal of Interactive Marketing*, Vol. 17, N. 1, pp. 41–55.

Krokos, Eric, Plaisant, Catherina, and Varshney, Amitabh, (2019), "Virtual Memory Palaces: Immersion Aids Recall", *Virtual Reality*, Vol. 23, N. 1, pp. 1–15.

Kyaw, B. Myint, et al., (2019), "Virtual Reality for Health Professions Education: Systematic Review and Meta-analysis by the Digital Health Education Collaboration", *Journal of Medical Internet Research*, Vol. 21, N. 1, pp. 1–13.

Lee, T. Han, Hsu, K. Shu, and Yeh, L. Jyi, (2011), "Design and Application of the Augmented Reality with Digital Museum and Digital Heritage", in *International Conference on Technologies for E-Learning and Digital Entertainment*, Vol.6872, pp. 25–26.

Linowes, Jonathan, (2015), *Unity Virtual Reality Projects*, UK: Packt Publishing.

Lombard, Matthew and Ditton, Theresa, (1997), "At the Heart of It All: The Concept of Telepresence", https://doi.org/10.1111/j.1083-6101.1997.tb00072.x, (19.05.2020).

Makransky, Guido and Lilleholt, Lau, (2018), "A Structural Equation Modeling Investigation of the Emotional Value of Immersive Virtual Reality in Education", *Educational Technology Research and Development*, Vol. 66, N. 5, pp. 1141–1164.

Merchant, Zahira, Goetz, T. Ernest, Cifuentes, Lauren, Keeney-Kennicutt, Wendy, and Davis, J. Trina, (2014), "Effectiveness of Virtual Reality-Based Instruction on Students' Learning Outcomes in K-12 and Higher Education: A Meta-Analysis", *Computers & Education*, Vol. 70, pp. 29–40.

Meyer, A. Oliver, Omdahl, K. Magnus, and Makransky, Guido, (2019), "Investigating the Effect of Pre-Training When Learning Through Immersive Virtual Reality and Video: A Media and Methods Experiment", *Computers & Education*, Vol. 140, N. 103603, pp. 1–17.

Mihelj, Matiaz, Novak, Domen, and Beguš, Samo, (2014), *Virtual Reality Technology and Applications*, Dordrecht: Springer.

Mikropoulos, A. Tansos and Natsis, Antonis, (2011), "Educational Virtual Environments: A Ten-Year Review of Empirical Research (1999–2009)", *Computers & Education*, Vol. 56, N. 3, pp. 769–780.

Moon, Jewoong and Ke, Fengfeng, (2019), "Exploring the Treatment Integrity of Virtual Reality-Based Social Skills Training for Children with High-functioning Autism", doi: 10.1080/10494820.2019.1613665, (15.05.2020).

Muhanna, A. Muhanna, (2015), "Virtual Reality and the CAVE: Taxonomy, Interaction Challenges and Research Directions", *Journal of King Saud University-Computer and Information Sciences*, Vol. 27, N. 3, pp. 344–361.

Murray, W. Jeff, (2017). *Building Virtual Reality with Unity and Steam VR*, USA: CRC Press.

Non-Immersive SG, (2020), "Scenario-based Learning Game Lets Students Design Forest-friendly Virtual Worlds", http://www.fao.org/forestry/news/94282/en/, (19.06.2020).

Oculus Quest, (2020), "Oculus Quest", https://www.oculus.com/quest/, (10.06.2020).

Passig, David, Tzuriel, David, and Eshel-Kedmi, Ganit, (2016), "Improving Children's Cognitive Modifiability by Dynamic Assessment in 3D Immersive Virtual Reality Environments", *Computers & Education*, Vol. 95, pp. 296–308.

Pham, H. Chien, et al., (2018), "Virtual Field Trip for Mobile Construction Safety Education Using 360-Degree Panoramic Virtual Reality", *International Journal of Engineering Education*, Vol. 34, N. 4, pp. 1174–1191.

Radianti, Jaziar, Majchrzak, A. Tim, Fromm, Jennifer, and Wohlgenannt, Isabell, (2020), "A Systematic Review of Immersive Reality Applications for Higher Education: Design Elements, Lessons Learned, and Research Agenda", *Computers & Education*, Vol. 147, N. 103778, pp. 1–29.

Rogers, Sol, (2019), "Virtual Reality: The Learning Aid of the 21st Century", https://www.forbes.com/sites/solrogers/2019/03/15/virtual-reality-the-learning-aid-of-the-21st-century/#2e10146c139b, (10.06.2020).

Rojas-Sola, I. Jose and Aguilera-García, Ángel, (2018), "Virtual and Augmented Reality: Applications for the Learning of Technical Historical Heritage", *Computer Applications in Engineering Education*, Vol. 26, N. 5, pp. 1725–1733.

Santos, B. Sousa, Dias, Paulo, Pimentel, Angela, Baggerman, Jan-Willem, Ferreira, Carlos, Silva, Samuel, and Madeira, Joaquim, (2009), "Head-Mounted Display Versus Desktop for 3D Navigation in Virtual Reality: A User Study", *Multimedia Tools and Applications*, Vol. 41, N. 1, pp. 161–181.

Sayre Glove, (2020), "Historique", http://www.viper-computing.be/2016/11/16/historique/, (24.05.2020).

Sensorama, (2020), "Inventor in the Field of Virtual Reality", http://uschefnerarchive.com/morton-heilig-inventor-vr/, (23.05.2020).

Sherman, R. William and Craig, B. Alan, (2002), *Understanding Virtual Reality: Interface, Application, and Design*. US: Elsevier.

Singh, R. Pratap, Javaid, Mohd, Kataria, Ravinder, Tyagi, Mohit, Haleem, Abid, and Suman, Rajiv, (2020), "Significant Applications of Virtual Reality for COVID-19 Pandemic", *Diabetes & Metabolic Syndrome: Clinical Research & Reviews*, Vol. 14, N. 4, pp. 661–664.

Slater, Mel, (1999), "Measuring Presence: A Response to the Witmer and Singer Presence Questionnaire", http://www0.cs.ucl.ac.uk/staff/m.slater/Papers/pq.pdf, (10.05.2020).

Slater, Mel, (2018), "Immersion and the Illusion of Presence in Virtual Reality", *British Journal of Psychology*, Vol. 109, N. 3, pp. 431–433.

Sood, K. Sandeep and Singh, D. Kiran, (2018), "An Optical-Fog Assisted EEG-Based Virtual Reality Framework for Enhancing E-Learning Through Educational Games", *Computer Applications in Engineering Education*, Vol. 26, N. 5, pp. 1565–1576.

Sopher, Hadas, Gewirtzman, D. Fisher, and Kalay, E. Yehuda, (2019), "Going Immersive in a Community of Learners? Assessment of Design Processes in a Multi-setting Architecture Atudio", *British Journal of Educational Technology*, Vol. 50, N. 5, pp. 2109–2128.

Statista VR, (2020), "Forecast Augmented (AR) and Virtual reality (VR) Market Size Worldwide from 2016 to 2022 (in Billion U.S. Dollars)", https://www.statista.com/statistics/591181/global-augmented-virtual-reality-market-size/, (18.05.2020).

Su, Chung-Hu and Cheng, Ching-Hsue, (2016), "Developing and Evaluating Creativity Gamification Rehabilitation System: The Application of PCA-ANFIS Based Emotions Model", *Eurasia Journal of Mathematics Science and Technology Education*, Vol. 12, N. 5, pp. 1443–1468.

Tarng, Wernhuar, Lee, Chi-Young, Lin, Chih-Ming, and Chen, Wei-Hun, (2018), "Applications of Virtual Reality in Learning the Photoelectric Effect of Liquid Crystal Display", *Computer Applications in Engineering Education*, Vol. 26, N. 6, pp. 1956–1967.

Tarng, Wernhuar, Liu, Chia-Lin, Lee, Chi-Young, Lin, Chih-Ming, and Lu, Yun-Chen, (2019), "A Virtual Laboratory for Learning Fullerene Production and Nanostructure Analysis", *Computer Applications in Engineering Education*, Vol. 27, N. 2, pp. 472–484.

van Ginkel, Stan, et al., (2019), "Fostering Oral Presentation Competence Through a Virtual Reality-Based Task for Delivering Feedback", *Computers & Education*, Vol. 134, pp. 78–97.

Watson, Paul and Livingstone, Dan, (2018), "Using Mixed Reality Displays for Observational Learning of Motor Skills: A Design Research Approach Enhancing Memory Recall and Usability", *Research in Learning Technology*, Vol. 26, pp. 1–18.

Webster, Rustin, (2016), "Declarative Knowledge Acquisition in Immersive Virtual Learning Environments", *Interactive Learning Environments*, Vol. 24, N. 6, pp. 1319–1333.

Webster, Rustin and Clark, Alex, (2015), "Turn-Key Solutions: Virtual Reality", https://asmedigitalcollection.asme.org/IDETC-CIE/proceedings-abstract/IDETC-CIE2015/57052/V01BT02A052/256961, (10.06.2020).

Wu, Jialing, (2018), “A Space Design Teaching Model Using Virtual Simulation Technology”, *International Journal of Emerging Technologies in Learning*, Vol. 13, N. 6, pp. 163–175.

Yang, Yongzhong and Meng, Lingfeng, (2019), “Physical Education Motion Correction System Based on Virtual Reality Technology”, *International Journal of Emerging Technologies in Learning*, Vol. 14, N. 13, pp. 105–116.

Yilmaz, Rabia, Topu, Fatma-Burcu, Goktas, Yuksel, and Coban, Murat, (2013), “Social Presence and Motivation in A Three-Dimensional Virtual World: An Explanatory Study”, *Australasian Journal of Educational Technology*, Vol. 29, N. 6. pp. 823–839.

Zhang, Lingling, (2019), “Illustration Design Teaching Mode Based on Virtual Wall Painting Technology”, *International Journal of Emerging Technologies in Learning*, Vol. 14, N. 3, pp.190–200.

Mehmet Akif BAL

Marketing of Sports and Communication in Digital Area: E-Sports and the Case of Turkey E-Sports Football League

1 Introduction

In our daily life, sports have emerged as an area that has the power to bring many people together and is an important and popular area in social life. That is why many companies have noticed this power of sports in the recent past and used marketing related to sports. Thus, the share of sports in marketing has increased gradually.

Recently, just like the two sides of a coin, a different side of sports has emerged based on technological developments. The e-sports (Warr, 2014), where amateur or professional players compete with each other through computer or game consoles in the electronic environment, formed the other side of the coin. The concept of electronic sports (e-sports), mediated by computer systems, has become a growing area with the development of games and online systems. Therefore, marketing of e-sports and e-sports products has also become important, as with sports marketing. In the study, the type and applications of e-sports that have reached a corporate identity for the first time in 2018 in Turkey are investigated in the literature review. The aim is to reveal the change in sports and sports marketing in digital media, to explain e-sports, and the development and applications of e-sports, which are becoming popular today. In this respect, the activities of Turkey E-sports Federation, which is the representative and manager of Turkish e-sports, were addressed in the national development process of e-sports. The sample of Turkey e-sports football league organized for the first time in the year 2019 was examined.

2 Sports Marketing and Electronic Sports in Digital Media

In today's sports world, the sports market has reached a budget expressed in billions of dollars. In the last 30 years, sports marketing has improved with the sports industry developing all over the world. Sports-related businesses, companies producing sports goods, the fields of sports, stadiums, players, sports teams, and professional leagues presented in the sports industry contributed to the growth of the sector (Terekli vd., 2000: 410). The concept of sports marketing was first used in the Advertising Age magazine in the United States in 1978. From that day to this, it has taken its place in the marketing literature. In Turkey, the concept of sports marketing came to the fore for the first time in the "The Seminar of Marketing in Sports" in 1991 (Mediacat, 2002: 18). The development of sports marketing in recent years in Turkey was realized with the following (Altunbaş, 2007: 94):

- Federations begin to increase the importance they attach to marketing.
- The enactment of the sponsorship law.
- The emergence of publications on sports marketing.
- Increasing sponsorship activities in other sports than football.
- The increasing number of firms sponsoring teams, leagues and federations.

Sports marketing is all activities created to meet the changing demands and expectations of sports consumers (Mullin et al., 1993: 7). Depending on the two factors that arise in sports marketing, direct marketing of sports takes place in terms of increasing the consumption of various sports activities and encouraging the audience to participate in these activities. The involvement of sports in marketing action is defined as the use of sports by commercial firms as a marketing and communication tool for the promotion of their products (Mediacat, 2002: 18). The first of two factors that arise in the marketing of sports is the marketing of sports, sports goods and services, and the second is the marketing of other goods and services, thanks to sports. While a professional team or club is concerned with the marketing of sports, a beverage company is interested in marketing through sports. The sponsorship comes to mind when it comes to marketing, thanks to sports. (Argan and Katırcı, 2002: 23).Therefore, sponsorship is an important function that directs sports.

Sports clubs, licensed sportsmen, national and international federations, and companies form the basis of modern sports. The healthy relationship between these four in today's sports prepares an environment for qualified athletes to be trained and for sports shows to attract the attention of the audiences. An important function of sports clubs is to train sportsman; nonetheless, another function is to make the sports they represent popular with their production. For this, many sports clubs have emerged that are affiliated with federations in the modern sports concept. Sports clubs are among the leading actors of sports communication and marketing in connecting the audiences to the clubs (Çetin, 2015: 14–19). While the federations are responsible for sports branches today, licensed sportsmen who want to sport regularly started to gather under the roof of sports clubs. Sportsmen must be licensed for healthy and measurable growth of sports branches. For a sportsman to be deemed licensed, he must be part of a club, and that club must get a license for that sportsman from federations for a fee. Thanks to this license, sportsmen can participate in competitions of both national and international federations (Çetin, 2015: 20). Federations are organizations that prepare the rules of the games of the sports branches they represent. Federations that ensure the standardization of the game rules provide the environment for national qualification and equal competition (Çetin, 2015: 26).

These mentioned processes, which form the basis of sports and sports marketing, have brought the concept of e-sports today. With the introduction of computers and the Internet into our lives, games that we can access from mobile devices today are also included in our lives. With the increase of the games, the tournaments organized in the last decade related to the games have also increased. This led to the birth of a new sector called e-sports. Factors such as clubs, federations, licensed players, sponsors, audience, sports product sales, advertising activities, and promotion channels, which have been handled in traditional sports, have played a major role also in the promotion and development of e-sports.

E-sports can be defined as a new and competitive type of sport that can be played individually or as a team, online or offline, on computers, consoles, and mobile phones, and that can be played with various modes of digital games (Youth Research Newsletter, 2018: 6). In other words, e-sports is a sport that brings two people together to play from one end of the world to the other end, and that requires more than most sports do, both physically and mentally, via the Internet. (Argan et al., 2006). According to Turkey Electronic Sports Federation, e-sports is a type of sport that includes all kinds of activities being participated both individually and as a team, both online and offline via a device.

E-sports have branches such as MOBA, FPS, RTS, CCG, and MRPG, just like in traditional sports. Besides, virtual versions of traditional sports games are made as e-sports under the name of sports games (Aktuna and Ünlüönen, 2017: 3). In the report prepared by the Ministry of Youth and Sports for e-sports, these branches are explained as follows:

1. ***MOBA (Massively Online Battle Arena):*** There are two competing teams generally consisting of five people. These teams try to destroy each other's energy sources on a map consisting of three main roads.
2. ***FPS (First Person Shot):*** The player sees the digital game environment through the eyes of the virtual character. Features such as reflex, quick decision making, and team management are necessary for success in this type of game.
3. ***RTS (Real-Time Strategy):*** In such games, the player tries to properly manage the resources at hand and sets up an army and then tries to defeat the opponent's armies.
4. ***Fighter:*** The player tries to prevail over the other virtual characters chosen by other players with the virtual characters of his choice representing a certain fighting discipline. It is commonly played as a player versus a player.
5. ***Sports:*** They are traditional sports branches adapted to the virtual environment.
6. ***MMORPG (Massively Multiplayer Online Role-Playing Game):*** These are role-playing games played by connecting to any server.

In e-sports, which has developed more in the world, players can receive fees up to $2 million today. The components that are effective in the development and spread of e-sports are video game producers, sponsors, spectators, teams, leagues, players, and broadcasters.

The components of digital sports marketing on the e-sports side and their roles can be summarized as follows:

1. ***Video Game Producers:*** According to the holistic perspective of digital marketing in the production of games, factors such as the players' enjoyment of the game, the viewing status of the game, and the connection between the players are taken into account. Due to the increasing transaction volume of computer games, video game producers see e-sports as an area that will turn their shares into big winnings.
2. ***Sponsors:*** Sponsors have always been present in almost all tournaments in the world examples of e-sports. This contributed greatly to the development of e-sports. Since e-sports is a sports branch using electronic devices, it was seen that the first sponsors were technology companies.
3. ***Televisions:*** The appearance of e-sports in traditional media started in the 2000s. South Korea, where e-sports was played intensively for the first time, has allowed for sports broadcasts from satellite channels, whereas the American Sports Media Organization has broadcast the "Dota 2 E-Sports Championship" live. In Turkey, channels like Bein Sports and S Sports show tournaments and activities related to e-sports.
4. ***Online Broadcasting Institutions:*** Companies such as Facebook, YouTube, Google, Steam, and PlayStation have also created their online broadcast systems.
5. ***Players and Teams:*** Today, thanks to the electronic systems and online environments used, everyone can easily form their team and struggle.
6. ***Leagues and Tournaments:*** Leagues are usually an environment where a particular game and main sponsor work together, as in traditional sports, where teams and sportsmen fight against each other. ESL (Electronic Sports League) is one of the most established leagues.

Effective management of all these components will facilitate faster marketing of sports in digital media and will create an opportunity for the audience to recognize it. However, there are opinions in the world that argue that the concept of e-sports is not a sport. The main reason for this is the view that e-sports are not based on physical activity. The idea that there should be physical activity in the sport revealed a biased perspective on e-sports. However, success in e-sports depends on brain functions and good use of hand-eye coordination. Whether e-sports is a sport is still controversial (Witkowski, 2012). This controversial issue between sports and e-sports is also a different topic to be examined. Therefore, in this study, their structural similarities and differences are not included.

3 E-Sports Applications in Turkey

E-sports, which has come a long way in reaching the target audience and has recently increased its transaction volume, emerges as an area in which many people do not know much. Given the number of followers of the e-sport, which allows an individual or team to struggle with the help of computers and mobile devices, it is useful to support the development of e-sports.

The team called Dark Passage, founded in 2003, made the beginning of e-sports in Turkey. This team participated in the Balkan tournament on behalf of Turkey; in the final, it beat Serbia 2-1 in the League of Legends eSports game and became the Balkan champion. Also, there have been positive developments in the name of e-sports from professional sports clubs in our country. In Turkey, Besiktas, Galatasaray, and Fenerbahce sports clubs have established their own e-sports teams and transferred the sportsmen involved in these sports to their cadres. These sports clubs, which have started to operate in the field of e-sports, have even started to transfer players from abroad. This situation shows once again how a serious sports branch e-port is. These three teams represent our country in the world (Aktuna and Ünlüönen, 2017: 6).

The Beşiktaş Gymnastics Club has created teams by the name of *Besiktas e-Sports* that struggle in games such as League of Legends, Playerunkown's Battlegrounds, and CS: GO. Fenerbahçe has a team called *1907 Fenerbahçe e-Sports.* It struggles in League of Legends and Wolfteam games. Galatasaray competes in games such as League of Legends, Wolfteam, and Zula with its team called *Galatasaray e-Sports.* Additionally, the Turkey Digital Games Federation was founded for e-sports in Turkey in 2011. With the establishment of the Turkey Electronic Sports Federation, it has become possible to issue an e-Sportsman license from the Ministry of Youth and Sports.

The Turkey Electronic Sports Federation was founded under the General Directorate with the approval of the Ministry of Youth and Sports of No. 277144 of 24/04/2018, based on the decision taken at the meeting of the Youth and Sports Center Advisory Board on March 22, 2018. According to the provisions of the Election Regulation of the Federation Presidency of the General Directorate of Sports, the first Federation President was elected in the Ordinary General Assembly, held on May 15, 2018. The aim of the Federation has been defined as ensuring the growth of Turkish e-sports in the national and international platforms and improving the brand of Turkey e-sports Federation. The Federation's organizational structure includes (www.tefsed.gov.tr):

1. Chairman and Board of Manager
2. Advisory Board
3. Central Referee Board
4. Technical and Education Board
5. Board of Advertising, Marketing, Sponsorship, and Organization
6. Board of Health
7. Provincial Representatives

In the first stage, the Federation prepared instructions for various topics in the area of e-sports. "Turkey E-Sports Federation Sportsman License, Visa, Registration, Transfer Instructions," "Instructions for Referees," and "Special E-Sports Qualification Certificate Instructions" were issued. As can be understood from here, one of the most important duties of the Federation is to issue visas to sportsmen and to register these visas. Then, the Federation, which signed cooperation protocols with some game companies, cooperated with the Ministry of National Education to direct the youth correctly. In this context, they have started preparations for projects to hold inter-high school e-sports tournaments and to put e-sports courses in the curriculum.

The Federation held the "First E-Sports Workshop" on September 25, 2019 for all its stakeholders. Local and international game companies, e-sports clubs, professional e-sportsmen, and representatives from universities attended the workshop. The workshop discussed the future of e-sports and mostly expectations from the Federation. In the workshop report, the Federation stated that it has 70 registered clubs; 63 women and 844 men, in total 907 licensed e-sportsmen; and has assigned provincial representations in 48 cities (Tefsed 1st Workshop Report, 2019).The Federation organized the first contest for TEFSED Vodafone Freezone Turkey Cup and was involved in the activity report for 2019. The contests have realized with the participation of contestants who are over 16 years old under the rules of the game. The sportsmen have participated in the contests with their visa licenses.

The Vodafone Freezone Turkey Cup, which was carried out under the coordination of the Turkey Electronic Sports Federation, consisted of six games, for five teams and one individual. The games that TEFSED included in the section of branches of 2019 are classified as follows.

1. League of Legends (PC – at least 5-people team)
2. Counter-Strike: Global Offensive (PC – at least 5-people team)
3. Player Unknown's Battle Grounds (PC – at least 4-people team)
4. Dota 2 –The Last Champion (PC – at least 5-people team)
5. Zula (PC – at least 5-people team)
6. FIFA 2020 (PS4 – individual participation)

For 2020, TEFSED has changed the e-sports branches in the following branches of the game. Accordingly, the e-sports game branches determined by the Federation for 2020 are as follows (www.tefsed.gov.tr):

1. MOBA (Massively Online Battle Arena)
2. FPS (First Person Shot)
3. RTS (Real-Time Strategy)
4. Battle Royale
5. Sport: They are traditional sports branches adapted to the virtual environment.
6. MMORPG (Massively Multiplayer Online Role-Playing Game)
7. Fighter

E-sports, which mostly include young people as their target audience and are growing rapidly day by day, are a type of sports that realizes what traditional sports do, but in a virtual environment. Traditional clubs do not see e-sports, which can also be referred to as "the new against the old," as a rival; instead they are integrated with e-sports activities. Therefore, both Turkey's young population as well as considering the world population, developing, spreading, and marketing of e-sports are the issues that should be emphasized. Thus, the establishment of the e-sports Federation in Turkey speeded up its work. Nowadays, every home has Internet, computers, tablets, and more than one mobile device; so e-sports extending and becoming a recognized market will also provide significant support to the digital transformation process. A few more important topics are as follows.

4 Example of Turkey Electronic Football League (TEFL)

It is useful to explain two important football games in e-sports before explaining the example of the Turkey Electronic Football League. One of the digital football games is Fédération Internationale de Football (FIFA), while the other is Pro Evolution Soccer (PES). These games are presented as two rivals that fully dominate the virtual football market in the field of e-sports. The games allow players to use their original names or club outfits. So these kinds of games are similar to real football matches. FIFA is a video sports game released every year since 1993. Its name is derived directly from the name of the World Football Federation (FIFA). The publisher of the game is a company called Electronic Arts. Due to its reality, it is a football simulator that is very preferred today (Stein, Mitgutsch and Consalvo, 2013). The game uses advanced technology, which is why the game and the appearances of the players are quite similar to reality (Markovits and Green, 2017). Its most important feature is that it requires an online mode, an Internet connection.

Each player plays his football with the name determined by himself, in his club and with virtual players in a virtual environment. The game is given with a widescreen so that the target audience can see it. The tournaments, which were small and local in the beginning have turned into international tournaments over time (Zagala and Strzelecki, 2019).

"Turkey Electronic Sports Football League" is an organization that is considered to be created based on the popularity of football games in digital media, and it can be addressed under e-sports activities in Turkey. The Turkey e-sports football league includes the FIFA Pro Clubs teams of football clubs and provides them an environment where they can compete. The Pro Clubs mode is described as EA Sports FIFA online game mode, where each footballer is led by a single player, and in total, 11 real players match against 11 real players. Today, licensed e-sports clubs that are members of the Turkey e-sports Football League are as follows (www.tefligi.co):

1. Antalyaspor
2. Bursaspor
3. Çaykur Rize Spor
4. Gençlerbirliği
5. Istanbulspor
6. Kocaelispor
7. Ankaragücü
8. Sakaryaspor,
9. Trabzonspor

The matches organized for these clubs were played on Sunday, February 23, 2020, this year. The matches were shown live on YouTube. These matches, which is an example for e-sports to establish and manage leagues in the field of football, were organized for the first time in 2019. Since it was the first, it attracted attention with both the inadequacy of sponsors and the lack of participation. It cannot be said that the audience of this organization, which is thought to have not been adequately promoted from social media, is sufficient. Although the participation of licensed e-sports clubs is ensured, it can be said that it has an amateur league structure. Consequently, increasing such an exemplary activity, supported by the state, may be discussed. Players of some major teams from Turkey participated for the first time in 2019 in the e-sports team FIFA World Cup events. Increasing this number professionally, the participation of many clubs in the football leagues and tournaments of e-sports in the international arena can be taken into account for the development of this market. In the e-sports branch, adequate participation in FIFA world competitions is not yet achieved in Turkey.

5 Conclusion

E-sports, which constitutes another side of traditional sports under the name of sports, can be expressed as the activity of playing virtual games interactively by one or more people in digital media. Therefore, the question of whether there is a sport is a matter to be discussed at first before the definition of e-sports due to the virtual game emphasis. However, in this study, what kind of sport is e-sports, its development, what takes place in it, and examples from Turkey were described. Especially in the field of marketing of sports in the electronic environment, e-sports components were emphasized. In light of the information obtained from the literature review, the results of the study and the recommendations and criticisms for e-sports can be listed as follows:

- **i.** One of the most important problems for e-sports is whether this sport is a real sport or not. In the development process of e-sports, it can be discussed supporting the development of e-sports in Turkey, instead of such debates. Because, if the definition of sports is considered in different dimensions, not seeing sports as just a physical activity can eliminate this mess.
- **ii.** The players and teams can be supported for the growth of the e-sports economy of Turkey.
- **iii.** Since there are game companies that determine each league due to the structure of e-sports, the rules are determined by these companies. Therefore, it seems difficult to create rules and regulations covering the whole area. Despite this, efforts to establish rules and pedestals that can be managed from a single center can be done with the support of the state.
- **iv.** The sponsorship mechanism is still among the unresolved issues. It can be provided to explain e-sports to the public to solve the sponsor problems and increase the sponsors as soon as possible.

v. It is very important to raise awareness of parents, firms, and sponsor companies, especially the state, to pave the way for e-sports and accelerate its development.
vi. One of the biggest shortcomings of this area is that the e-sportsman agreements are made with very low fees and the e-sportsmen are not satisfied enough. It may be a matter of increasing the awards given and escalating the licensed player fees.
vii. Clubs and players should be ensured to take place under the Federation.
viii. In view of the fact that that the target audience of e-sports is comprised of young people, it is necessary to make parents conscious and heighten the awareness of them.

As a result, it can be said that e-sports became popular at the national and international level when all the information addressed in the study was compiled. Notably, the world, in Turkey, gaining acceleration rapidly of developing e-sports applications depends on the marketing of this sport as well. In our country, especially taking into account all the e-sports stakeholders and correctly introducing the function of e-sports will accelerate the awareness of this sport branch. In this study, the e-sports ecosystem in Turkey and the world has been explained in general terms. From this point of view, it is seen that necessary steps have been taken in our country with the support of both state and federation. It is now accepted in the corporate sense that e-sports are also a branch of sports and that strategies for e-sports should be developed. For this, the Electronic Sports Federation of Turkey has started to spread awareness on e-sports. At the point reached today, recognition and awareness of e-sports have started to accelerate the spread of e-sports.

Bibliography

Aktuna Halil Can, Ünlüönen Kurban, (2017), "Yeni Bir Turizm Çeşidi Olarak Elektronik Spor Turizmi", *Gazi Üniversitesi Turizm Fakültesi Dergisi,* Vol. 2, pp. 1–15.

Hüseyin Altunbaş, (2007), "Sporun Pazarlaması ve Pazarlama İletişiminde Spor", *Selçuk İletişim,* Vol. 5, N. 1, pp. 93–101.

Argan Metin, Katırcı Hakan, (2002), *Spor Pazarlaması,* Ankara: Nobel Yayın.

Cem Çetin, (2015), *Spor İletişimi,* Ankara: Nobel Yayın.

E-Spor Raporu, (2018), *Gençlik Araştırmaları Haber Bülteni,* N. 5, pp. 1–72.

Emma Witkowski, (2012), "On the Digital Playing Field: How we "Do Sport" With Networked Computer Games", *Games and Culture,* Vol. 7, N. 5, pp. 349–374.

Evren Tuğberk, Kargün Mehmet, Pala Adem, Yazarer İlkay, (2019), "Spora Yenilikçi Yaklaşım: E-Spor", *Uluslararası Sosyal Araştırmalar Dergisi,* Vol. 12, N. 66, pp. 1424–1434.

Gençlik ve Spor Bakanlığı, (2019), http://yayinlar.gsb.gov.tr/Public/Files/2018.05.14_16.57.28_espor-raporu-say%C4%B15.pdf, (06.02.2020).

Kacper Zagala, Strzelecki Artur, (2019), "eSports Evolution in Football Game Series Physical Culture and Sport", *Physical Culture and Sport Studies and Research*, Vol.83, N.1, pp. 50–62.

Kayalı Fares, Purgathofer Peter, (2008), "Two Halves of Play – Simulation Versus Abstraction and Transformation in Sports Videogames Design", *Journal for Computer Game Culture*, Vol. 2, N. 2, pp. 105–127.

Markovits Andrei, Green Adam, (2017), "FIFA, The Video Game: A Major Vehicle For Soccer's Popularization in the United States", *Sport in Society*, Vol. 20, N. 5–6, pp. 716–734. https://doi.org/10.1080/17430437.2016.1158473.

Mullin Bernars James, Sutton William, Hardy Stephan, (1993), *Sport Marketing*, Human Kinetics Publisher.

Philippa Warr, (2014), "eSports in Numbers: Five Mind-Blowing Stats", http://www.redbull.com/en/esports/stories/1331644628389/esports-in-numbers-five-mind (06.02.2020).

Samet Taşlı, (2015), https://dijitalsporlar.com/makaleler/geleneksel-spor-ve-espor-arasindaki-iliski, (06.02.2020).

Spor Pazarlama, (2002), *Mediacat Aylık Pazarlama İletişim Dergisi*, N. 89, pp. 18–27.

Stein Abraham, Mitgutsch Konstantin, Consalvo Mia, (2013), "Who Are Sports Gamers? A Large Scale Study of Sports Video Game Players", *Convergence: The International Journal of Research into New Media Technologies*, Vol. 19, N. 13, pp. 345–363. https://doi.org/10.1177/1354856512459840.

Terekli Mustafa Serdar, Katırcı Hakan, Mert Erkan, (2000), "Sporda Çağdaş Pazarlama Anlayışı", *1. Gazi Üniversitesi Beden Eğitimi ve Spor Bilimleri Kongresi Bildiriler Kitabı*, Ankara.

Muhammed Fatih AKSARAYLI

Transformation of Higher Education in Electronic Environment: The Sample of Digital Higher Education Project

1 Introduction

The development of internet and mobile technologies affects the education and training of the university and necessitates its transformation and differentiation. According to today's understanding what is technological and digital is quality. Essentially, the concept of quality is closely linked to valuing people, innovation, continuous change, development, and renewal. At this point, the adoption of a quality-oriented approach by universities will move these institutions to a more advantageous and reputable position in the future.

The quality expectation of higher education students includes many concrete and abstract elements, including courses, teaching tools and materials, campus, dining hall, library and information centers, means of transport, knowledge and attitudes of academic and administrative staff, and technologies used. How these elements are perceived by students reveals the quality of service in higher education.

With the developing technology and widespread internet, face-to-face education, which is in the center of universities, is replaced by distance education, online courses, and mobile learning. However, this digitalization in higher education has not only been limited to teaching models but has also made necessary the use of information systems and stakeholders to provide information online in business processes in some universities (Tosyalı, 2019: 172). Therefore, the expansion of the range of services and determination of the aspects that integrate with digitalization have become very important issues to be investigated in the field.

This study is prepared based on The Council of Higher Education of Turkey's (YÖK) quality-oriented Digital Transformation Project. It aims to be a guide for young universities in Turkey to create quality awareness in technology-based service delivery. The digitization process in higher education has been discussed on the quality axis.

2 Quality in Higher Education and University Transformation in Electronic Environment

Quality in the service sector is related to perceptions and is the evaluations and judgments about what is presented (Gürbüz and Ergülen, 2008: 24). Two main factors affecting the quality of the service are the tools and equipment that provide it and the people offering it (Karahan, 2006: 38). The combination of the concepts of quality and service can actually be expressed as the behavior showed and the process applied accordingly. Therefore, the common ground of the two concepts is to determine the needs and expectations and to provide services in better conditions than their counterparts in that sector. The development, tastes, and habits of the society change the concept of quality according to the expectations and needs of consumers. For this reason, the main purpose is to determine the needs and expectations of the consumers well and to offer better products and services than the equivalent products and services in the market (Bozkurt, 1995: 175).

Today's university, which is the leverage of economic development, works with the business world with its research and development (R&D) activities and supports innovation processes (Altbach, Reisberg and Rumbley, 2009: xii; Oosterlinck, 2001: 4). It has a growing influence on the competitiveness of the national economy and gives people useful abilities to adapt to the economic environment (Dewey, 1996: 52; Singh, 2012: 8). The university is highly aware that high-quality education and training that works for the training of the highly qualified workforce needed will not only provide an advantage in the international competition but also become synonymous with a sustainable healthy economy (Korkut, 1996: 24; Jonathan, 1997: 73). In this stage, quality has great importance for the university, which closely affects instructors, families, other institutions, and the public especially students.

With globalization in the twentieth century, universities were included in competitions worldwide. In addition to being able to realize its objectives and serve, it has gained great importance to provide these services in the best way since this field has widened a lot. Therefore, the understanding of service quality in higher education has emerged. The concept of a customer associated with the quality of service has given its place in higher education primarily to students, academics, and other institutions – in other words, internal and external customers, which we can call stakeholders. Internal customers can be specified as students, academic, and administrative staff whereas external customers can be specified as society, families, public, and private institutions (Çavdar, 2009: 105). According to the scales developed in researches toward service quality in higher education, quality elements whose reliability researchers test and agree on can be listed as follows: (Owlia and Aspinwall, 1996: 19);

i. Concrete Elements: Adequate and modern equipment facilities, physical infrastructure, information center capacity, and resource capacity
ii. Competence: Competence, qualification, and communication skills of instructors
iii. Attitude: Understanding student needs, showing interest, and positive approach
iv. Content: Content and validity of educational programs, computer use, and teamwork
v. Presentation: Presentation of content and feedback
vi. Reliability: Keeping a promise, evaluating complaints, and making efforts to resolve them

Considering the above elements, quality in higher education institutions appears multilayered in education and training, physical and academic infrastructure, academic programs, curriculum and exam evaluation systems, distribution of titles and promotions, research and publications, and relations with society (Gürbüz, 2006: 26). However, the content offered to higher education stakeholders and their presentation methods has become different from traditional methods with the digital age. While it is easy to evaluate the perceptions of the party receiving service in face-to-face education, it has become difficult to determine the quality of these processes with digitalization, which is the window to the abstract world. Implementations in the fields of resource planning and management, sustainable effective management and leadership, determining the perceptions of learners and stakeholders, designing, developing, and implementing active learning-based and user-friendly programs, developing and increasing the capacity of the physical infrastructure, establishing effective digital management and feedback systems help universities to improve their quality in the digitalization process (Gümüşoğlu, 2017: 37).

3 Digital Transformation in Higher Education

Digital transformation is a process that involves the progress of humans, processes, and technology dimensions together and harmoniously. Thus, institutions aim to provide more effective, quality, and fast service (Karahan, 2006: 46). According to the World Economic Forum, it is calculated that the impact of digitalization on world economy for the next 10 years will amount to approximately 100 trillion dollars. This situation agitates companies that strive to keep their profits at the top every year and universities that are constantly struggling to increase resources.

As a result of the integration of today's universities with information and communication technologies, a digital transformation is actualized with the processes of e-management in administrative terms and e-learning in academic terms. Education and training can now be done over the internet, regardless of time and place, and is supported by virtual reality, three-dimensional simulations, and videos. In this way, every academic has become a digital designer of his/her own course. Its interlocutors are Z generation students who are fond of freedom, learn and live with the internet, are active and excited in digital life, and passive and reluctant in real life and were born 1995 or later.

The rapid change process in information technologies has changed the expectations and tendencies of students, who are considered as the most important customers of universities in a different way. The idea of how education can be more effective and efficient has been realized by both instructors and university administrations. In this respect, examination of the samples in the world in Turkey and regulation of the relevant legislation is important for higher education in order to keep up with the digital era.

Interactive courses in digital media in the United States and Europe today are the most important example of the integration of technology into learning processes. Massive open online courses, public educational resources, artificial intelligence, cloud technologies, and data mining studies support the process. The modules on Learning Management Systems (LMS) can transfer real-time information to instructors, students, and supervisors through learning analytics.

Some universities have started large-budget projects in order to reduce the costs of education and training with information and communication technologies to address different student profiles from all corners of the world and to increase the quality. In addition to being a campus, the first thing that comes to mind when it comes to a university (Cronje, 2016: 138), California Berkeley University is now running a digital transformation project to offer noncampus lessons to cope with the increasing number of students. In this project, started in the digital transformation and adaptation process of the university, case studies were conducted and perceptions of instructors and students, cost impact analyses, and infrastructure adequacy of faculties were examined.

Higher education institutions have initiated the distance education process with the introduction of open online courses. . With the support of digital production, courses enriched with a variety of images, sounds, pictures, languages, and graphics have been designed. These online courses are supported by actors who learn and teach based on mutual interaction. In addition, these courses provide students with modules that enable them to get a certification. For example, The Open University in the United Kingdom has adapted to this change and has started to implement the programs of start-ups and spin-offs (Taşkıran, 2017: 101–102). Today nearly a third of students in the United States take at least one course online and universities such as MIT, Stanford, and Harvard give some lectures through only distance education.

Courses that have vital dangers, such as deepwater drilling simulations with new technologies used within the framework of digital education, or that which are costly, such as cockpit simulations, can be moved to the computer environment. In this way, students can do their courses and experiments in a safe environment. These educations with very low costs appeal to a larger number of students and provide equality of opportunity. Additionally, these courses are provided to support hearing-impaired students (Tosyalı, 2018: 179). In the field of applied sciences, the learning processes of the students are supported with visual elements, thanks to augmented reality. With these realities, for example, the final shape of an unfinished building can be seen (Coşkun, 2017: 63). In digital media, communities learning interactively are created and are provided open learning resources that offer education opportunities for everyone.

4 The Sample of Digital Transformation Project of YÖK

The studies for Turkey become competitive in the field of digital transformation, started with the Digital Transformation Project by the Council of Higher Education of Turkey (YÖK) in 2019. This project is a study that enables instructors and students to communicate interactively via the LMS network and continue their courses and projects in also digital media besides face-to-face education. The process began with the introductory meeting of the project named Turkey's Window of Opportunity: Turkish Higher Education, organized in Ağrı İbrahim Çeçen University. It was held with the participation of Prof. Dr. Yekta SARAÇ, who is the President of YÖK; Council members; the governor; the mayor; the rectors of 26 university; and academicians. At the meeting, emphasis was placed on the digital transformation and literacy issues that YÖK has set as the new target in the education of the young population of the country, studies in the field of vocational education, new undergraduate/postgraduate programs, and the roles of universities in regional development. As a result of the ceremony held within the scope of the Digital Transformation Project, Ağrı İbrahim Çeçen University, Bayburt University, Bingöl University, Iğdır University, Muş Alparslan University, Munzur University, Siirt University, and Şırnak University signed a protocol.

With the project prepared under the New YÖK title, it was stated that universities will contribute to the sustainable development process of the country. Also it has been emphasized that such projects will install new meaning to the universities in Turkey's future vision, and it is an important step in the field of education and training. Within the scope of this project, students can access all educational resources such as books, assignments, articles, questions, videos, and animations related to their courses from their computers and mobile phones at any time. In addition, it is planned for instructors to easily see which digital material is used and how long with course follow-up and learning analytics in the virtual environment.

Anadolu University also makes important contributions to the project. One of these is that it supports pilot universities with online courses. In the 2018–2019 academic year, all instructors of universities determined within the scope of this project were taught Learning and Teaching in Higher Education in the Digital Age and Digital Literacy courses for their students. Both courses are taught via distance education. Anadolu University presented the contents of the Digital Literacy course for students as the exams were carried out by the project universities. Exams were conducted according to each university's own regulations. We can summarize the stages of the Digital Transformation in Higher Education project in terms of lecturers and students as follows:

i. Orientation (Guidance) Process: This process is a weekly time period that includes introducing, analyzing, and experiencing the system before classes begin for both instructors and students.
ii. Inclusion of Instructors in the Process: It is also the process that includes the registration of instructors to the course opened in the name of Learning and Teaching in Higher Education, completing the course, and issuing certificates to those who meet the necessary requirements. This course includes contents about the roles and skills expected from instructors in digital transformation, new learning approaches, gaining skills about digital media, and its implementations.

iii. Inclusion of Students in the Process: It is the process that includes the registration of students to the course opened with the name of Digital Literacy. Its content provides a road map for adaptation to online education and training world, and that is about providing the ability to find, understand, and analyze information through smart devices.

Each university that signs the protocol within the scope of the project chooses the course that is compulsory for its students and conducts the course process online throughout the semester. Each student has his own private area on this system and follows all digital course materials from there. The contents of the course are internet technologies, portable technologies, social networks, technology-society and humans, ethics of informatics, technology and lifelong learning, cloud computing, and technologies of the future.

An important point on which the Digital Transformation Project is based is that such projects are also given great importance by the Turkish Presidency. In the Presidential 2019–2023 Strategic Plan Report, digitalization is also given a place. In the section of technology and informatics infrastructure analysis in the plan, the realization of e-transformation projects and the effective use of information and communication technologies has been mentioned (Presidency of the Republic of Turkey Strategic Plan Report 2019–2023: 32).

The universities included in the Digital Transformation Project showed great interest. In the first stage of the Higher Education Learning and Teaching course, 3,112 instructors enrolled and entitled the certificate. In addition, approximately 30,000 students have taken the course on Digital Literacy. Gümüşhane University, which is involved in the project, as the process progresses has put the Digital Transformation Coordinator into service and aims to provide planning and coordination of all kinds of digital services and to establish digital peer services of all existing face to face and document-based university services.

5 Conclusion

Within the scope of the study, a group of students from different universities that have taken the Digital Literacy course, which is studied in eight pilot universities and is made obligatory within the Digital Transformation Project, has been reached. Although this group does not reflect the characteristics of the entire universe, some common experiences and focused ideas have been reached. The students were asked the following questions: "Did you learn more easily in digital media? Were you able to concentrate on your lesson? Are you satisfied with the education you received? Would you prefer an education at school or distance education?" In the feedback received, dissatisfaction has come to the forefront at a high rate. Along with student feedback, some instructors who taught this course in distance mode were also reached, and they were asked to make a general evaluation. In the light of all the feedback and evaluations received, the following basic conclusions were reached:

i. A harmonious effort and motivation that covers the whole of the university are required for the success of the digital transformation. It should not be considered that this transformation process will be completed without any problems only with the technical efforts of the information processing department staff. It should be remembered that the most important success factor is human.
ii. Objective and clear criteria should be established for instructors and students to evaluate distance education processes and outcomes and to measure the perception of quality in service. The level of satisfaction and efficiency should be freely discussed.
iii. It should be taken into consideration that physical and digital spaces are separate in distance education but the boundary between them has become blurred. This environment in which interest shifts to other directions in terms of both narrator and listener requires a high focus effort. It should also be taken into consideration that not everyone has the discipline to manage their own education and to learn independently. In this context, education of keeping the relevant person alive and concentration in a virtual environment should be given.
iv. In education and training, abstract concepts are the most difficult to teach and understand. Digital pedagogy should also be considered along with traditional pedagogical approaches for the virtual environment, which is an abstract world by its nature.
v. The expectations, perceptions and behaviors of the Z generation, which is at the center of education and training, should be managed well. People who are passive with traditional media, especially this generation, have become more participatory and active with social media. This energy of young people should be included in the distance education process.
vi. It should be noted that the message will have different effects when transmitted through different communication channels. For example, just as face-to-face talking, talking on the phone, or texting has different effects, the difference of influence and efficiency between the digital courses and the courses at school should be taken into consideration. In this framework, the exam and homework success bar in a virtual environment should be revised.
vii. Programs that are difficult to develop and use should not be selected. The software and hardware infrastructure should be established solidly, and education and training should not be interrupted with continuous technical problems. The value-cost-impact analysis should be done well.
viii. Digital education materials should have content enriched in both the instructor and learner dimensions.
ix. The digital age is also an age of surveillance and journal. It should be noted that virtual data, words, and behavior can always be recorded.

The university is an exceptional school. It is not just a place established for education and has practical goals. It is a place that gives intellectual discipline to its students, arouses interest in science, instills a sense of responsibility and conscience toward homeland and humanity and nurtures morally and emotionally (Jaspers, 1959: 1; Schwartz, 2007: 474; Fikret, 1930: 2). These dimensions of the university can only be realized through their experienced, respected, and wise academicians who act as role models for their students. The provider of this is the human interaction environment created by face-to-face education between the instructor and the student. In this framework, face-to-face education should not be disabled completely; hybrid models for learning should be developed, as recommended in the report named Transforming Higher Education: How to Teach in The Digital Age, prepared by the European Commission in 2017.

Bibliography

Altbach, Philip, G., Reisberg, Liz, and Rumbley, Laura E., (2009), "Trends in Global Higher Education: Tracking an Academic Revolution", *UNESCO* 2009 *World Conference on Higher Education Report,* Paris: UNESCO.

Bozkurt, Rıdvan, (1995), "Hizmet Endüstrilerinde Kalite", *Verimlilik Dergisi,* Vol. 4, pp. 175–186.

Coşkun, Cumhur, (2017), "Bir Sergileme Yöntemi Olarak Artırılmış Gerçeklik", *Sanat ve Tasarım Dergisi,* Vol. 20, pp. 61–75.

Cronje, Johannes, (2016), "Learning Technology in Higher Education", (Nick Rushby and Daniel W. Surry, Eds.), *The Wiley Handbook of Learning Technology,* UK: Wiley Blackwell, pp. 131–144.

Çavdar, Ertuğrul, (2009), "Yükseköğretimde Hizmet Kalitesi Unsurları ve Bir Uygulama", *Niğde Üniversitesi İİBF Dergisi,* Vol. 2, N. 2, pp. 100–115.

Dewey, John, (1996), *Demokrasi ve Eğitim: Eğitim Felsefesine Giriş,* (T. Yılmaz, Trans.), İzmir: Ege Üniversitesi Edebiyat Fakültesi Yayınları.

Fikret, Halil, (1930), *Terbiye ve Tedris Tarihi,* İstanbul: Devlet Matbaası.

Gümüşoğlu, Eylem Koral, (2017), "Yüseköğretimde Dijital Dönüşüm", *Açıköğretim Uygulamaları ve Araştırmaları Dergisi,* Vol. 3, N. 4, pp. 30–42.

Gürbüz, Esen, (2008), *Yükseköğretim Kurumlarında Hizmet Kalitesi Ölçü ve Modelleri,* Ankara: Detay Yayıncılık.

Gürbüz, Esen and Ergülen, Ahmet, (2006), "Hizmet Kalitesinin Ölçümü ve Grönroos Modeli Üzerine Bir Araştırma", *İstanbul Üniversitesi Siyasal Bilgiler Fakültesi Dergisi,* N. 35, pp. 173–190.

Jaspers, Karl, (1959), *The Idea of the University,* USA: Beacon Press.

Jonathan, Ruth, (1997), "Educational Goods: Value and Benefit", *Journal of Philosophy of Education,* Vol. 31, N. 1, pp. 59–82.

Karahan, Kasım, (2006), *Hizmet Pazarlaması,* İstanbul: Beta Basım Yayım.

Korkut, Hüseyin, (1996), "Üniversitelerde Kalite ve Akreditasyon", *Türk Yurdu Dergisi*, Vol. 16, N. 111, pp. 24–26.

Oosterlinck, André, (2001), *University/Industry Knowledge Management: A University Perspective.* Retrieved from https://www.oecd.org/education/innovation-education/2668232.pdf, (01.02.2020).

Owlia, Mohammed S. and Aspinwall, Elaine M. (1996), "A Framework for the Dimensions of the Quality Higher Education", *Quality Assurance in Education*, Vol. 4, N. 2, pp. 12–20.

Presidency of the Republic of Turkey Strategic Plan Report, (2019–2023), Retrieved from http://www.sbb.gov.tr/wp-content/uploads/2019/11/Stratejik-Plan-2019-2023.pdf, (01.02.2020).

Schwartz, Philipp, (2007), "İstanbul Üniversitesinin Bugünkü Durumu ve İstikbali", (Namık Kemal Aras, Emre Dölen and Osman Bahadır, Eds.), *Türkiye'de Üniversite Anlayışının Gelişimi 1861–1961*, Ankara: TÜBA Yayınları, pp. 474–497.

Singh, Mala, (2012), "Re-inserting the Public Good Into Higher Education Transformation", (Brenda Leibowitz, Ed.), *Higher Education for the Public Good: Views From the South*, USA: Trentham Books, pp. 1–15.

Taşkıran, Ayşe, (2017), "Dijital Çağda Yükseköğretim", *Açıköğretim Uygulamaları ve Araştırmaları Dergisi*, Vol. 3, N. 1, pp. 96–109.

Tosyalı, Hikmet, (2018), "Dijitalleşmenin Yükseköğretime Etkileri" (Gonca Telli Yamamoto and Mustafa Özhan Kalaç, Eds.), *Kamuda Dijital Dönüşüm*, İstanbul: Kriter Yayınevi, pp. 172–188.

Andreas KAPLAN

Your Attention, Please: You've Got 15 Seconds! TikTok and How Organizations Can Make Use of It

1 Inbox: Contrast to the Past

Social media giants such as Facebook and Twitter increasingly have to battle declining user statistics, particularly from users born between 1995/2000 and 2010, with one notable exception: TikTok, the entertainment-based platform launched in 2018, exhibits exponential growth rates within this age group, the so-called Generation Z (Fromm 2020).

At first glance, TikTok does not look different from other social media applications (Kaplan and Haenlein 2010c). It can be described as a video-sharing mobile app that lets its users watch, create, and share clips 3–15 seconds in length. Usually, these clips make use of a soundtrack, often current top-charting songs, and offer a variety of filters, augmented reality options, and other gimmicks to refine the clips before uploading. TikTok hosts several challenges, nearly always positive and fun, similar to Facebook's 2013/2014 Ice Bucket Challenge (Lafferty 2014).

However, TikTok is quite distinct from other social media: it makes ample use of artificial intelligence (AI; Kaplan and Haenlein 2019; Kaplan and Haenlein 2020) and algorithms to decide which content will be shown to which users in order to hold their attention as long as possible. In contrast to other social media applications, TikTok's algorithms are disconnected from users' friends and social connections: They depend solely upon what content the user has seen, liked, and shared. TikTok primarily is about watching and filming videos. It is about social entertainment. It is not about building and fostering social relationships.

In this chapter, we first give an overview of social media, AI, and the combining of both. In the second part, we survey TikTok's characteristics and how it differs from other social media applications. Then we analyze how corporations employ TikTok for their branding, communications, and marketing purposes. In the last part, we offer advice to help organizations use TikTok optimally.

2 Home: The Connection Between Social Media and Artificial Intelligence

In this section, we survey social media and AI. For each, we provide a definition as well as a classification taxonomy. We also offer a brief historical overview, which shows the origins of both social media and AI, as well as their interconnections.

2.1 Social Media

Social media, commonly defined as "a group of Internet-based applications that build on the ideological and technological foundations of Web 2.0, and that allow the creation and exchange of user-generated content" (Kaplan and Haenlein 2010c, p. 61; Kaplan 2012), can be classified into collaborative projects (e.g., Wikipedia; Kaplan and Haenlein 2014), microblogs/blogs (e.g., Twitter), content communities (e.g., YouTube), social networks (e.g., Facebook), virtual game worlds or MMORPGs (massively multiplayer online role-playing games) (e.g., World of Warcraft), and virtual social worlds (e.g., Second Life; Kaplan 2009; Kaplan and Haenlein 2009; Kaplan and Haenlein 2010a).

Viewed historically, the rise of social media can be interpreted as the Internet "going back to its roots" (Kaplan and Haenlein 2012a). In its beginnings, the Internet mainly consisted of newsgroups, in their approach similar to today's online forums, most of which would be categorized as content communities. On such newsgroups, primarily focused around high tech and science fiction, individual users shared content and information. Concomitantly, the first MMORPGs popped up, which could definitively be considered the precursors of today's virtual game worlds (Kaplan 2018b; Kaplan and Haenlein 2010b; Kaplan and Haenlein 2011).

In the mid-1990s, big companies such as Amazon and eBay made their digital debut, gradually gaining control. Corporate activity rapidly overtook individual user activity, rendering the latter largely invisible to the broad public. The reconquering of the cybersphere by individual users then slowly began with the launch of Wikipedia in 2001, often considered the advent of modern social media applications. The beginnings of Facebook in 2004, YouTube in 2005, and Twitter in 2006 reinforced this reappropriation of the Internet by individuals. TikTok, launched in 2018, might be considered a new milestone in the history of social media, as it mainly makes use of AI.

2.2 Artificial Intelligence

Artificial intelligence, commonly defined as "a system's ability to correctly interpret external data, to learn from such data, and to use those learnings to achieve specific goals and tasks through flexible adaptation" (Kaplan and Haenlein 2019, p. 17), can be classified into three types: analytical, human-inspired, and humanized AI (Kaplan and Haenlein 2019). Analytical AI features only characteristics of cognitive intelligence. Human-inspired AI, alongside cognitive elements, features elements of emotional intelligence. Finally, humanized AI is characterized by all three types of competencies (i.e., cognitive, emotional, and social intelligence).

AI's birth, also called "AI's Spring," occurred in 1950 with the publication of Alan Turing's paper titled "Computing Machinery and Intelligence." Therein, Turing (1950) gives the "recipe" for how a machine's intelligence can be tested. The term "artificial intelligence," however, was officially coined only in 1956 at a conference organized by Marvin Minsky and John McCarthy, the now-famous

Dartmouth Summer Research Project on Artificial Intelligence (DSRPAI). Several AI "summers" and "winters" followed, i.e., terms metaphorically used to describe the iterating ups and downs of AI's progress. While winters were characterized by pessimism and scarce funding for AI projects, the opposite was the case during AI summers, defined by high optimism and even higher investments (Haenlein and Kaplan 2019, Kaplan 2020a, b).

The results of these high investments are now being harvested in AI's autumn period. This last season of AI is usually considered to have begun in 2015 when Google's AlphaGo AI system successfully played Go against the human champion at the time. As this board game is higher in complexity than, for example, chess, until that time, a machine playing Go was considered impossible. Therefore, AlphaGo constitutes a milestone in the history of AI (Haenlein and Kaplan 2019).

2.3 Social Media and Artificial Intelligence

Social media and AI are both based on algorithms, or finite sequences of clearly defined specifications and instructions for automated reasoning, calculation, data processing, and so forth. A further link between social media and AI is that with the help of the latter, hypertargeting on social media becomes possible. Ultimately, users are fed only content that they like. For example, users who are more "cat people" than they are "dog people" will more likely enjoy images of cats than dogs on, say, Facebook. The algorithm consequently will learn this about them and accordingly expose them to images of cute cats only. At some point, such users will believe that there are only cats on Facebook and that everybody is a cat person, dog persons having disappeared entirely (Kaplan 2020c).

3 Discover: Characteristics of TikTok

What follows are three characteristics typical of TikTok, that are rather atypical for other mainstream social media applications. TikTok is centered on positivity, happiness, and entertaining content, curated by algorithms only. Other social media increasingly face user envy and negativity, with a given user's social contacts and friends playing an important role (Baccarella 2018) therein.

3.1 Algorithms Versus Friends

In contrast to social networking sites, TikTok is independent of an individual's social network. In fact, at no time does TikTok ask for contacts, either during onboarding or thereafter. Imagine Facebook showing a feed without knowing any of the user's contacts or friends: this is TikTok. Instead, TikTok's focus is on content production and content delivery customized to the individual user. The objective is to catch its user's utmost and maximum attention. When on the app, TikTokers do not want to catch up with their social network; they want to watch and enjoy endless clips of maximum 15 seconds in duration, perfectly tailored to their taste and consumption behavior.

To know which content to show on its "For you" feed, TikTok analyzes a user's online behavior. Examples include which videos a given user views until the end, at what time, and with ones they are most likely to interact in the various options that the platform provides for this purpose. Moreover, the content is tagged in order to know as precisely as possible what is displayed therein. This information helps to provide users with a "perfect match" based on continually improved algorithms.

Obviously, on TikTok, users also can follow specific people and see only their content on the "Following" feature and talk to them privately via the messaging function. They also can be followed and, if they have enough followers, become a TikTok star and influencer. Also Facebook, Twitter, and YouTube do not deprive themselves of using algorithms. It is simply a question of where the cursor lies: on TikTok, it lies entirely on the side of the algorithm. As soon as the user opens the app, TikTok begins making assumptions about what content they might want to watch.

3.2 Entertainment Versus News

When on TikTok, users do not want to be informed of current events, breaking news, or anything else. They want to be entertained. In this respect, TikTok can be best compared to Instagram with its feed of thousands of images and pictures. As such, hashtags and trends on TikTok have almost nothing in common with what one would watch on the evening news. In the event that a news item would trend on TikTok, it is augmented and distorted in a humorous, fun way that appeals to Generation Z.

It is almost as if Generation Z tries to escape on TikTok from the (fake) news and (mass) media to which they are exposed day in and day out (West 2017). Moreover, Gen Z wants to escape data privacy risks and decided to increasingly share personal content on WhatsApp and Facebook Messenger (Wilson 2020) instead of publicly sharing it on social media. Finally, they might also just want to escape boredom. Generation Z differs from Gen X, millennials, or baby boomers, who mostly use social media to get updated on friends and family, post their opinions, and share (very) personal data about their activities and lives.

3.3 Happiness Versus Envy

Entertainment goes hand in hand with happiness. While TikTok is similar to Instagram in the sense that it is not about news or any current information, there is a difference between the two social media applications. Instagram has become a place for users to show off their augmented pictures and "perfect lives," fomenting envy and judgment. On TikTok, there's no room for jealousy; it is about happiness and about entertaining others by doing funny, cool things.

Although there are plenty of beautiful people on TikTok, it seems that TikTokers simply want to escape, at least to a certain extent, the dictate of keeping up the perfect lifestyle and appearance "required" on other social media applications (Tuten, Solomon, Kaplan 2019). In that sense, TikTok is similar to Snapchat, which all but eliminates the pressure to look perfect, as its videos disappear after having been viewed by the recipient(s). Likewise, TikTok also reduces stress, as it provides users with the option of lip-syncing to the hit songs of the moment that everybody's listening to. This by itself is already cool. Moreover, on TikTok, a user's friends do not constitute her primary audience, but rather people whom they do not know at all. Any ordinary user who posts a video on TikTok will immediately find herself an audience of likeminded people who enjoy the same challenges, hashtags, and music.

On TikTok, one will find people dancing (or some form thereof), streaking with friends (note: TikTok is PG13), or admitting to bad behavior (in front of their Mum). Generation Z'ers are not fans of jealousy and negativity, preferring happiness and positivity. The striving for positivity is one of the reasons why Z'ers represent a mass exodus from social networking sites. Edison Research (Edison Research 2019) showed that approximately 60 % of US users aged 13 and above do not use Facebook, citing too much negativity thereon. TikTok, in contrast, promotes happiness and positivity, highlighting and pushing positive trends and hashtags only.

4 Templates: Corporations' Utilization of TikTok

Although one might question TikTok's usefulness for corporate purposes with its clips of a maximum length of 15 seconds, it actually contains much potential for branding, communications, and marketing. As aforementioned, TikTok appeals to Generation Z, an age group that is extremely difficult for marketers to reach. In what follows, the three different types of marketing that can be found on TikTok are: content marketing, influencer marketing, and the more traditional paid-for advertising.

4.1 Content Marketing

TikTok is a video-sharing platform that is compatible with content marketing, which augments a brand's consumer awareness and value. Content marketing is an online type of marketing focused on creating, publishing, and distributing content for a very precise and targeted audience. The fact that users can see a brand's content in their feeds without even following them is of particular interest to brands. Users will potentially be exposed to a brand's content on TikTok's "Discover" function, or if they are searching for hashtags related to their and your preferred topics. Moreover, TikTok enables marketers to hypertarget users according to their content preferences, which throws the doors wide to psychographic targeting.

So-called (hashtag) challenges are often used by brands to engage with TikTokers, encouraging them to create or recreate content, adding the soliciting brand or branded hashtag thereto. Guess, for example, created the #InMyJeans challenge, encouraging TikTokers to film themselves wearing the brand's latest denim line. McDonald's #bigmactiktok challenge asked TikTok users to produce and share videos showing them dancing to a specific "Big Mac sound." Chipotle likewise organized its #GuacDance challenge, encouraging fans of the restaurant chain to present their avocado-themed dance moves to Dr. Jean's Guacamole Song.

Thus, hashtags on TikTok play a vital role, as they are used as a functional organizing tool for its seemingly endless challenges. Hashtags on TikTok actually work the way Twitter wanted them to: twitter users group around certain hashtags, combining and compiling individual tweets on specific topics, "funneling" into a more exhaustive discussion thereon. TikTok regularly sends out a newsletter to brands informing them of trending hashtags, which are then promoted by the respective companies. The newsletter advises that "If you're interested in participating, make sure to upload your video no earlier than one day before the hashtag's launch."

4.2 Influencer Marketing

In some cases, brands decide to collaborate with an influencer for their marketing activities. This is called *influencer marketing*, defined as a form of social media marketing involving endorsements and product placement from influencers or individuals and organizations that have a purported expert level of knowledge or social influence in their fields. Such collaboration is particularly used in the case of a brand's not yet being well-known. In such cases, influencers help to promote a brand's hashtag challenge, causing it more likely to go viral. For example, take Chipotle, which collaborated with several TikTok influencers such as Brittany Broski or Zach King to give more visibility to their #boorito challenge. They encouraged TikTokers to post a video featuring their apparel before and after their Halloween costumes, with the winner getting free burritos for an entire year. This challenge resulted in an impressive four billion views (Frier and Wagner 2020).

TikTok provides brands with several KPIs, which are all compiled on TikTok's Creator Marketplace. On this dashboard, brands are enabled to select and compare various influencers according to their number of followers, region, average viewership of videos, and topics of interest. Moreover, the dashboard also precisely shows both the influencer's followers' engagement with the video, hashtag challenge, and the like and the precise demographics of users engaging with the content. Further data shows how effective a challenge was in increasing a brand's awareness on consumers' parts, taking, for example, real-time viewership as a proxy. With the data at its disposal, a brand can potentially calculate the ROI (return-on-investment) of the respective campaign.

4.3 Advertising Marketing

Finally, traditional online advertising campaigns are also available on TikTok. Their KPIs are likewise collected on TikTok's Creator Marketplace, displaying audience breakdowns, engagement rates, number of views, and so forth of paid advertising. Mainly three different ad-for-pay campaigns are to be found on TikTok:

First, banner ads, which one often finds on TikTok's Discover page and which are mostly hypertargeted. Banners usually help users click through to information on the brand's challenge and its regulations and guidelines. Second, so-called in-feed native ads, which allow brands to place links or action buttons that transfer TikTokers to landing pages or e-commerce sites. Third, brand takeover ads, which enable brands to transfer TikTokers to landing pages or the particular challenge via videos, images, or GIFs. However, only one brand can take over a specific brand category per day.

5 For You: What You Should Know About Using TikTok

Three essential precautions should be taken concerning how to optimally make use of TikTok. First, TikTok is about the audience, not about the brand. Second, one should aim for authenticity, not perfection. Finally, TikTok is more about increasing brand awareness and less about directly making high numbers of sales.

5.1 The Audience – Not You

By now, we have made clear that TikTok is all about entertaining the audience and not about showing off one's image and brand. By doing the former, a brand's content *might* go viral, but this is decided quite quickly by TikTok's algorithm. Usually, TikTok displays the video first on the For You page to only a tiny number of users. If they engage with the content, then the video will be exposed to a wider audience. With an increasing number of viewers, at some point, the video will go viral with an exponentially increasing exposure and growth. Thus, the playing field is level. TikTok gives everybody a second chance: even if your video garners only a few views, the algorithm restarts the just-described process a couple of weeks later.

Thinking of the *Washington Post*'s TikTok presence, one might imagine the serious content for which this high-quality daily newspaper is known. However, the brand has completely adjusted its content to TikTokers, who are fed entertaining, behind-the-scenes videos and satire of current events and news. As such, instead of videos reporting on the Presidential election, the daily's staff instead analyzes and reports on the latest episode of The Bachelorette. The *Washington Post*'s success rests not on presenting itself but rather on its appeal to TikTok's audience, who simply want to be entertained – no more, no less. Thus, the brand's TikTok presence is fun and entertaining, yet, it is still consistent with the *Washington Post*'s brand identity, which showcases their investigative journalists both relatably and authentically.

As aforementioned, TikTok's audience at the moment mainly consists of Generation Z'ers. Thus, the average TikToker is 16–24 years old, slightly more female than male, and most likely uses the app from their mobile phones (Talbot 2020). It is noteworthy that many minors use TikTok. As always on social media, brands lose control of their message to a certain extent and should be ready to do so. That having been said, we need to be careful not to encourage minors to mimic or engage with a brand's videos in age-inappropriate or even illegal ways, as among other consequences, their parents might blame the brand rather than the medium.

5.2 Authenticity Versus Perfection

On TikTok, authenticity largely overtakes perfection, as Gen Z'ers are skeptical of the "perfect lifestyle" shown on other social media such as Instagram. TikTok is not about beautiful people who show off their best angle: it is precisely about amateurism, imperfection, and spontaneity.

Brands on TikTok definitely should go for a somehow less professional, more amateur approach concerning their videos. Such an approach will render the brand more relatable and approachable to the TikTok community. Accordingly, mistakes are allowed but originality and creativity are a must. The phrase "Nobody is perfect" completely applies here, and is even celebrated, as neither should a company's brand be perfect on TikTok.

As an example, take the NBA's TikTok presence featuring pep talks from players, workout sessions of NBA athletes playing sports to highly dramatic music, and players or their mascots dancing and "hamming it up" on the basketball court. Due to their authenticity, the NBA's videos blend perfectly into a TikToker's feed without being considered more professional than, or superior to, videos uploaded by any other user. Moreover, the NBA's videos are rather amateur in form, but also content-wise, they are done in a way that appeals to both basketball fans and nonfans.

5.3 Awareness Versus Sales

We stress here that TikTok should not be considered a sales channel but rather a means to develop brand awareness and engagement. When organizing hashtag challenges and the like, a brand should keep this in mind, encouraging TikTokers to engage with it and animate comments, shares, reactions, and much more. Sales will, at some point, follow an increase in awareness and engagement, but it should not be the primary objective of using TikTok.

Take Crocs, for example, which promoted the opening of its TikTok account with a challenge called #ThousandDollarCrocs. This challenge was linked to a song by Post Malone, a Crocs brand ambassador. Just a week after launching its TikTok account, Crocs already had 100,000 followers. The number of views of the aforementioned hashtag is by now in the billions. Although it is difficult to prove a link from this challenge to actual sales, Post Malone Crocs sold out in just 1.5 hours post TikTok launch (Wolfe and Ries 2019).

Within the scope of this discussion are also musicians and recording artists. Given the importance of music and background sounds, TikTok has an entire team dedicated to its relationships with recording artists. The artists are not paid but increase their brand awareness. Music sales will take place elsewhere online, with some TikTok videos promoting a link to streaming platforms where users can purchase the artist's song.

6 Following: Consequences for the Future

This chapter began with an overview of social media, AI, and their interdependencies. Consequently, we profiled TikTok, focusing on the differences between it and social media applications. How to make use of TikTok as a tool for branding, communications, and marketing was discussed in the third section of this chapter. In the final part, three precautions were suggested toward organizations making optimal use of TikTok.

We showed that TikTok is a new social medium, based entirely on AI-driven algorithms. TikTok most likely will have further impact on traditional social media such as Facebook and the like, which have seen a radical decrease in users from the Generation Z age group. TikTok might also influence the offline world; it could, for example, influence the composition of future hit songs toward a shorter format than is the current norm. American singer-songwriter and producer Adam Friedman uses TikTok as a test platform for as-yet-unproduced songs. Only if the 15-second clip performs well on TikTok is it produced in a longer format. Moreover, Friedman began producing music directly for TikTok influencers, recording them for maximum success on the video-sharing platform. Moreover, stars are born on TikTok. As such, 19-year-old Sub Urban aka Danny Maisonneuve signed a record deal with Warner following millions of TikTok users dancing to his song "Cradles." Stunna Girl, aka Suzanne, a 21-year-old rap artist from Sacramento, signed with Capitol for the same reasons.

Although TikTok is about entertainment and fun, one last issue – more serious in nature and with potential consequences – is worth discussing. TikTok is a Chinese company, wholly owned by ByteDance, with Chinese corporations legally bound to share (big) data such as users' locations, images, and biometrics with the Chinese government and authorities. The Peterson Institute for International Economics (Yang 2019), a US think tank, described TikTok as a national security threat to the West and as a "Huawei-sized problem." It might not come as a surprise, then, that hashtag searches on #HongKong will lead to several videos on a variety of topics related to Hong Kong, but none related to the ongoing protests. The same happens if one searches for the hashtag #protest, which yields plenty of them in London, Melbourne, and elsewhere, but none in Hongkong. As aforementioned, TikTok is about fun and entertainment only, which might explain this lacuna.

Bibliography

Baccarella, Christian V., Wagner, Timm F., Kietzmann, Jan H., and McCarthy, Ian P., (2018), "Social Media? It's Serious! Understanding the Dark Side of Social Media", *European Management Journal*, Vol. 36, N. 4, pp. 431–438.

Edison Research, (2019), *The Infinite Dial®2019*, Somerville, NJ: Edison Research.

Frier, Sarah and Wagner, Kurt, (2020), "TikTok Marketers Chase Billions of Views in Uncharted Terrain: Popular Creators Are Paid to Promote Hashtags as Mobile Video App Develops More Targeting Tools for Advertisers", *Bloomberg News*, February 27, Retrieved from https://www.bloomberg.com/news/articles/2020-02-27/tiktok-marketers-chase-billions-of-views-in-uncharted-terrain, (24.07.2020).

Fromm, Jeff, (2020), "TikTok Doesn't Stop: How to Reach Gen Z Consumers with Influencer Campaigns", *Forbes*, March 5, Retrieved from https://www.forbes.com/sites/jefffromm/2020/03/05/tiktok-doesnt-stop-how-to-reach-gen-z-consumers-with-influencer-campaigns/#dfa872b168af, (24.07.2020).

Haenlein, Michael and Kaplan, Andreas M., (2019), "A Brief History of AI: On the Past, Present, And Future of Artificial Intelligence", *California Management Review*, Vol. 61, N. 4, pp. 5–14.

Kaplan, Andreas M., (2009), "Virtual Worlds and Business Schools: The Case of INSEAD" in Wankel, C. and Kingsley, J. (eds.) *Higher Education in Virtual Worlds: Teaching and Learning in Second Life*, Bingley, UK: Emerald Group Publishing, pp. 83–100.

Kaplan, Andreas M., (2012), "If You Love Something, Let It Go Mobile: Mobile Marketing and Mobile Social Media 4x4", *Business Horizons*, Vol. 55, N. 2, pp. 129–139.

Kaplan, Andreas M., (2018b), "Social Media: Definition and History", (Alhajj, R. and Rokne, J. Eds.) *Encyclopedia of Social Network Analysis and Mining*, 2nd edition, Springer, pp. 1825–1827.

Kaplan, Andreas M., (2020a), "Artificial Intelligence, Marketing, and the Fourth Industrial Revolution: Clarifications, Challenges, Concerns" in Christiansen B. and Škrinjarić T. (eds.), *Handbook of Research on Applied AI for International Business and Marketing Applications*, IGI, forthcoming.

Kaplan, Andreas M., (2020b), "Artificial Intelligence, Social Media, And Fake News: Is This the End of Democracy?" in Gül, A. A., Ertürk, Y. D., and Elmer, P. (eds.), *Digital Transformation in Communication and Media Studies*. Istanbul University Press Books, forthcoming.

Kaplan, Andreas M., (2020c), "Social Media Powered by Artificial Intelligence, Violence, and Nonviolence", (Kurtz, L. Ed.), *Encyclopedia of Violence, Peace & Conflict*, 3rd edition. Elsevier, forthcoming.

Kaplan, Andreas M. and Haenlein, Michael, (2009), "The Fairyland of Second Life: About Virtual Social Worlds and How to Use Them", *Business Horizons*, Vol. 52, N. 6, pp. 563–572.

Kaplan, Andreas M. and Haenlein, Michael, (2010a), "Mondes virtuels: Retour au réalisme", *Expansion Management Review*, Vol. 138, N. 9, pp. 90–102.

Kaplan, Andreas M. and Haenlein, Michael, (2010b), "Uitdagingen en kansen rond social media", *Management Executive*, Vol. 8, N. 3, pp. 18–19.

Kaplan, Andreas M. and Haenlein, Michael, (2010c), "Users of The World, Unite! The Challenges and Opportunities of Social Media", *Business Horizons*, Vol. 53, N. 1, pp. 59–68.

Kaplan, Andreas M. and Haenlein, Michael, (2011), "Two Hearts in 3/4 Time: How to Waltz the Social Media-Viral Marketing Dance", *Business Horizons*, Vol. 54, N. 3, pp. 253–263.

Kaplan, Andreas M. and Haenlein, Michael, (2012a), "Social Media: Back to The Roots and Back to The Future", *Journal of Systems and Information Technology*, Vol. 14, N. 2, pp. 101–104.

Kaplan, Andreas M. and Haenlein, Michael, (2014), "Collaborative Projects (Social Media Application): About Wikipedia, The Free Encyclopaedia", *Business Horizons*, Vol. 57, N. 5, pp. 617–626.

Kaplan, Andreas M. and Haenlein, Michael, (2019), "Siri, Siri in My Hand, Who Is the Fairest in The Land? On the Interpretations, Illustrations, And Implications of Artificial Intelligence", *Business Horizons*, Vol. 62, N. 1, pp. 15–25.

Kaplan, Andreas M. and Haenlein, Michael, (2020), "Rulers of The World, Unite! The Challenges and Opportunities of Artificial Intelligence", *Business Horizons*, Vol. 63, N. 1, pp. 37–50.

Lafferty, Justin, (2014), "Facebook: 15M People Talking About ALS Ice Bucket Challenge, 1.2M videos posted", *Adweek*, August 15, Retrieved from https://www.adweek.com/digital/facebook-15m-people-talking-about-als-ice-bucket-challenge-1-2m-videos-posted/, (24.07.2020).

Talbot, Kate, (2020), "3 Reasons Why Families Love TikTok", *Forbes*, April 28, Retrieved from https://www.forbes.com/sites/katetalbot/2020/04/28/3-reasons-why-families-love-tiktok/#6d7bb5644e9f, (24.07.2020).

Turing, Alan, (1950), "Computing Machinery and Intelligence", *Mind*, Vol. 59, pp. 433–460.

Tuten, Tracy, Solomon, Michael, and Kaplan, Andreas M., (2019), Marketing des médias sociaux, Paris: Pearson.

West, Darrell, (2017), "How to Combat Fake News and Disinformation", Brookings Institution, December 18, Retrieved from https://www.brookings.edu/research/how-to-combat-fake-news-and-disinformation/, (24.07.2020).

Wilson, Sara, (2020), "The Era of Antisocial Social Media", *Harvard Business Review*, February 5, Retrieved from https://hbr.org/2020/02/the-era-of-antisocial-social-media, (24.07.2020).

Wolfe, Elisabeth and Ries, Brian, (2019), "Post Malone's Special Crocs Sold Out in Under Two Hours", CNN, December 11, Retrieved from https://edition.cnn.com/style/article/post-malone-crocs-sold-out-trnd/index.html, (24.07.2020).

Yang, Sophia, (2019), "Report Warns of Security Concerns Amid Sudden Rise of TikTok", *Taiwan News*, January 14, Retrieved from https://www.taiwannews.com.tw/en/news/3616776, (24.07.2020).

Hema MANDAPANDA APPACHU and Biddappa MANDAPANDA APPACHU

Impact of Digitalisation on the Education System in India: A Critical Perspective

1 Introduction

Education is a process of laying a foundation for a strong and widespread historical, social, cultural, political, psychological and scientific knowledge base to build an assertive, responsible and informed society. It is an investment which aims at further steering the prospect of collective growth and holistic development of a nation. The system of education of a country is distinguished by its ideal in the life of a nation. In this regard, 'Education' serves as a custodian, permitting and facilitating an upward mobility in the society (Cheney, Ruzzi & Murulidharan, 2005).

2 Value-Based Education

Theodore Roosevelt states, "To educate a man in mind and not in morals is to educate a menace to society" (Srivastava, 2018). 'Values' are defined as moral beliefs and principles that are accepted by the majority so as to ensure the continuity of a society (Ergil, 1984). Values are practised moral standards and gained tendencies to act humanely in various situations of life, thus forming the core of human existence. Since they are not intrinsic in nature, it requires a process of responsible transfer, which, in the context of the society, is obliged onto the older generation (Lakota, Sirca & Dermol, 2016). Therefore, values are the strengths that have to be infused in the personality of the younger generation, thus developing in them the ability to empathise in order to stimulate their potential that would further impact their performance. Based on several parameters, values can be categorised in the form of content (aesthetic, scientific, educational, economic, political and social values), extensiveness (societal, national and global values) and process – modern and traditional values (Türkkahraman, 2014).

Education serves three primary functions (Macionis, 1997): protective function, by transmitting values and culture for the ease of collaboration and coherence in a society; commutative function, focussed at behaviour modification; and formative function, facilitating cultural innovation and scientific temperament. The functions served by value education include: guiding the ideal thought process and behaviour, helping in judgement and preservation of important cultural objects, facilitating the execution of social roles and acting as a social control mechanism and thus enhancing solidarity (Fichter, 1990). Therefore, since values are social standards and fundamental beliefs that help to distinguish right and wrong for individuals, variables such as learner, school settings, family, society, environment etc. define areas of socialisation that contribute significantly in the acquisition of values. In this context, this chapter aims at focussing on the transitioning roles of education systems from the traditional patterns to digital platforms and its impact on imparting values.

3 Indian Education System

The education system of ancient India is considered to be the most unique and distinguishing system in the world (Rai, 2018). People in ancient India strongly believed that the fundamentals of education should be based on self-progression in order to achieve inner transformation. Such a development was thought to inculcate the values of social responsibility in the people so that they may contribute to the socio-economic development of the country. The ancient education system in India emphasised on physical, moral, intellectual, religious and spiritual growth as important stages in the development of personality of the individual (Mustafa, 2017). Until the colonial rule, education in India was grounded in indigenous systems of learning and scholarship with the ultimate aim of conservation of human values and culture throughout the nation (Basu, 1982). However, a formal system of education following the English medium of instruction was introduced by the British, which further continued even after the country's independence in 1947. Over the years, the importance of education was ingrained in the societal transformation to such an extent that Sri. Pandit Jawaharlal Nehru, the first prime minister of independent India, enormously contributed to the development of higher education and wished to achieve world-class proficiency in research and learning (Gautam, Singh, Fartyal, Tiwari & Arya, 2016). Furthermore, a great number of governmental policies and programmes were executed to significantly improve the quality of education imparted and ensure that educational opportunities are available to all segments of the society. More so, with liberalisation, privatisation and globalisation, systematic and advanced deliberations were involved in achieving the vision of building a competent and informed society. In this endeavour, efforts were initiated to embrace 'digitalisation' – the superlative product of the twenty-first century, into the Indian education system. Successful digital transformation requires efficient capacity building in response to the context and needs of its recipients (Carcary, Doherty & Conway, 2016). Thus, digitalisation is the 'ability to turn existing products or services into digital variants, and thus offer advantages over tangible products' (Gassmann, Frankenberger & Csik, 2014).

4 Digital Education in India

Digitalisation of education in India is a process of implementation of modern methodologies in the field of education, helping the students in various fields to embrace the advancements in technology and face the level to which the competition currently exists on the world stage. In addition to this, the quality of knowledge imparted is considered to be top-notch with the induction of artificial intelligence where the scope for imparting knowledge increases by a considerable margin (Subrahmanyam & Swathi, 2018).

India is a country which is very proud of its culture, heritage, values and history (Srivastava, 2015). The deep-rootedness of these core ideals is evident in the day-to-day activities of its citizens and the sense of nationalism one experiences in the face of any grave adversity. Traditionally, Indians have placed their teachers on a very tall pedestal, comparing them to the God of wisdom and knowledge on various contexts. The teachers' quest for knowledge and their proficiency in imparting information has gained them the love and respect of their students. This is an evidence to the privilege the teachers enjoy in the Indian society. The different pedagogy of teaching employed by various teachers based on the needs of the individual is what sets apart a good teacher from the rest. However, there have been evidences of recent advancements which seem to aim at replacement of the human involvement in education by artificial intelligence. One such instance is the introduction of completely automated robots for primary education in one of the schools in Bengaluru, Karnataka, India (Balaji, 2019). In such a scenario, complete digitalisation of the education system, sidelining the whole cluster of talented, innovative and experienced teachers to be replaced by artificial intelligence for its vast knowledge base and flawless delivery of information is the question being probed into in this chapter.

The emphasis of this chapter is on the digitalisation of content, specifically in terms of availability of multiple sources of information on the Internet, as well as the ability to comprehend the available material. The applications of digitalisation in academics is undoubtedly manifold. For example, video conferencing applications have simplified the procedures of admission to institutions globally; tracking systems have eased the burden of the applicants; e-library facilities have enabled comfortable readership. However, when looked into from a holistic perspective, while on one hand, digitalisation has broadened the scope of connecting people in academic platforms through innovations like virtual classrooms, webinars etc. on the other hand, the moderation required in terms of delivering age-appropriate content and its related interpretations has been a constant challenge. Therefore, while the focus is on the role of digital platforms in the process of encoding and decoding the education delivered, attention is concentrated on the digital spaces taking over the traditional systems of imparting education and its impact on the value base of the society.

5 Methodology

This research is designed based on the social constructivist paradigm, which focuses on the role of social interaction in the construction of knowledge and human development. In this regard, a questionnaire was developed to understand the perception of people about digital transformation of the Indian education system. Further, emphasis was placed on the merits and demerits of the changing scenario of the educational setup dedicated towards imparting quality education. The data was collected from participants representing different age parameters, gender and socio-cultural backgrounds through surveys, individual interactions and secondary sources of literature and subjected to thematic analysis. Perspectives on the preparedness of the country for complete digitalisation of its education system is discussed further on the basis of Stuart Hall's encoding/decoding model of communication.

The rationale for analysing the content of digital transformation of education based on the tenets of Stuart Hall's encoding/decoding model of communication has great significance in this context. Education is a two-way process between the producer of content and the consumer of knowledge. The successful creation of an informed and responsible society is certainly based on the intent of production being received at the same level, which otherwise might lead to destructive impact on the society. In this regard, the contemporary, flexible and dynamic nature of the Reception theory proposed by Stuart Hall is of utmost relevance in terms of understanding the implications of being a participant to adopting such a paradigm shift in the context of education.

6 Analysis and Discussion

Stuart Hall is a critical and cultural theorist who proposed the encoding/decoding model of communication (Bodker, 2016). In the approach, the encoding model encapsulates the conception of an idea, the related content and means used for execution. The process of decoding is based on the social and political context in which the content is produced and interpreted. When there is a discrepancy in the socio-political context of encoding and decoding of a message, it leads to miscommunication. Thus, the entire process of encoding and decoding follows the four steps of production, circulation, distribution/consumption and reproduction (Yousmann, 2013).

Production: The process of producing a content is greatly influenced by the ideological and cultural hegemony of the time. The ideology, belief and practices dominantly adopted by the society combined with the core set of ideals of the creator of knowledge is reflected in the content produced. While understanding the traditional pattern of knowledge production in India, the arrival of the British brought along with them the cultural hegemony creating a dominant educational system. The novel system suppressed the existent Indian way of education, which focussed around imparting knowledge based on locally practised cultural and traditional way of life. For example, Sanskrit, which was the dominant language of pre-colonial India, also called the 'Language of the Learned', was the direct result of the upheld Brahminical culture of those times (Mane & Hirve, 2013). Thus, it was the means and basis for imparting education. Macaulay (1835) in his 'Minute on Education' mentions about the superiority of Western literature over any of the vernacular languages of India, in the intellectual prosperity of the citizens. Therefore, the Anglicisation of the education system was a way of taking over the production line of education, thus stamping the hegemony of the colonisers in India.

Circulation: Based on the cultural system, pre-existing knowledge and pre-conceived notion of a particular context, the information is received and interpreted by the individual. The knowledge imparted by the modern education system of the British was circulated with the addition of an element of 'perspective'. This element which was determined by the social and cultural structure of the time, along with the transcending power structure and politics of the era, influenced the absorption and interpretation of knowledge by the receivers, in this context, the Indians.

Distribution/consumption: Any message which is in circulation will be distributed only after its interpretation as a meaningful discourse by the receivers. In the context of historical background of the Indian education system, the meaningful analysis of the knowledge imparted through the anglicised education system impacted the wider distribution of the system, which further effected popularity.

Reproduction: It refers to the actions advanced based on the knowledge consumed in the previous step. The consumption of knowledge through the anglicised education system led to the development of the modern methodology of education in India, which is a culmination of Western and indigenous concepts. This left the Indian education system at a juncture which strives to maintain a balance between traditional and modern approaches.

In the context of decoding information, Stuart Hall proposed three positions (Tomaselli, 2016): dominant/hegemonic, negotiated and oppositional positions. In the dominant position, the intent of the information encoded and decoded are congruent with each other, thus creating a hegemonic stand among the receivers. The position where congruency between the encoded and decoded messages are achieved but the information is customised to adapt to the socio-political and cultural context is called the negotiated position. The latter position is based on the critical understanding of the encoded message to be decoded in a global contrary manner. Contextualising and decoding the standpoint of the Indian education system, it is best represented through the negotiated position of decoding. This chapter will follow the structure of discussing the encoding themes, followed by the decoding themes, in the context of education and its transformation into a digitalised forum.

7 Encoding Network in Indian Education System

Education is a process which aims at biological, psychological, social and moral development of a human being, who is a member of the society (Thornburg, 1984). It is an encoding process made easy through the efficient role of facilitators in the form of professionals, digital media or any other form of information which is fed to the decoder. The encoding process is of utmost importance due to its responsibility of transferring knowledge related to facts, traditions, culture, linguistics etc. to the decoder, who will form the next society. In this context, the data received from an empirical investigation recognises an array of themes in the encoding pattern of Indian education system, as discussed further:

Structural framework to the Indian-ness in thought: There are different ways and methods of transfer of knowledge. In India, with the prevalence of a multitude of languages, several languages lack scripts through which information can be recorded for the benefit of the future generations. Hence, transfer of knowledge takes the oral route. Thus, the flow of traditional knowledge of culture, deeper understanding of tradition, survival skills as well as valuable inputs of life have been passed down from one generation to the other through verbal discourse. A culture that elevates a Guru (Teacher) to the pedestal of God has a teacher in all walks of life because anyone who offers knowledge is bestowed the same respect as that of a proficient teacher. Therefore, with these set of values and ideals forming the structural framework of the Indian society, there is easy transference of knowledge along with maintenance of a reverential position for the provider, thus forming a hierarchy for the movement of valuable information.

Democratic approach: India is the largest democracy in the world (Jakobsen, Nielsen, Nilsen & Vaidya, 2018). This system of governance is the basis of most of the administrative machineries of the country. Education is considered to be one of the critical dimensions of democratic governance (Gawas, 2015). Indeed so, the nature of education system in India is perceived to be functioning on the same model of democracy. The primary encoders to the production of education are: the importance given to regional languages and inexpensive facilitation of knowledge. The country functions on the basic principle of 'Unity in Diversity'. With inhabitants practicing and professing different faiths, the country is also a homeland to a range of languages. Therefore, the eighth schedule of the Constitution of the Republic of India identifies twenty-two languages for its empowerment and growth (Bakshi, 2002). This process of officiating has reflected in promoting regional languages as the medium of instruction along with English. Another constitutional right bestowed to the citizens is the Right to Education Act (Ramachandran, 2009). Based on this endowment, government provides free of cost primary education to all children. Further, several scholarships and reservations are provided with an aim to reduce the financial burden of the people, together with dismantling the barrier of financial constraints from effecting the literariness of the citizens (Cho, 2014). These initiatives have been successful in creating a sense of inclusiveness among the citizens with efforts reaching out until the lowest strata of the society (Kirschner, 2015). It is a structure which is helping in widening the base for those who would want to reap the benefits of education.

Another prominent feature of the education system in India is a combination of traditional and modern techniques in imparting knowledge. While the traditional methods involve interaction and moral-based education as an inseparable part of the system, modernity is professed especially in the technical aids to delivery. This rich combination contributes to the high quality of education. The richness in standard created by the value-based education has also created a common moral ground for the students, thus helping them blend into the pre-set morality clauses of the society, overall forming an inclusive niche (Bacanli, 2011).

These features contribute to setting a high standard and trying to reach and make the people of all the strata knowledgeable enough to access pre-set quality standards of education enhanced by the democratic facilitation of resources. However, in the context of understanding digitalisation as a method of encoding information, the data from participants' express certain concerns and admonitions, which are as discussed further:

Restricts methodical and structured approach to education: The role of a facilitator in the system of education is perceived to be more than just delivering lectures on specified topics. An age old Sanskrit verse goes, 'Acharya Devo Bhava', meaning the teacher is assigned the position of God, which is akin to the position of parents (Joshi, 1991). Hence, with a primary focus on imparting knowledge with a personalised touch, the facilitator is a person who interacts with the participants, recognises individual potentials and promotes creativity. The enabler is also available to discuss the tiniest of distress experienced by the pupil, ranging from academics, co-curricular, extra-curricular to a personal front. Sometimes, the nurturing quality of the facilitator also makes the student comfortable enough to confide in him/ her, thus shouldering a position of responsibility by the facilitator. Hence, the role of the propagator of knowledge is much more, but not limited to personality development of individuals. A digital platform to education restricts the personal connect (Kaur, 2019) which in turn impacts the all-round development of the recipients.

Another predicament to the digital medium of knowledge production is related to the content of ideas published. It is important to note that information encoded in the digital platform is a result of subjective decoding, which might be restricted to the experiences and abilities of the specific decoder. While on one hand, an overload of information is likely to foster confusion among the receivers, on the other hand, there is a greater likelihood of manipulation through propagation of false information. The chances of misleading are greater with regard to young audience where age-inappropriate content act as potential distractors (Chassiakos, Radesky, Christakis, Moreno & Cross, 2016). Such situations demand a formal monitoring system aimed at supporting the well-being of individuals belonging to all age groups. This necessitates to fulfil the requirements of trained professionals to aid the reasoning and interpretative abilities of the recipients by inculcating in them the capacity to identify and assess the authenticity of the available content and respond accordingly. Furthermore, a major structural challenge is the lack of appropriate resources in terms of electricity and data services in the remotest areas of the country (Kaur, 2019). This may also be applicable to accessing abilities of the underprivileged sections of the society.

These factors collectively result in a resistance to transition, which calls for efficient work directed towards behaviour modification. By all means, the digital forum of knowledge generation calls for right choice of information made available, supported with appropriate guidance, facilitated by human intervention.

Lack of preparedness: While assessing the current structure available for digitalisation, the demand and supply analysis show an imbalance with the demands overpowering the supply chain. The demand chain requires fulfilment in terms of adequate training for professionals to use the digital platforms without compromise on the personal connect, setting up a team to troubleshoot technical glitches that can hinder the supply and accessibility of information. A systematic approach would also be required to address the concerns related to adequate supply of power through electrical transmission, inaccessibility to network services due to geography of the land, poor maintenance of the installed digital equipment and their upgradation. A major hurdle may however be the implementation of the digital services at the grassroots level. The availability of adequately upgraded instruments and gadgets amongst all the participants belonging to the different strata of the society is also a concern that suggests the lack of preparedness for complete digitalisation in the Indian context. These further augments the need for a systematic and inclusive form of upgradation of technology for the benefit of its recipients.

Need for delegated and collective effort: The process of evolution involves a step by step mechanism involving undo and redo options. Likewise, the developmental activities of a progressive country also work on a trial and error method of identification and problem-solving. In this course, considering the inhibitions and resistance to change being the norm of any behaviour modification process, it becomes the duty of the facilitators to initiate the desired change in a periodic manner. Considering that identification of the problem is half the solution, a regular review and monitoring process should be initiated to identify the problems at every stage of implementation. Further, a strong administrative and executive machinery would have to be set up to keenly observe and plan a method of implementation of the desired change. The importance of such a machinery arises to assure the smooth transition of the traditional method of education into a partially digitalised system (Johnson, Jacovina, Russell & Soto, 2016) wherein the facilitators play an important role in bridging the gap between the supply and demand chain of production, which in this context form the encoding process of information.

Need for a balance: While driving steadily on a progressive path, it is quintessential to be prepared for roadblocks on the way. Complete digitalisation is likely to promote individualistic thought process along with diluted mentor–mentee relationship, thus reducing social interaction and weakening the crux of the Indian society, that is collectivism (Sinha, Sinha, Verma & Sinha, 2004). Development is certainly the need of the hour. However, it is also imperative to deliberate on the nature of compromise demanded by the situation. Though a dilemma, an integrated and balanced approach would be the way forward to achieve the goals of the modern society, while firmly being grounded in the social fabric that consolidates the Indian identity (Ray, 2017).

These master themes which have emerged from the observations, experiences and introspections of the participants encapsulate the intervention required to improve the efficiency of encoding process in digital education in the Indian context. However, along with concerted efforts directed at plugging the structural difficulties in imparting digital education, attention should also be focussed on the content delivered through the platform. One of the primary goals of education is to provide confidence and the consciousness of being an individual. This may be imparted through an integration of value-based education in addition to academic content, thus raising individuals who can convert them into behaviour. The identity of an individual is defined when values are experienced, internalised and practised with a feeling. Thus, the pre-condition for imparting value education is to prepare an environment for the person to implement free will in actions, which should be the goal of encoding process.

8 Decoding Network in Indian Education System

Learning is defined as a relatively permanent change in human performance or potential as a result of change in experience (Driscoll, 1994). The process of decoding involves receiving the encoded information, deciphering, understanding and interpreting the content based on pre-existing knowledge and social roles in order to enhance the database of the individual. Overall, the process involves comprehension and acceptance of different forms of information made available in various platforms to further apply and utilise in appropriate situations. In the context of the Indian education system, the following themes are discussed based on the responses of the participants:

Growth potential: The capacity of education is perceived as having enormous potential in terms of developing the skill set of individuals and shouldering the greater responsibility of shaping the personality. The participants on the receiving end look up to the process of education and interpret the delivery as a preparation to compete and excel in this modern competitive world, which encourages students to grow to become the best version of themselves and ultimately lead a happy and successful life. In this regard, digital platforms widen the scope of knowledge base pertaining to various domains of life. However, the unfiltered nature of information encoded is a double-edged sword that requires a keen, informed and rational eye during the process of decoding. In the pursuit of identifying such individuals with a balanced potential for sensible decoding of information, Huitt (2011) enumerated the following steps in the identification of potential capacities: investigate human capacities, investigate if the identified capacities can be transformed into competence through guided learning experiences, and investigate the applicability of the competence. In the context of formal education systems, the role of a facilitator (encoder) is driven by knowledge delivery coupled with identifying and developing potential capacities of the recipient (decoder). Alternately, in digital platforms, the process is limited to encoding information, with complete absence in the second phase, which may be a deterrent to the growth potential of an individual/society/nation.

Nevertheless, the steps of identifying potential capacities would be at its performing best with the help of technology assisting human judgement, thus helping in delivering the maximum output, thereby building a stronger society. Therefore, the role of digital education is viewed in the context of complementing the human efforts in identifying, developing and executing the strengths, rather than replacing the human intervention.

Empowerment: The concept of empowerment can be understood with the difference between 'power' and 'powerlessness' (Lord & Hutchison, 1993). Education being a source of power, the Indian education system offers a variety of subjects to the students. This creates a broader knowledge base further strengthening analytical abilities through multiple perspectives. The system further reinforces conceptual clarity with application in real-life settings. Therefore, a combination of theoretical and practical orientation is successful in promoting better reasoning abilities among its participants. The Doctrine of Formal Reasoning proposes that education disciplines the faculties of the mind (Bunge & Leib, 2019). It helps individuals to compare, understand analogies and derive logical conclusion, in application to social contexts, thus impacting the outcomes of life. These abilities are assumed to hone reasoning skills required to solve novel problems. With this backdrop, the role of digital education is interpreted as fulfilling the objective of facilitating informed and responsible human beings. For example, digital platforms have successfully broadened the exposure of a common man to his/her individual rights enabling appropriate understanding and response to situations, in all, empowering a society by creating awareness about their right to legitimate actions. Work opportunities in e-platforms have also helped in developing a prudent workforce, thus enhancing the employability of the citizens, in turn assisting in stabilising the welfare of the country. Another instance referred to is the e-tendering provisions that have introduced transparency in the system. This has also brought in the need for accountability and responsible deliverance of services, indirectly influencing the value system of the citizens. As education is also deciphered as a process of applying scientific and moral concepts along with acquisition of values and ethics, the role of digitalisation seems to have a subtle presence in developing conscience, awareness and sense of right or wrong, therefore encouraging the construction of a virtuous and empowered society.

Enables sustainability: The Indian education system functions strongly on the principle of morality, ethics, social relevance, nurturance, strengthening historical knowledge and conceptual knowledge, further being embedded in the Constitution as a basic right. Article 21A of the Constitution of India promulgates free and compulsory education to children from 6 to 14 years of age making education a Fundamental Right (Gawas, 2018). In this way, the Indian education system strives to provide affordable and quality education to the remotest sections of the society. It aims to cross the barriers of gender, class and creed, empowering all its citizens. Universal access to education is the basic motto with an aim to achieve non-discrimination, solidarity, and equality of opportunity, thus enabling social justice. Based on the awareness of rights given to a person, he/she moulds self and others, thus developing individually and working to contribute to nation building. This in turn is hopeful to create a cycle of sustainability. While literature supports that digital literacy fosters digital rights and aids social equality, thus increasing the participation of citizens in public domain and contributing to sustainable development of the country (Hargittai & Walejko, 2008), digital media is certainly not immune to compromised confidentiality along with inaccurate, inappropriate and unsafe content (Chassiakos, Radesky, Christakis, Moreno & Cross, 2016). Therefore, for a country which has the basic foundation for equality and education in its Constitution, information and communication technology would serve as an advantage in wheeling the cycle of sustainability.

Moulds thinking to enable pragmatism: The participants describe education to be an 'enlightening experience'. The endeavour of learning is observed as a means of acquisition of knowledge, skills and habits, develop an individual intellectually and also inspire creativity, thus strengthening the cognitive abilities of an individual. Education is described to be a framework that acts as basics to further understand, develop and innovate, thus igniting novelty in thought process. Engagement in idea-generation tasks are a means of stimulating originality and inventive abilities that stem from creative thought processes (Pang, 2015). Sternberg (2000) proposed the Triarchic theory of intelligence and identified three categories of cognitive abilities: practical, analytical and creative intelligence. According to this theory, practical intelligence is related to overall success in living. Two other concepts interrelated with practical intelligence were referred to: analytic intelligence, defined by the ability to observe, compare and analyse situations; creative intelligence, associated with generation of novel ideas. He further elaborated that though some of these capacities could be inherited, it involves a lot of developmental processes to improve the skills related to the different components of successful intellectual abilities (Feldman, 2011).

In this endeavour, the process places great importance to practicality-fostering critical thinking abilities, realistic thinking and cognitive development to deal with the real world. This requires a combination of human deliberations and technological advancements aimed at catering to different learning styles and supplying to the need-based demands of individuals. This amalgamation would also help in stimulating the abilities of the facilitators to customise the content based on their experiences and observations regarding the needs of the decoders. Such a combination of cognition, novelty and practicality is thus looked upon as a robust force in stimulating a pragmatic society.

Perspective: Socialisation is a multidirectional process which provides an experience, allows reproduction of human relations and promotes self-assertion of a concrete personality (Zvoleyko, Kalashnikova & Klimenko, 2016). One of the significant aspects of social integrity is the transfer of value systems through generations. Based on the renowned proverb, 'Each one, teach one', an efficient mechanism of teaching values is identified to be the realistic solution in tackling the changes propelled by socio-economic and technological advancements on a global scale (Turkkahraman, 2014). Thus, values and morals of an individual are developed largely during the process of socialisation. Digitalisation, in a way, limits the much needed personal and social connect, which is essential for transference of values that are very important in the overall moulding of a personality. One of the reasons cited for the collapse of value education system in the present times is the influence of creative modification in the media. Unfortunately, the effect of such an influence is found to negatively impact the perspectives of the recipients, due to immature thought processes and training in decoding. The inability of young minds to understand the gravity of the morals gives rise to flawed and confused perspectives, which is not a desirable outcome.

Digitalisation as a facilitator of learning: The advancement of digitalisation creates a learning space characterised by dynamism, flexibility and maximum exposure to content. Digitalisation of education offers choice-based learning systems taking into consideration the vast interest of the individual. Therefore, the limitations of the Indian education system in terms of a pre-defined set of courses can be mitigated by enrolment in various online learning platforms. It allows pace-appropriate learning to individuals following different learning styles: visual, aural and kinaesthetic. Further, easy access to wider collection of data and the confidence that the data can be revisited promotes self-learning among the participants. Therefore, digitalisation allows customisation based on individual priority. Digitalisation is also viewed to be a solution to overcome the structural challenges of geographical restrictions, quality of educators and paucity of time, thus expanding the reach of education to all.

The likelihood of outcomes expected of an integration of digitalisation in the Indian education system is analysed at the individual/social level as follows:

Better informed and empowered self: Digitalisation as a medium of communication has aided the promotion of psychological attributes like creativity and personality development. It is also perceived to be a means to enhance the employment potential through skill development for expanding the employability of participants. It is further credited to be an inclusive forum providing equal opportunities to people with disabilities too. Another significant contribution of the digital world is the facilitation of governmental schemes and policies to individuals, thus creating equality in opportunity. Availability of content through virtual platforms have supplemented the utilisation abilities of receivers ranging from students getting exposed to global marketing through e-markets to recipients of governmental and other services, thus nurturing a better informed and empowered self.

This understanding is in tandem with the four 'Pillars of Learning' in holistic education as indicated by the UNESCO (1996): learning to 'Learn' by asking – which is professed as a natural act of gaining knowledge; learning to 'Do' by acquiring more of behavioural skills than intellectual skills; learning to 'Live together' by developing qualities of self-awareness, inter-dependence, mutual respect, conflict resolution and competency to work towards common objectives; and finally, learning to 'Be' by working towards discovering the inner wisdom through introspection.

Vasudaiva Kutumbakam: Digitalisation has supplemented the existing opportunities with access to provisions in global platforms. Hence, the world has become a single community principled on the idea of 'Vasudaiva Kutumbakam', where there is understanding and inter-relatability with different cultures, languages, study patterns, pedagogies etc. With this idea, there has been a plethora of attempts for cross-cultural studies through student exchange programmes as well as interdisciplinary studies aimed at understanding and accommodating multiple and differently angled opinions and perspectives. In the Indian context, the concept of scientific temper is referred to as 'a way of life, a process of thinking, a method of acting and associating with our fellowmen' (Nehru, 1946). This tradition of enquiry in the Indian intellectual tradition dates back to the preaching of Jainism, Sankya and Buddist traditions, which gradually ingrained itself in the structure and function of the Indian ethos (Palampur Declaration, 1981 as cited in Mahanti, 2013). This has further been facilitated through the enhancement of opportunities made possible by digital platforms. It has also been productive in reducing discrimination on various fronts, thus making the society more inclusive, promoting harmony, along with scientific temperament amongst the masses.

However, extreme freedom, anonymity and pervasiveness that characterise cyberspace, concerns about values and moral education have now shifted into overdrive. Cyber culture is understood to simultaneously lead to marginalisation of traditional form of mediation of scientific knowledge with its vast sources producing scientific information in abundance. As a result, diversified means of communication have engaged both professional scientists and passionate amateurs in dispensing information. This has resulted in constricted flow of actual knowledge through the mediation process by scientists and mediators, as they represent the smaller community in the present times (Schiele, 2012).

9 Conclusion

Digital convenience has penetrated in all walks of life, thus making it a basic necessity in today's world. It is an automated process which unfortunately limits a personal connect between the encoder and the decoder. Therefore, the role of virtual podiums have to be perceived and considered in the context of connecting different cultures and pedagogies of teaching. The content available in the internet should be interpreted in a manner which is constructive in the development of a society. Since it is a technology-driven process, there is a need for human facilitation to filter the contents and induce a clause of values and morality. Such moral grounds have been the focus of imparting value-based education, hence the crux of this chapter.

The role of education in building a society is recognised with pride. This careful contribution to the societal construction is due to an acknowledgement of the role of the society in inculcating a set of moral values in a person, principled on the idea that morality is inculcated and not inherent. In this context, two systems are envisioned to function in parallel, where the decoder of a particular information becomes the encoder for the following chain of decoders. Likewise, the encoding society leaves a significant imprint on the decoding citizens because the latter adopts the encoding roles thereafter. In such a deeply structured endeavour, complete digitalisation hinders the availability of subjectivity in thought process, thus generating a predictable and mechanised mindset for interpretation.

Another concern is related to the risk of producing professionals with incomplete or misinterpreted knowledge, which might be a greater roadblock to the establishment of a progressive society. On the other hand, the advanced technologies of virtual interface maybe to a certain extent, working towards achieving a personal connect as in any human interaction. However, the lack of necessary infrastructure in all parts of India, due to reasons varying from geography, finances and administrative machineries, may impact the intent of production and transfer of knowledge, values, ideals and morals.

In response to the recognised lacunae in the digital form of education, the chapter proposes partial digitalisation in the Indian context, with a systematic and methodical integration of traditional and modern methods for imparting education, thus creating opportunities to accommodate newer perspectives.

Bibliography

Bacanli, H. (2011), "Value is Valuable", *Egitime Bakis*, 19, pp.18 – 21.

Bakshi, P.M. (2002), *The Constitution of India*, Delhi: Universal Law Publishing.

Balaji, F. (2019, September 01), "Robot Turns Teachers in Bengaluru School, Thanks to AI", Livemint. Retrieved from https:// www.livemint.com/ technology/tech-news/robot-turns-teachers-in-Bengaluru-school-thanks-to-ai-1567336339210.html.

Basu, A. (1982), "The Origins and Operations of the Indian Education System 1757–1947". In *Essays in the History of Indian Education*, New Delhi: Concept.

Bodker, H. (2016), "Stuart Hall's Encoding/Decoding Model and The Circulation of Journalism in the Digital Landscape", *Critical Studies in Media Communication*, Vol. 33, N. 5, pp. 409 – 423.

Bunge, S.A., and Leib, E.R. (2019), "How Does Education Hone Reasoning Ability?" *Current Directions in Psychological Science*, pp.1–7.

Carcary, M., Doherty, E., Conway, G. (2016), *A Dynamic Capability Approach to Digital Transformation – A Focus on Key Foundational Themes.* In: 10th European Conference on Information Systems Management. Academic Conferences and Publishing limited, pp. 20–28.

Chassiakos, Y.R., Radesky, J., Christakis, D., Moreno, M.A., and Cross, C. (2016), "Children and Adolescents and Digital Media", *Paediatrics*, Vol. 138, N. 5, pp. 1–18.

Cho, E. (2014), *The Impact of Reservation Policy on Education in India: An Evaluation of Karnataka's Gram Panchayat Presidency Reservation and Children's Education Attainment for Scheduled Caste and Scheduled Tribe Households* (Unpublished Dissertation), Duke University, England.

Driscoll, M.P. (1994), *Psychology of Learning for Instruction*, Needham, MA: Allyn & Bacon.

Ergil, D. (1984), *Society and Human*, Ankara: Turhan Kitabevi.

Feldman, R.S. (2011), *Understanding Psychology*, New York: McGraw-Hill.

Fichter, J. (1990), *What is Sociology?* (N. Çelebi, Trans.), Ankara: Attila Bookstore.

Gassmann, O., Frankenberger, K., and Csik, M. (2014), *The St. Gallen Business Model Navigator* [Online], Available:

Gautam, M., Singh, S., Fartyal, G., Tiwari, A., and Arya, K.S. (2016), "Education System in Modern India", *International Journal of Scientific Research and Education*, Vol. 4, N. 1, pp. 4876–4879.

Gawas, V.M. (2015), "Political Economy of Tribal Development in Goa", *Journal for Interdisciplinary Studies*, Vol. 4, N. 14, pp. 2319–4766.

Gawas, V.M. (2018), "Right to Education Under Constitution of India and Development among the Scheduled Tribes: A Socio-Legal Study", *International Journal of Law*, Vol. 4, N. 3, pp. 63–72. http://www.im.ethz.ch/education/HS13/MIS13/Business_Model_Navigator.pdf[8] E.

Huitt, W. (2011), *A Holistic View of Education and Schooling: Guiding Students to Develop Capacities, Acquire Virtues, and Provide Service*, Revision of paper presented at the 12th Annual International Conference sponsored by the Athens Institute for Education and Research (ATINER), May 24–27, Athens, Greece.

Inkeles, A. (1964), *What is Sociology?*, USA Englewood Cliffs: Prentice-Hall.

Jakobsen, J., Nielsen, K.B., Nilsen, A.F., and Vaidya, A. (2018), *Mapping the World's Largest Democracy (1947–2017)*, Forum for Development Studies, doi:10.1080/08039410.2018.1465461

Johnson, A.M., Jacovina, M.E., Russell, D.E., and Soto, C.M. (2016), "Challenges and Solutions When Using Technologies in Classroom", (S.A. Crossley and D.S. McNamara Eds.) *Adaptive Educational Technologies for Literacy Instruction*, New York: Taylor & Francis, pp. 13–29.

Joshi, K. (1991), *The Veda and Indian Culture: An Introductory Essay*, New Delhi: Motilal Banarsidass Publishers.

Kaur, H. (2019), "Digitalization of Education: Advantages and Disadvantages", *International Journal of Applied Sciences*, Vol. 4, pp. 286–288.

Kirschner, S.R. (2015), "Inclusive Education", (W. George Scarlett Ed.), *Sage Encyclopedia of Classroom Management*, Sage, pp. 1–16.

Lakota, A.B., Sirca, N.T., and Dermol, V. (2016), "Virtues – The Centre of Quality Education System – For Successful Integration in the International Society", *Procedia – Social and Behavioral Sciences*, Vol. 221, pp. 302–307.

Lord, J., and Hutchison, P. (1993), "The Process of Empowerment: Implications for Theory and Practice", *Canadian Journal of Community Mental Health*, Vol. 12, N. 1, pp. 5 – 22.

Macaulay, T.B. (1835), Macaulay's Minute on Education. February 2, 1835.

Macionis, J. (1997), *Sociology,* New Jersey: Prentice-Hall International.

Mahanti, S. (2013). "A Perspective on Scientific Temper in India", *NISCAIR Publications,* Vol. 1, N. 1–2, pp. 46–62.

Mane, D., and Hirve, A. (2013), "Study of Various Approaches ın Machine Translation For Sanskrit Language", *International Journal of Advancements in Research & Technology,* Vol. 2, N. 4, pp. 383–387.

Mustafa, S. (2017), *Historical Development of Teacher Education in India,* Srinagar: Institute of Advanced Studies in Education.

Nehru, J. (1946), *The Discovery of India,* Calcutta: The Signet Press.

Pang, W. (2015), "Promoting Creativity in the Classroom: A Generic View", *Psychology of Aesthetics, Creativity and the Arts,* Vol. 9, N. 2, pp. 122–127.

Rai, S. (2018), "Indian Education System: A Comprehensive Analysis by Sister Nivedita", *PEOPLE: International Journal of Social Sciences,* Vol. 4, N. 2, pp. 17–26.

Ramachandran, V. (2009), "Right to Education Act: A Comment", *Economic & Political Weekly,* Vol. XLIV, N. 28, pp. 155–157.

Ray, S. (2017), "Indian National Identity: Post Independence Journey in the Light of Miller's Theory", *Asia Pacific Journal of Research,* Vol. 1, N. 1, pp. 67–71.

Schiele, B. (2012), "Science Communication: New World, New Challenges", (H.J. Khan, G. Raza and S. Singh Ed.), *Quest for Scientific Temper,* New Delhi: National Institute of Science Communication and Information Resources (NISCAIR), Council of Scientific & Industrial Research (CSIR).

Sinha, J.B.P., Sinha, T., Verma, J., and Sinha, R.B.N. (2004), "Collectivism Coexisting with Individualism: An Indian Scenario", *Asian Journal of Social Psychology,* Vol. 4, N. 2, pp. 133–145.

Srivastava, P. (2018, January 1), "The Need for Moral Education", *New Delhi Times.* Retrieved from www.newdelhitimes.com/the-need-for-moral-education/.

Srivastava, S. (2015), "A Study of Awareness of Cultural Heritage among the Teachers of University Level", *Universal Journal of Educational Research,* Vol. 3, N. 5, pp. 336–344.

Subrahmanyam, V.V., and Swathi,K. (2018), Artificial Intelligence and Its Implications in Education, International Conference on Improved Access to Distance Higher Education Focus on Undeserved Communities and Uncovered Regions, Telangana, 11–12 August. Warangal: Kakatiya University.

Swami Nirvedananda (2011), *Our Education* (6 ed.). India: Ramakrishna Mission Calcutta.

Thornburg, H.D. (1984). *Introduction to Educational Psychology,* New York: West Publishing Company.

Tomaselli, K. (2016), "Encoding/Decoding, the Transmission Model and a Court of Law", *International Journal of Cultural Studies*, Vol. 19, N. 1, pp. 59–70.

Turkkahraman, M. (2014), "Social Values and Value Education", *Procedia – Social and Behavioral Sciences*, Vol. 116, pp. 633–638.

UNESCO (1996), "The Four Pillars of Education", in *Leaning: The Treasure Within: Report to UNESCO of the International Commission on Education for the Twenty First Century*, Paris: UNESCO.

Yousmann, B. (2013), "Revisiting Hall's Encoding/Decoding Model: Ex-Prisoners Respond to Television Representations of Incarceration", *Review of Education, Pedagogy, and Cultural Studies*, Vol. 35, N. 3, pp. 197–216.

Zvoleyko, E.V., Kalashnikova, S.A., and Klimenko, T.K. (2016), "Socialization of Students with Disabilities in an Inclusive Educational Environment", *International Journal of Environmental & Science Education*, Vol. 11, N. 14, pp. 6469–6481.

www.ingramcontent.com/pod-product-compliance
Ingram Content Group UK Ltd.
Pitfield, Milton Keynes, MK11 3LW, UK
UKHW021921120926
8556IPUK00042BA/1108

9 783631 834893